QUANTUM CHEMISTRY RESEARCH TRENDS

QUANTUM CHEMISTRY RESEARCH TRENDS

MIKAS P. KAISAS
EDITOR

Nova Science Publishers, Inc.
New York

For permission to use material from this book please contact us:
Telephone 631-231-7269; Fax 631-231-8175
Web Site: http://www.novapublishers.com

Library of Congress Cataloging-in-Publication Data

Quantum chemistry research trends / editor, Mikas P. Kaisas.
p. cm.
Includes index.
ISBN-13: 978-1-60021-620-6 (hardcover)
ISBN-10: 1-60021-620-X (hardcover)
1. Quantum chemistry. I. Kaisas, Mikas P.
QD462.Q346 2006
541'.28--dc22
2007002393

Published by Nova Science Publishers, Inc. ✢ New York

CONTENTS

PREFACE

Quantum chemistry is a branch of theoretical chemistry, which applies quantum mechanics and quantum field theory to address issues and problems in chemistry. The description of the electronic behavior of atoms and molecules as pertaining to their reactivity is one of the applications of quantum chemistry. Quantum chemistry lies on the border between chemistry and physics, and significant contributions have been made by scientists from both fields. It has a strong and active overlap with the field of atomic physics and molecular physics, as well as physical chemistry. This new book presents leading research in the field.

Chapter 1 - The size of 'molecules' varies from small systems up to huge complexes. In order to assess intermolecular interactions, the electronic structure of the subsystems should be described to a considerable extent. In many cases, however, the size of molecules exceeds what sophisticated methods such as quantum chemistry approaches have access to. In addition, the type of interactions among the components is usually unknown in detail. Using theoretical (such as ab initio) methods one can study the interaction of only two individual molecules to a high degree of accuracy.

A promising procedure to overcome these difficulties is the application of localization. In the localization method a given mathematical representation to describe the electronic structure of each of the systems participating in the interaction is transformed into another one. The localized representation allows thus to study the near surrounding of the interactions to a significant extent – if relevant geometry data are available. Nowadays, experimental measurements (NMR, crystallography) are usually able to provide data for the starting geometry.

Various quantum chemical program packages (GAUSSIAN, GAMESS) have been elaborated for studying molecular structures at the ab initio level. The software GAMESS-UK has already been proved especially useful for treating interacting molecular complexes. The van der Waals (vdW) bondings are of importance in human biomolecules, they are often responsible for stabilizing bondings between molecules. The study of these molecular interactions should contribute significantly to get a deeper understanding of important (bio)molecular processes.

It has already been pointed out that several one-electron properties are advantageous for studying vdW interacting structures. Some useful characteristics (including transferability) can be exhausted in these cases. The electric moments (such as dipoles) derived using one-electron operators have also a transferable feature. The dipole and higher order moments can be calculated in localized representation, too. The results obtained on the determination of electric moments in localized representation are promising. This is supported by various investigations performed in the authors groups. The subject of the authors contribution is to summarize the available references on intermolecular interactions and to present the various aspects of applying localized representation in this field. Computational results obtained in the authors works on interacting molecular structures are also given.

Chapter 2 - A brief overview of the author's line of scientific inquiry has been presented. Discussed are results of the establishment of predictive structure - property relationships, the study of interrelation between the substances reactivity and the molecules electronic structure, the elucidation of mechanisms of chemical reactions and the allowance for medium effects on their proceeding, the solving of other interpretational tasks of chemistry. The methods of different hierarchy and theory level, adequate to the problems to be solved, are used, from semiempirical to modern sophisticated ab initio and DFT methods, along with comprehensive approaches and techniques including the NBO, energetic deletion, AIM, orbital interactions, electrostatic potentials analyses, calculation of vibrational, electronic absorption and NMR spectra, scaling of vibrational frequencies, separation and analysis of normal vibrations, ZPVE, thermal and BSSE corrections, localization of transition states, IRC, SCRF, et al. The capabilities and potentialities of the current quantum chemistry have been considered. The problems of systematization and generalization of the information resources on natural sciences, on ecology are also involved.

Chapter 3 - The electronic structure problem is introduced and briefly reviewed within the context of density-functional theory. The use of numerical grid based schemes, circumventing the use of a basis set, will be discussed. The authors will briefly introduce their numerical grid methodology for use in electronic structure theory. This method, which is based upon a combination of the Smolyak cubature construction and a novel non-linear transformation, has been shown to be efficacious when applied to cubature of physically relevent functions. In this paper the authors extend these results to interpolation and differentiation; these being the other fundamental operations essential for a numerical solution of the electronic structure problem. The authors shall investigate the performance of their method as a function of spatial dimension, number of grid points and, across a wide range of relevant functions.

Chapter 4 - With the increased sophistication of semiconductor technology, it is now possible to create designer nanostructures named as quantum dots which display many of the characteristic properties normally associated with atoms and molecules. The ability to manipulate the size and shape of these quantum dots has opened up a wide range of possibilities and areas for examination. Transitions never observed in the spectra of natural atoms can be obtained from the artificial ones. Consequently, quantum dots may be used to build more efficient and precisely controlled lasers with otherwise inaccessible wavelengths, as well as vital components of nanoelectronic devices. It is also hoped that quantum dots may one day be able to help realize the dream of quantum computing.

Since these structures are made in the nano-meter regime, conventional classical or semi-classical theories and methods no longer apply. A pressing challenge to theoreticians is to provide an accurate description of the electronic properties of such nanostructures based on fully quantum mechanical theories. This chapter describes a newly developed method, the spin adapted configuration interaction method, for calculating multi-electron quantum dot structures. This method takes into account all perceivable interactions and full spin correlation effects, giving a level of accuracy not available in alternative approaches such as the density functional theory or the self-consistent Hartree-Fock method. It also has an advantage over the standard configuration interaction method using Slater determinants, in that a much smaller basis is needed since the spin eigenfunctions already have the required symmetry. The results presented in this paper, where the matrix elements are evaluated analytically and thus exactly, provide a benchmark against which one can test other approximate methodologies developed to study more complex systems.

Chapter 5 - The aim of this work is to show that it is possible to obtain a new approach of relativistic quantum mechanics based on relativistic classical field theory to study atomic and molecular systems. Recently, relativistic effects have been shown to be important to study several chemical systems. In this work, starting from the basic principles of Special Relativity Theory (SRT), without using the usual concepts of Quantum Mechanics (QM), the authors obtain the equation of motion for the studied systems. The interaction between two particles is introduced through their respective metric functions. This method is accomplished by taking the interaction among particles moving with the standing wave radiated by them in the four dimensional space-time. According to the classical theory of fields, the geometrical properties of the solutions are governed by a classical-like equation of motion. The eigenvalue equation is derived from an improvement of the relativistic variational method introduced by Schrödinger in 1926. As this work is developed in a classical formalism, the authors use real action functions and thus they can obtain well-defined trajectories, which can be useful to study chemistry reactions. As examples, this formalism is applied to one, two and three-body problems, yielding solutions of the equation of motion by integrals of action or by solving the associated differential eigenvalue equation. Both set of solutions reproduce the correct structure of the energy levels.

Chapter 6 - A molecular dynamics simulation of aqueous solutions at infinite dilution for several molecules was performed using Lennard-Jones 12-6-1 potentials to describe the solute-solvent interactions, and TIP3P to describe the water-water interactions. The Morokuma decomposition scheme of ab initio interaction energies at the SCF level and the ESIE solute atomic charges were used to reproduce the molecular parameters of the solute-water interaction potential. Some improvements, such as including the dispersion component evaluated at MP2 level in the potential function and the extension of this procedure to reactive systems, have been added in order to improve the molecular properties results (solvation energies, dielectric constant, relaxation time, hydration shell, hydrogen bond lifetimes, reaction and activation free energies, etc.). The results showed that the functions that use the EX-PL-DIS-ES interaction model lead to good values of the structural, dielectric, and energy properties when they are compared with those from using other potentials, and with the available theoretical and experimental data.

Chapter 7 - In this work, the authors report a theoretical study on the interactions between the 1,3-butanediol and the chiral hydrogen peroxide. The complexes formed between two isolate chiral hydrogen peroxide (R and S) and 1,3-butanediol have been investigated by using B3LYP method at different basis set levels. Four pairs of equilibrium structures have been determined, one of which are one-hydrogen bonded structure and others are cyclic double–hydrogen bonded structures at the largest basis level. The optimized geometric parameters, interaction energies and chirodiastatic energy for various isomers at different levels are estimated. The infrared spectrum frequencies and IR intensities for various complexes are reported.

Chapter 8 - The present chapter is a study in the evaluation of the Coulomb Potential (CP) generated by the product of two Slater Type Orbitals (STO). This study is motivated by the recent developements carried out in the authors research group, see refs. [1,2,3]. The methos presented in these references open several possibilities. Here, the efforts are directed to obtain approximated values of a parameter called Q, similar to that defined in [1], and also, to improve the evaluations in the intermediate auxiliary integrals which are analogous to those defined in [2]. The CP is represented as a series whose terms involve integrals which are functions of the approximated value for Q. This parameter is deermined in two differents ways: as an appropriated average and considering a model of the error of truncation of the series expansion. In both cases the value of CP corresponds to the truncated series evaluated at these particular values of the parameter. The interest in this approach resides in the fact that it is possible to perform all the necessary integrals in the complex plane, similarly as it is done in ref. [3]. This step is performed by using rational functions to approximate the nonoscilating parts of the integrands. The error observed in the obtained CP values can be explained from the errors involved in the rational functions used. The test calculations are performed for the ethylene molecule.

Expert Commentary

In: Quantum Chemistry Research Trends
Editor: Mikas P. Kaisas, pp. 3-5
ISBN: 978-1-60021-620-6

EXPERT COMMENTARY: CAN QUANTUM-MECHANICAL DESCRIPTION OF CHEMICAL BOND BE CONSIDERED COMPLETE?

Mihai V. Putz[*]
Laboratory of Computational and Structural Physical Chemistry, Chemistry Department, West University of Timişoara, Str. Pestalozzi No. 16, Timişoara, RO-300115, Romania

Following the famous papers bearing the name from which the above title was inspired [1, 2], there is time to address the question as to whether the chemical bond can be considered sufficient and necessary described by the quantum-mechanical paradigm of matter.

In this respect, firstly let's note that usually the chemical bond is referred to as the covalent bond where electrons are shared by the encountered atoms. However, we can generalize this picture to assume that the chemical bond is generated by a special field of the neighbouring electrons and nucleus such that the field geometry can be optimized. In this respect, the optimization procedure includes variational principles of energy [3], of chemical potential (i.e. electronegativity) [4], of chemical force (i.e. the chemical hardness) [5], or, more recently, of chemical action (i.e. averaged quantum fluctuation field) [6].

Certainly, electrons as well as atomic particles and atoms themselves behave in a complementary frame of wave and particles: they are wave through their influential field and particles by caring their mass. In these conditions, the nature of the chemical bond is compulsory quantum.

This is the necessary condition.

What about the sufficiency?

This part is approached by recalling that, indeed, the chemical bond implies a variety of situations that span from the small molecules in the gaseous phase up to the chemical reactions in lattice or the affinity among biomolecules.

Hopefully, what is usually considered a weakness of the quantum mechanics – namely the impossibility to asses a complete analytical solution for whatever potential governing

[*] E-mail address: mvputz@cbg.uvt.ro, mv_putz@yahoo.com. Tel:+40-256-592633, Fax: +40-256-592620

whatever collection of electrons and nuclei, becomes here the argument due to different levels of approximations that are suited for assaying various chemical situations. It is the same that when, having to describe a stone falling from an arbitrary altitude on Earth, we solve the movement within Newtonian rather than with special relativity theory, thus appealing to an approximate theory, that in no way departs from the observed reality.

As such, mutatis mutandis, combining the different approximation to the electronic density or wave function with the proper combination of the preferred optimization procedures one can arrive at the averaged reactivity path of electrons in bonding space. Moreover, we have to note that the recent emphasised quantum inequality principles of chemistry, namely the inequality of the chemical potential [7] and of the chemical force (i.e. the maximum hardness principle) [8], take further account on the quantum fluctuations that a chemical bonding process imply.

The impressive experimental evidence of the quantum fluctuations of electrons and electronic potential when the time of interaction practically freezes, through the so called ultra fast femto-spectroscopy [9], indicates that the quantum nature of the chemical bond is manifested both in space and time.

Consequently, based on the Wick transformation between time and temperature, $\tau = \hbar\beta$ [10], we can equally admit that at the very short times in the bonding process the equivalent local temperature becomes asymptotically high.

It is this the proof that since quantum fluctuations in bonding are accompanied by the rise of the local temperature, as the fluctuations are smeared out by the optimum reactive path through variational principles the bond is achieved through a stationary state in the low temperature limit.

This way, the quantum paradigm of physical reality furnishes a sufficient frame in which all relevant steps in bonding can be properly described and of which averaged quantum effects can be also evaluated, observed and measured. It is therefore the occasion to answer the issue at hand in the positive.

However, the remaining theme is to construct, as in physics there is the complete set of observations that describe the quantum motion in the Hilbert space, the minimum complete set of reactivity descriptors and of their variational principles to closely assist the quantum nature of bond and bonding.

With this goal, the quantum scenario based on electronegativity and hardness indices and principles stands as a natural first step in developing the quantum chemical space of reactivity [11].

Acknowledgement

This work belongs to the project "Quantum Unification of The Chemical Bond through Density Functional Electronegativity and Electronic Localization Function with Applications at Reactivity of The Complex Natural Nano-Systems" kindly supported by Romanian National Council of Scientific Research in Universities (CNCSIS) through the Grant AT/54/2006-2007.

References

[1] Einstein, A.; Podolsky, B.; Rosen, N. *Phys. Rev.* 1935, 47, 777-780.

[2] Bohr, N. *Phys. Rev.* 1935, 48, 696-702.

[3] Hohenberg, P.; Kohn, W.; *Phys. Rev.* 1964, 136, B684-B871.

[4] Parr, R. G.; Donnelly, R. A.; Levy, M.; Palke, W. E. *J. Chem. Phys.* 1978, 68, 3801-3807.

[5] Chattaraj, P. K.; Lee, H.; Parr, R. G. *J. Am. Chem. Soc.* 1991, 113, 1855-1856.

[6] Putz, M. V. Contributions within Density Functional Theory with Applications in Chemical Reactivity Theory and Electronegativity; Dissertation.com: Parkland, FL, 2003.

[7] Tachibana, A. *Theor. Chem. Acc.* 1999, 102, 188-195.

[8] Parr, R. G.; Chattaraj, P. K. *J. Am. Chem. Soc.* 1991, 113, 1854-1855.

[9] Felker, P. M.; Zewail, A.H. *Phys. Rev. Lett.* 1984, 53, 501-504.

[10] Kleinert, H.; Pelster, A.; Putz, M. V. *Phys. Rev. E* 2002, 65, 066128/1-7.

[11] Putz, M. V. in Progress in Quantum Chemistry Research, Erik O. Hoffman (Ed.), Nova Science Publishers, Inc., NY, 2007.

RESEARCH AND REVIEW STUDIES

In: Quantum Chemistry Research Trends
Editor: Mikas P. Kaisas, pp. 9-55
ISBN: 978-1-60021-620-6

Chapter 1

APPLYING LOCALIZATION: A TOOL FOR TREATING INTERMOLECULAR INTERACTIONS

C. Kozmutza[1], J. H. van Lenthe[2], L. Udvardi[1] and I. Varga[1]
[1]Department of Theoretical Physics, Budapest University of Technology and Economics, Budafoki ut 8. H-1111, Budapest, Hungary,
[2]Theoretical Chemistry Group, Chemistry Department, Utrecht University

Abstract

The size of 'molecules' varies from small systems up to huge complexes. In order to assess intermolecular interactions, the electronic structure of the subsystems should be described to a considerable extent. In many cases, however, the size of molecules exceeds what sophisticated methods such as quantum chemistry approaches have access to. In addition, the type of interactions among the components is usually unknown in detail. Using theoretical (such as *ab initio*) methods one can study the interaction of only two individual molecules to a high degree of accuracy.

A promising procedure to overcome these difficulties is the application of localization. In the localization method a given mathematical representation to describe the electronic structure of each of the systems participating in the interaction is transformed into another one. The localized representation allows thus to study the near surrounding of the interactions to a significant extent – if relevant geometry data are available. Nowadays, experimental measurements (NMR, crystallography) are usually able to provide data for the starting geometry.

Various quantum chemical program packages (GAUSSIAN, GAMESS) have been elaborated for studying molecular structures at the *ab initio* level. The software GAMESS-UK has already been proved especially useful for treating interacting molecular complexes. The van der Waals (vdW) bondings are of importance in human biomolecules, they are often responsible for stabilizing bondings between molecules. The study of these molecular interactions should contribute significantly to get a deeper understanding of important (bio)molecular processes.

It has already been pointed out that several one-electron properties are advantageous for studying vdW interacting structures. Some useful characteristics (including transferability) can be exhausted in these cases. The electric moments (such as dipoles) derived using one-electron operators have also a transferable feature. The dipole and higher order moments can be calculated in localized representation, too. The results obtained on the determination of electric moments in localized representation are promising. This is supported by various investigations performed in our groups. The subject of our contribution is to summarize the available references on intermolecular interactions and to present the various aspects of applying localized representation in this field. Computational results obtained in our works on interacting molecular structures are also given.

I. Introduction

Recently the experimental technology has been advanced to a great extent. No better evidence should be provided but the Nobel prizes such on NMR technique. The experimental methods lean on two auxiliary pillars. One of them involves the application of an advanced level of the statistical analysis. The other pillar emerged through the application of quantum mechanics invoking the calculation of the electronic structure. The main aim of the calculations of quantum chemistry is to supplement the experiments. Moreover, the theoretical predictions are supposed to give information in cases when the experimental technology is not reliable enough or is burdened with unacceptable inaccuracy. The goal of prediction of the theoretical calculations has only been reached for smaller molecules. On the other hand, the interpretation of a number of biological processes has been performed considering theoretical investigations.

The interaction between closed shell molecules plays an important role in biophysical and biochemical processes. These intermolecular interactions affect the molecular conformation of several very important molecules as well as their functioning (e.g. DNA). These systems serve as prototypes of the demand that not only the constituents should be known but also the weak interactions between them. The consideration of H-bonding in molecular systems with weak interactions of van der Waals type is unavoidable. The description of H-bonds is by all means essential in order to theoretically supplement the experimental data obtained for these systems. Nowadays using medium size basis sets one can account for the properties of the H-bonds at Hartree-Fock level with satisfactory precision. The reason for this is the relatively large electrostatic effect of the H-bond. With the most up-to-date software, e.g. Gamess-UK, running on hardware with the largest capacity and most advanced architecture, e.g. GRID, the electron number to be considered in order to deal with the intermolecular interactions is still beyond the present possibilities. The tendency will be the same in the future, too, even if the limits shift gradually. Therefore further ideas ought to be brought in based on the combination of the variety below:

a) Transferability and additivity
b) Definition of well-defined and transferable quantities
c) Collection, systematization and application of appropriate (measured and calculated) data

As far as the a) point is concerned, this is a viable possibility according to historical survey, especially when a well-founded method is used. One of them is the application of localized molecular orbitals (LMO) based on localization techniques. According to b) it is suggested to introduce one-electron contributions of the LMOs, which can be calculated at Hartree-Fock level with acceptable precision even with medium size basis set. The last point, c), involves the construction of databases using the results of a) and b) together with experimental data with the possibility of constant renewal and upgrade.

Our present work would contribute to this field. The theoretical aspects to points a) and b) will be monitored. The tasks related to point c) will be summarized in another contribution.

II. Historical Overview

The theoretical investigation of atomic and small molecular systems consisting of 2-4 atoms has been started right at the cradle of quantum mechanics. The fast development of the mathematical techniques and the inclusion of computational systems made it possible to theoretically study larger and larger molecular and interacting systems. Nowadays *ab initio* techniques are already available for the calculation of properties of the electronic structure of relatively large molecules. It is obvious that most calculations of large scale molecules is performed using the Hartree-Fock approximation. The technological advance of the computer hardware does not allow for the application of the HF technique for the calculation of large, weakly interacting molecular systems. The description of intermolecular, van der Waals (vdW) type, bonding must always be supplemented with additional techniques, i.e. the electronic correlations must be taken into account. The investigation of H-bonds appearing in weakly interacting systems can be done at a satisfactory level when the electrostatic contributions are significant. In the case of extended systems (this holds for single molecules or vdW systems) two major steps should be considered with great care:

a) The choice of the applied basis set and the *a priori* (or *a posteriori*) correction of this set when applicable.
b) Rational reduction of the numerical work which strongly depends on the problem at hand and the available technical background.

The one-particle Hartree-Fock states in molecules and interacting molecules are usually constructed as the linear combination of analytical functions, the LCAO-MO-SCF method. The coefficients are determined using a variational technique. This method requires the careful selection of the basis functions. Several different types of basis sets have been introduced and put forward during the past decades, the best-known of them are combinations of Slater- type and Gauss-type basis sets [1-13]. One of the most significant sets was introduced by Huzinaga, Pople and Dunning [14-24]. A particular basis set is the universal set [25-28]. Several atomic basis sets have been used and tested for the description of a great variety of molecules [29-35]. Various works dealt with the mathematical foundations of the complete basis set problem in order to calculate the Hartree-Fock energy [36-44]. Many times the so-called virial coefficients help to analyze basis sets. The quality of the basis set can be connected with the virial ratio [45].

The *ab initio* investigation of the interaction between atoms and molecules has contributed to the understanding of the intermolecular forces. The applied approximations can be cast into two major groups. One of them considers the interaction as a perturbation to the constituting parties. The other method, also known as the supermolecule approach, calculates the total energy of the whole system and the interaction energy is obtained as the difference between that of the constituents and the total energy. The latter method operates with the small difference of large numbers, hence imposes severe difficulties. One of these difficulties is the appearance of the basis set superposition error (BSSE). Several basis sets have been worked out especially applied for the description of intermolecular interactions [46-52]. The BSSE problem has been dealt with and the possibilities of its elimination have been analyzed [53-67]. Molecular systems including H-bond dimers have been investigated together with the effect of the BSSE using different types and quality of methods [68-82]. Several attempts have been performed in order to investigate interactions of biological importance. The number of *ab initio* calculations in this field is constantly increasing [83-85]. However, many times lower level methods, e.g. molecular modeling, is applied for the study of such systems.

The *ab initio* methods require the calculation of one- and two-electron integrals between the given basis functions. At the first generation of such calculations the computational time was determined by the number of two-electron integrals. For the case of a set of N basis functions the essential number is N^4, i.e. this is the scaling factor. Several attempts have tried to reduce this large number; a major breakthrough was the introduction of the so-called direct SCF method. The number of integrals has not been reduced, however, the input/output operation has been eliminated completely. Several attempts aimed at the reduction of computational time and memory requirements. Beyond the calculation of integrals [86-94], the applications of the direct SCF technique [95-100] and the application of symmetries were helpful, as well [101-105]. Two remarkable methods should be mentioned [106-109] and [110, 111] together with three programs developed [112-114]. Some applications were supplemented by computer usage data [115, 116].

The molecular orbitals determined using the canonical LCAO method extends over the whole molecule or the supermolecular system. Other types of molecular orbitals may be defined as well, and that could bear some advantages. In particular the so-called localized molecular orbitals seem to be favorable as it has been confirmed by several research projects. Various criteria allow for a definition of localized molecular orbitals (LMOs). The pioneering works of Boys and Ruedenberg have been applied [117-130]. Both intrinsic and external criteria can be defined [131-140]. Several methods are in particular notable: some are interesting, some are useful [141-143], [144-148], [149-152], [153-155]. Many researchers have been dealing with the localization of virtual orbitals, especially for the application of the CI [156-164]. Other methods have been worked out for the investigation of molecules and their properties especially for atoms in molecules [165-171]. In condensed matter localized (Wannier) representation was used in [172, 173]. The localized methods utilized for a large variety of molecular and interacting molecular structures are discussed in detail in a forthcoming chapter of this work.

The classical work of Heitler and London about the H_2 has been published in 1927. During subsequent decades the H-bond in two- and many-atomic molecules have been investigated intensively. Great attention has been paid for the theoretical study of the water dimer. Several works deal especially with the understanding of the chemical bonding, with the partitioning of the bonding energy. The energy components appearing in the formation of the vdW bonds have been investigated, as well.

A nůmber of theoretical approximations have been worked out in order to understand the interaction between molecules and atomic systems. The numerical determination of the potential energy of the interaction between atoms and molecules at an acceptable level requires the application of methods beyond the precision of the Hartree-Fock technique. In order to accomplish this task several researchers have investigated the contribution of electronic correlations to the intermolecular forces. The so-called vdW minima, i.e. the equilibrium distances have been calculated for diatomic molecules, first for H_2, He_2 [174-180]. The interaction between noble gas atoms have been calculated by several researchers [181-183], likewise for diatomic molecules [184-188]. The effect of alkaline metallic atoms was especially treated for some special properties [189-190]. A number of calculations are known for the water dimer interaction [191-206]. The intermolecular interactions in clusters of small ($n<8$) water molecules have been performed in [207-225].

The water dimer as a prototype has been investigated in detail and also mainly the different ten-electron dimer structures. Several basis sets have been applied for dimers in order to find out if single or double H-bonds are formed. Many works have been devoted for the investigation of hydrogen-fluorid [226-234]. The ammonia dimer shows peculiar characteristics. Therefore much work has been devoted to understand in the $(NH_3)_2$ dimer the proton's donor and acceptor properties [235-239]. The interaction between the methane molecules has been investigated, as well, together with the interaction between the CH_4 and other molecules [240-244]. Many works deal with the theoretical understanding of the H-bonds. The literature considering the formation of H-bonds in smaller and larger molecules is exceedingly large, we may mention only a part of it [245-253].

The investigation of weakly interacting molecular systems leads to the study of vdW bonds. Such bonds are related to the conventional bond-theories with properties that can be transferred to the case of H-bonds. After the elementary theories [254, 255] a major improvement was the applicability of quantum mechanics for the analysis of molecular bonds [256-261]. Many work focused on the determination of the properties of hydrogen bonds and the intermolecular interactions [262-270]. The energetic aspect of the intermolecular interactions was worked out in several papers [271-276] in terms of the factorization of the bonding energy into components. The DFT method, which has been worked out for the investigation of the weak interaction of molecular systems with large number of molecules[277-282], has failed many times. Attempts for the construction of intermolecular potentials were sometimes successful [283-285].

Beyond the investigation of vdW bonds in dimer structures trimer structures have been investigated, too. It is conceivable that beyond the two-body interactions three-body and higher order interactions could be as important. In order to understand these effects investigations have been performed for the profound understanding of the structure of liquid water and solid ice using clusters containing several numbers of monomers. With respect to biological importance several interesting molecules, mainly solvatation (water) shell was the topic of several researches.

Besides the noble gas dimers (homogeneous and heterogeneous) trimer structures were also investigated. The main goal for these studies was the understanding whether the so-called non-additive cooperation effects do influence and to what extent the total energy. The electronic structure of clusters consisting of three, four, or more atoms was determined [286-291]. A variety of molecular clusters was treated in [292-302]. Trimers formed by water molecules have been discussed, as well as, larger clusters in [303-308]. A number of studies focused on the understanding of the structure of the ice [309-314]. The non-additive effects were dealt with by several researchers, especially for the case of liquid water [315-323]. Interactions in one- and two-dimensional systems were treated in [324-326].

Molecular interactions play an essential role in biophysical and biochemical processes. For instance these interactions are responsible for the structural formation of the DNA. The structural properties of biomolecules can be investigated experimentally both in solid phase and in liquid (solved) phase. The theoretical extension of these data is in the frontline of research in order to widen our knowledge about these processes. Interactions of several biologically important molecules have been intensively studied [327-332]. The quest for the enzymes was performed in several works [333-337]. To take into account the solvent (primarily water) has been a research topic, first the environment of ions and small molecules were investigated in the presence of water [338-346]. The clarification of biological and biochemical processes intensive studies of the behavior of molecules, their properties and characteristics have been performed using diverse theoretical methods [347-359].

Several experimental techniques have been worked out for the investigation of intermolecular bonds. The knowledge of the parameters of the constituents facilitates the measurement of the characteristics of the compound system more precisely. Many works have been published about smaller (ten-, twenty-electron) molecules [360, 361], and furthermore about more complex systems [362-366]. Intensive experimental work has been performed for the structural analysis of ice especially in recent years [367-371]. Molecules with peptide bonds that bear biophysical and biochemical significance have been investigated experimentally [372-378]. Specific techniques allow for the reconstruction of molecular orbitals which makes it possible to complement theoretical and experimental research [379].

III. Methods for Localization

The introduction of localized molecular orbitals (LMOs) has given an important contribution to come closer to chemical intuition, to better understand the transferability. It has also lead to a convenient study of the additive properties of certain quantities derived from LMOs. The „localization" of electron density in atomic and molecular systems has mainly been carried out in the framework of the independent particle model. The significant part of calculations has been performed for closed shell structures.

In order to prepare LMOs, several methods have been elaborated. Depending on the criteria used, the methods can be divided into two main groups: internal and external localization methods can be distinguished. There are procedures, which obtain LMOs by a suitable transformation of the canonical orbitals (= internal localization criterion). In the methods using certain external localization criterion, some further restrictions are taken into consideration. The procedures which transform the Hartree-Fock operator are called direct (internal) methods.

One of the most famous methods is that of Edmiston and Ruedenberg, based on the generalization of equivalent orbitals, using the one-determinant approximation. The one-determinant wave function for closed-shell system is invariant to any unitary transformation of the $\{\varphi_i\}$ canonical molecular orbitals (CMOs). This holds to a phase factor of unity. Both the Coulomb and the exchange interaction energy terms contain the same quantity. A suitable **U** transformation can thus be chosen for finding a maximal value for $D\{\varphi\}$. The resulting localized orbitals applying this procedure, $\{\overleftarrow{\varphi_i}\}$, are called energy localized molecular orbitals. The $D\{\overleftarrow{\varphi}\} = \max$ relation can be interpreted as the „concentration" (to a maximal extent) of the electric charge distribution of atomic/molecular systems to certain spatial regions.

The transformation procedure implied by the Edmiston-Ruedenberg method requires the matrix transformation of N^4 elements (in a $2N$ electron system). In case of extended structures this requirement makes the practical utilization of this procedure tiresome and restricts its wide use.

The elaboration of the exclusive orbitals was originally proposed by Foster and Boys. This procedure for finding localized molecular orbitals using a convenient unitary transformation has been modified by Boys. The Boys method is often used for producing LMOs. The criterion applied in the procedure of Boys is defined as follows. The $\eta = \sum_{i \neq j} \left| \mathbf{r}_i - \mathbf{r}_j \right|$, i.e. the distances of the centroids of charge vectors of the orbitals should be the maximal. This requirement is equivalent to the criterion that the quadratic self-repulsion of the orbitals is also the maximal:

$$\sum_i \left\langle \overline{\varphi}_i^{\,2} \middle| r_{12}^2 \middle| \overline{\varphi}_i^{\,2} \right\rangle = \max .$$

Using the method of Boys the matrix of N^2 corresponding elements of the first order moment operator should be transformed. This is a rather attractive property of this method, which is, therefore, often used for preparing LMOs.

The method of charge- (or density-) localization is also based on a suitable unitary transformation of the canonical molecular orbitals. Its starting point is that the electron density corresponding to a one-determinant wave function has the following form

$$\rho = 2\sum_i |\varphi_i|^2 .$$

In cases of LMOs, the $\left|\overline{\varphi}_i\right|^2$ terms are the contributions of the orbital charge density. Presumably the „charge-overlap" is small between well-localized charge densities. The localization criterion is, therefore, formulated for minimizing the $\sum_{i \neq j} \left\langle |\varphi_i|^2 \middle\| \varphi_j|^2 \right\rangle$ expression. An alternative (equivalent) choice is to maximize the $\sum_i \left\langle \varphi_i^2 \middle| \varphi_i^2 \right\rangle$ term. The charge localized method has been applied successfully in the study of extended systems, too.

Localized orbitals can be obtained directly as the solutions of an eigenvalue problem, obtained by a suitable modification of the HF operator. Adams direct localization method is based on the assumption that the basis functions are not necessarily and mutually orthogonal. Although the basis functions are linearly independent and normalizable, for obtaining localized orbitals, the previous degree of liberty can be exhausted.

The direct method of Gilbert leads to a special eigenvalue problem by the introduction of a localization potential. The solutions of the equation corresponding to the above problem provide directly localized orbitals. The formalism used in this method has, furthermore, an additional theoretical value. Applying that formalism every localization method can be interpreted in a common way. This is due to the fact that the various localization criteria correspond to functionals of extreme value problems. These problems can be solved by using the variation method, and the solutions will lead to the „best" localized orbitals (according to the localization criterion actually used).

An external localization method was proposed by Magnasco and Perico. The localization criterion implies the maximization of the atomic population, obtained by using the Mulliken population analysis. The number of localized orbitals as well as their association to a given atom should be a priori defined.

The method of Peters starts with approximate orbitals, and then the localized orbitals are prepared in a consecutive way, using an iterative procedure.

An interesting idea was formulated by using statistical consideration (loge theory). The loge theory proposes to divide the spatial region into cells. These cells are determined by maximizing the probability of that a given number of electrons are in a given cell. One can also minimize the „missing information" on the system (according to the cells).

One of the most often used practical methods nowadays is that of proposed by Pipek and Mezey. This intrinsic localization procedure is based on a mathematical measure, and no external criteria are to be defined a priori for the orbitals. To one hand, in certain sense the method is similar to the Edmiston-Ruedenberg localization method. It works, on the other hand as economically as the Boys procedure does. The determination of atomic overlap integrals of the given molecular structure is sufficient for the application of the localization algorithm. This feature makes possible the adoption of the algorithm to both ab initio and semiempirical methods. The implementation of the procedure in existing program systems is, furthermore, quite easy. Due to this property, the Pipek-Mezey method is rather attractive for practical use in several cases.

One of the most useful methods is that introduced by Saebo and Pulay. The local (correlation) method has been elaborated for treating electron correlation in a straightforward manner. The method is designed as most of other localization methods, especially for closed shell systems, and it works in two steps. The internal, strongly occupied canonical orbitals are localized using a standard localization technique, by applying the Boys or the Pipek-Mezey procedure. As to the external, virtual orbitals, here atomic basis functions are used projected against the occupied canonical orbitals. One of the most attractive features of the localized correlation method is the significant computational saving. It was also pointed out, that the degree of arbitrariness (while neglecting certain contributions) is decreasing with increasing basis set.

The above methods were elaborated for unique molecules. In addition, an important role of the localization of the electronic structures has been established while investigating weakly interacting atomic and molecular systems. In these cases the transferability of localized units between the monomers and the complex (named also supermolecule) makes possible a straightforward study of the given electronic structure. The application of a localization procedure can be performed in two different ways, depending on the goal of the study: local effects or energetic (both total and contributing) terms are to be analyzed.

First, the LMOs can be determined in both the monomers as well as in the supersystems, and then the corresponding contributions are compared. The transferability of LMOs from the non-interacting monomers to the supermolecule can be performed in various ways. One of the most promising procedures is the comparison of some „local" quantities (these are derived from LMOs) in the corresponding systems. The one-electron contributions such as electric moments turned out to be especially useful for this purpose. A significant bias from transferability suggests some sensitivity (a site of interaction ability, e.g.) of the given region in the supermolecule. The variation of the bond and lone pair LMO densities can be analyzed as the function of the intermolecular distance. The consideration of the chemical intuition might be especially helpful in certain cases.

Second, the localization does not necessarily imply the localization of the electron density into small spatial regions, such as provided by LMOs. The localizability can also be interpreted as the separation/concentration of the charge density according to the „whole" spatial region of each interacting monomers. The regions in these cases are not as small as for localized molecular orbitals and LMO densities. The separation of the charge densities in weakly interacting systems according to the „whole region" of the contributing monomers opens a way to assess intermolecular interaction energy. Such a partition procedure makes, namely, possible the calculated electron densities of the supermolecule be directly compared to those resulted in the non-interacting monomers. The partitioned energy terms imply not only the total, but the various contributing terms as well. The partitioned one-electron potential and kinetic energy quantities can also be analyzed this way. Due to the suitable partitioning, the deviations from transferability are expected to reflect reasonably the polarization effects. The latter effects (formulated in terms of various energy quantities) allow estimating the intermolecular forces in the given supermolecule.

The electron density in weakly interacting atomic and molecular systems can be localized using the standard methods listed above. Those methods, however, do not ensure automatically a suitable comparison of LMO densities determined in the monomers and in the supermolecule. This is simply due to the fact that a calculation using a given localization procedure is carried out for each molecular structure separately. A unitary transformation, namely, does not restrict each parameter of the LMOs (it depends on the degree of liberty allowed by the spatial symmetry). In addition, the localization of the supermolecule often leads to convergence problems, especially when using extended basis sets. Both of these difficulties could be overcome by using a convenient localization procedure.

The essence of the method named Separation of Molecular Orbitals (SMOs) lies in obtaining the „closest possible" wave functions in the complex (supersystem) and in the monomers. The canonical representation is an excellent choice for this purpose, as offering a simple manner for a comparison between the CMOs of monomers as well as the CMOs of the supermolecule. This is possible, however, only in case when the CMOs of the supermolecule could be established in a form similar to those determined in the non-interacting monomers. To other word, the CMOs of the supersystem ought to be distributed according to the non-interacting CMOs. This can be reached by applying the SMO method reported by Kapuy and Kozmutza. The SMO method has been elaborated for studying weakly interacting molecules, i.e. for structures composed from units, whose wave functions do not alter significantly during the formation of the supersystem. This requirement implies that the deviation from transferability is attributed to the effects arising due to interaction. Although this assumption looks „sharp", it is expected that for related systems would certainly hold. This has been proved by calculations performed for related molecular structures and reported in literature. The results obtained allowed, that (at least) the magnitude of the interaction ability could be predicted to a rather good extent.

The procedure for obtaining SMOs starts from the occupied and virtual canonical molecular orbitals (CMOs) of the supersystem. It uses a bridge in order to transform these

CMOs towards a prescribed set of orbitals. The criterion is related to maximize the overlap between the corresponding CMOs. The criterion applied ensures that the overlap between the constructed orbitals SMO*i* of a supermolecule should be maximal with the initial, canonical orbitals XMO*i* of the non-interacting molecules. The bridge between the sets of CMOs implies an overlap criterion:

$$\sum_{i=1}^{N} \left| \langle \Phi_i | \mathrm{K}_i \rangle \right|^2 = \max$$

The set of resulting SMOs of the supermolecule can thus be compared - as the slightly 'modified' sets of monomer CMOs - to the original CMOs in the contributing monomers. On one hand, the extent of transferability of CMOs reflects the magnitude of the interaction. In order to describe the interaction ability in a quantitative way, on the other hand, it is straightforward to analyze the several energy terms as derived from the SMOs obtained.

At the Hartree-Fock SCF level the closed shell total energy has the following form:

$$E_{\mathrm{HF}} = E_{\mathrm{nuc}} + 2\sum_{i} (T_i + V_i) + \sum_{i,j} (2J_{ij} - K_{ij})$$

where

$$E_{\mathrm{nuc}} = \sum_{\substack{a,b \\ a>b}} Z_a Z_b R_{ab}^{-1} ,$$

$$T_i = -\frac{1}{2} \int \psi_i^*(\mathbf{r})(\Delta \psi)(\mathbf{r}) \mathrm{d}^3\mathbf{r} ,$$

$$V_i = -\sum_{a} Z_a \int \psi_i^*(\mathbf{r}) |\mathbf{r} - \mathbf{R}_a|^{-1} \psi_i(\mathbf{r}) \mathrm{d}^3\mathbf{r} ,$$

$$J_{ij} = \iint \psi_i^*(\mathbf{r}_1) \psi_j^*(\mathbf{r}_2) |\mathbf{r}_1 - \mathbf{r}_2|^{-1} \psi_i(\mathbf{r}_1) \psi_j(\mathbf{r}_2) \mathrm{d}^3\mathbf{r}_1 \mathrm{d}^3\mathbf{r}_2 ,$$

$$K_{ij} = \iint \psi_i^*(\mathbf{r}_1) \psi_j^*(\mathbf{r}_2) |\mathbf{r}_1 - \mathbf{r}_2|^{-1} \psi_j(\mathbf{r}_1) \psi_i(\mathbf{r}_2) \mathrm{d}^3\mathbf{r}_1 \mathrm{d}^3\mathbf{r}_2 .$$

Using SMOs this expression can be separated into one- and two-body terms by

$$E_{\mathrm{HF}} = \sum_{A} E_{\mathrm{HF}}^{A} + \sum_{\substack{A,B \\ A>B}} E_{\mathrm{HF}}^{AB} ,$$

where

$$E_{\mathrm{HF}}^{A} = E_{\mathrm{nuc}}^{A} + 2\sum_{i\in O_A}\left(T_i + V_i^A\right) + \sum_{i,j\in O_A}\left(2J_{ij} - K_{ij}\right)$$

and

$$E_{\mathrm{HF}}^{AB} = E_{\mathrm{nuc}}^{AB} + 2\left(\sum_{i\in O_A} V_i^B + \sum_{i\in O_B} V_i^A\right) + 2\sum_{\substack{i\in O_A\\ j\in O_B}}\left(2J_{ij} - K_{ij}\right).$$

A and B label the monomers, O-s are the sets of occupied orbitals according to monomer A or B, respectively, and

$$E_{\mathrm{nuc}}^{A} = \sum_{\substack{a,b\in N_A\\ a>b}} Z_a Z_b R_{ab}^{-1},$$

$$E_{\mathrm{nuc}}^{AB} = \sum_{\substack{a\in N_A\\ b\in N_B}} Z_a Z_b R_{ab}^{-1},$$

$$V_i^A = -\sum_{a\in N_A} Z_a \int \psi_i^*(\mathbf{r})\left|\mathbf{r} - \mathbf{R}_a\right|^{-1}\psi_i(\mathbf{r})\mathrm{d}^3\mathbf{r}.,$$

where N_A (N_B) denotes the indices of the nuclei for monomer A (B). The partitioned energy terms can be derived separately for each contributing monomer. The sum of the terms per each subunit corresponds to a partitioned total energy quantity for the given subsystem.

Using the monomers' contributions, at the Hartree-Fock SCF level one can thus describe the various partitioned energy terms. For a weakly interacting structure consisting from monomer A and B, the dimer (AB) will have the following terms:

$E_0^{A \text{ or } B}$	the nuclear-nuclear interaction on monomer A/B (N/N),
E_0^{AB}	the nuclear-nuclear interaction between A and B (N/N),
$2\sum_{i\in O_A \text{ or } O_B} T_i$	the kinetic energy on monomer A/B (EKIN),
$2\sum_{i\in O_A \text{ or } O_B} V_i^{A \text{ or } B}$	the electron-nuclear attraction on monomer A/B (E/N),
$2\sum_{i\in O_A} V_i^B$	the electron (on A)-nuclear (on B) attraction (E/N),
$2\sum_{i\in O_B} V_i^A$	the electron (on B)-nuclear (on A) attraction (E/N),
$\sum_{i,j\in O_A \text{ or } O_B}\left(2J_{ij} - K_{ij}\right)$	the Coulomb + exchange interaction on monomer A/B (E/E),

and

$2\sum_{\substack{i \in O_A \\ j \in O_B}} \left(2J_{ij} - K_{ij}\right)$ the Coulomb + exchange interaction between A and B (E/E).

The definitions show the decomposed terms are invariant in cases the SMOs are subjected to a unitary transformation such as U^A, U^B, ..

$$\psi_i' = \sum_{j \in O_A} U_{ji}^A \psi_j, \quad \text{if} \quad i \in O_A,$$

$$\psi_i' = \sum_{j \in O_B} U_{ji}^B \psi_j, \quad \text{if} \quad i \in O_B.$$

The decomposition scheme presented above can be simply extended to the inclusion of three, four, etc, monomers. The comparison of the corresponding terms obtained from the supermolecule calculation and the sum of those resulted for the non-interacting monomer allows a detailed, two-fold analysis. First, the change of the electronic structure of the given subunit when entering into weak interaction provides useful information on the role of the given monomer. Second, making the differences of the values calculated for the various energy terms (see above, such as kinetic, potential) for the monomers and for the supermolecule, estimation could be given for the interaction energy values. This prediction does not suffer significantly from the basis set superposition error (as it has already been reported).

The various localization methods, the procedures applied aim at a suitable study of the electronic structure of monomer (or dimer, trimer, etc,) systems considering usually the chemical intuition. In this way the investigation of more and more extended, related systems could be carried out. The applications of different localization procedures listed above on various systems are continuous. It should be mentioned, that in addition to the procedures discussed, recently the elaboration of new localization methods is rarely reported in literature.

IV. Electric Moments of Charge Densities

The charge distribution of a quantum chemical system can be expanded in terms of its multipole electric moments. The zeroth order moment equals the total charge Z

$$Z = \sum_i e_i \, .$$

The first order electric moment is the dipole moment $\boldsymbol{\mu}$

$$\boldsymbol{\mu} = \sum_i e_i \mathbf{r}_i \, .$$

These electric moments in general are not invariant under the shift of the origin of the coordinate system. However, the first non-vanishing moment is independent of the choice of the coordinates. The origin of coordinate system is thus often chosen as the center of gravity (or the center of the charge distribution).

In case of molecular structures the first order electric moments can be expressed in the following way (in atomic units):

$$\mu_u^{\mathrm{e}} = -\sum_{i=1}^{N} \langle \Phi | r_{ui} | \Phi \rangle \,, \; u = x, y, z, .$$

Here Φ denotes the eigenfunction of the system (N is the number of electrons). The contributions of the nuclei are

$$\mu_u^{\mathrm{n}} = \sum_{a=1}^{n} Z_a \delta(r_u - r_{ua}) r_u \,, \; u = x, y, z \,,$$

and they are determined by the position ($\mathbf{r}_a$) and the charge (Z_a) of nuclei. The total dipole moment of a molecular system is obtained by a summation here in each of three coordinate directions:

$$\boldsymbol{\mu} = \boldsymbol{\mu}^{\mathrm{e}} + \boldsymbol{\mu}^{\mathrm{n}} .$$

In the framework of the one-determinant approximation of the independent particle model the first order properties can be written by summing up the corresponding contributions. The electric moments can, therefore, be written as the sum of contributions from the individual orbitals. The first order moments will have the following form:

$$\mu_u^{\mathrm{e}} = \sum_{k=1}^{N} \mu_u^{\mathrm{e}}(k) = -\sum_{k=1}^{N} \langle \varphi_k | r_u | \varphi_k \rangle = -\sum_{k=1}^{N} r_u(k), \; u = x, y, z \,,$$

where φ_k denotes the k-th one-particle function. Subjecting the orbitals to a unitary transformation (symmetry transformation, e.g.):

$$\overline{\varphi}_j = \sum_{k=1}^{N} U_{jk} \varphi_k \text{ for all } j \,, \; j = 1, N \,,$$

the individual components $r_u(k)$ will change but their sum $\boldsymbol{\mu}^{\mathrm{e}}$ remains invariant. The first order moments are called the centroids of charge. To other word, the centroid length for $\overline{\varphi}_k$ is the scalar quantity $|\mathbf{r}(k)|$. By the name centroid the endpoint of the vector $\mathbf{r}(k)$ is meant.

The second and higher order electric moments can be defined in various ways.

In case of localized densities the centroid of the orbitals can be taken as the center of the quadrupole tensor. Due to the shifting the individual values of the components will change.

For a given localized orbital, $\bar{\varphi}_k$, the components $\bar{Q}_{uv}\ k$ can be determined in the following way:

$$\bar{Q}_{uv}(k) = \left\langle \bar{\varphi}_k \left| [r_u - r_u(k)][r_v - r_v(k)] \right| \bar{\varphi}_k \right\rangle,\ u,v = x,y,z,$$

where $r_u(k)$ and $r_v(k)$ are the components of the centroid vector of orbital $\bar{\varphi}_k$ (the index e is omitted). The shifted components define a quadrupole moment tensor **Q** for the individual localized orbital.

The moments of a charge distribution can be related to the statistical moments of random variables. The orbital centroid vector is, namely, identical to the central first moment $M(\xi)$ of a random variable ξ with density function $p(x)$:

$$M(\xi) = \int_{-\infty}^{\infty} xp(x)dx.$$

The definition of each diagonal element of the orbital second-moment tensor is formally identical to that of the statistical central second moment, m_2:

$$m_2 = D^2(\xi) = \int_{-\infty}^{\infty} p(x)[x - M(\xi)]^2 dx,$$

where $D(\xi)$ is the standard deviation of ξ. The centroid lengths and the elements of the diagonal quadrupole tensor are thus denoted by $\langle r \rangle$, $\langle x'^2 \rangle$, $\langle y'^2 \rangle$ and $\langle z'^2 \rangle$, respectively.

For the standard deviation, $D(\xi)$, the inequality holds (according to the Bienaymé-Chebyshev theorem):

$$P(|\xi - M(\xi)| > \lambda D) \le \frac{1}{\lambda^2},\ \lambda \ge 1.$$

The expression $P(|\xi - M(\xi)| > \lambda D)$ denotes the probability that $|\xi - M(\xi)|$ is larger than λ times the standard deviation. Taken $\lambda = 2$ as an example, then according to the above inequality, the probability that ξ lies within the range $-2D(\xi) \le \xi \le 2D(\xi)$ is at least 0.75. For $0 \le \lambda \le 1$ the inequality still remains valid. When the probability function is Gaussian,

$$P(|\xi - M(\xi)| > D) = 0.3174\ldots,$$

value can be obtained.

Although the above inequality is not „sharp", it gives some limiting value of the dispersion of the random variable ξ, even if its distribution function is not explicitly known. A similar relation applies to the eigenvalues $\langle x'^2 \rangle_i, \langle y'^2 \rangle_i, \langle z'^2 \rangle_i$ of the LMO second-moment tensor; hence, those are analogues of the statistical squared standard deviations. One thus can conclude that, in principle, the main features of the spatial extension of LMO's can be described by the first and second moments of their charge distribution. Hence the changes in the above moments are in relation with the change in the chemical environment of the given LMO.

One thus expects, that the main features of the spatial extension of LMOs can be described by the use of their electric moments. The first and second moments of the localized charge distribution provide some useful information about their 'region' i.e. where the given LMO is situated. From this it also follows, that the changes in the above moments might be in relation with the change in the chemical environment of the given LMO. In most cases useful values have been obtained, as one could draw the conclusions in the view of the calculations reported in this field.

A method based and elaborated on the use of the electric moments of LMO densities of structures in weak interaction such as van der Waals (vdW) systems thus should be a valuable procedure. It allows studying the change of the neighborhood of localized densities due to the effects caused by weak interaction. The stability of the method against the restricted size of the basis set one has to apply as the systems become larger and lager is ensured by the one-electron nature of the quantities investigated. The Hartree-Fock level, in addition, can nowadays be tackled using several thousands of basis functions. The software and hardware is intensively developing, the method on using LMO moments for studying vdW structures is thus ahead of a wide application in the near future.

V. Model Calculations on the Methane + Water Dimer

Based on the historical survey we may conclude that despite of considerable efforts present techniques do not allow for a precise calculation of the strength of a van der Waals bond at an arbitrary location. To advance a well-grounded and systematic work should be performed. We have to achieve to lean on the constantly developing computer technology. Simulations should be applied but only up to the level the precision can be reconciled with quantum chemistry. Such a technique is used when the starting model geometry (based on chemistry) is applied to investigate all the possible and realistic interactions. In order to achieve further development the methods of physics and mathematics (statistics) should approach each other. Several specific compounds and molecular systems will be used investigating their participation in some weak interactions to demonstrate the applicability and the potential of this method. For the exploitation of the three main points mentioned in the **Introduction**, the historical survey and the present technological capacities have assisted to a great extent.

For the application of the LMOs several methods have been worked out (see **Chapter III**). To investigate van der Waals bonds the SMO versions of the LMOs seem to be the most appropriate. On the other hand the construction of the SMOs has not been implemented in every standard software package, we will, therefore, apply the LMOs and their contribution using the Boys localization technique. Based on the demand assigned in the **Introduction**, those one-electron quantities seem to be acceptable that, besides having transferability properties can also be controlled experimentally. Henceforth we will consider the electric (dipole) moment, namely the LMO contributions of the first order electric moments. As we have previously shown in our other works, the Hartree-Fock localized representation of the supermolecule makes possible a good description of the van der Waals bonds. The deviations of the first order moments for different configurations allow, in addition, a direct comparison of the possible interaction effects. An analysis of such type is thus a challenging task. Based on some well-established theoretical works found in literature, the H-bond problem is approached using a very simple complex structure. This structure is the CH_4 + H_2O dimer, where the proton dimer/proton acceptor characteristics has been in the focus of a variety of theoretical works and continues to be at the frontline of research. Such a small complex is a very good example showing the deficient picture about the strength of the H-bond yielding in general results that are below the desirable level of accuracy. Summarizing three significant contributions from the literature [380-383] we may unambiguously establish that such small sized, theoretical molecular structures really test the potential of nowadays' research even using the full capacity of computational resources.

We present here our results obtained in localized representation. The numerical results for the LMO first order moments have been calculated using small and medium sized basis sets. The proton donor affinity for both systems, namely for the methane and the water molecule has been investigated separately. The intermolecular distance has been varied (in the range of 5.0-8.0 a.u. between atoms O and C). The actual values given in the Figures are obtained by the 6-31 G* basis set.

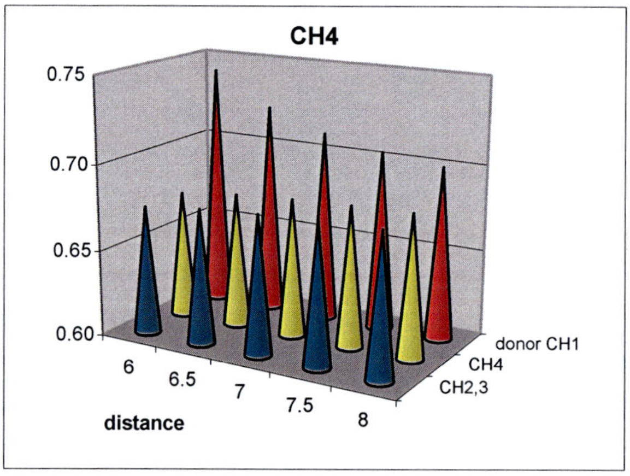

Figure 5.1.

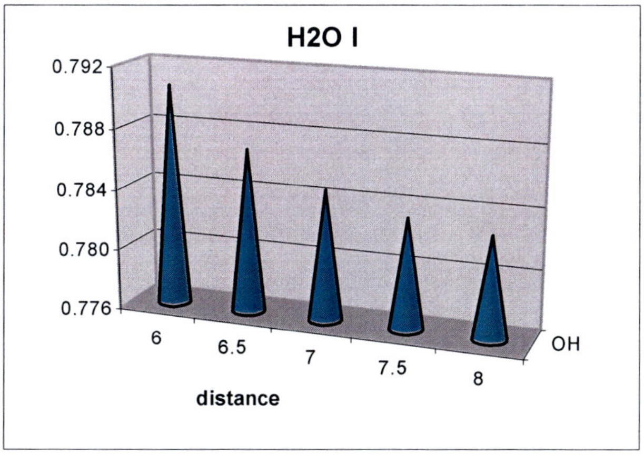

Figure 5.2.

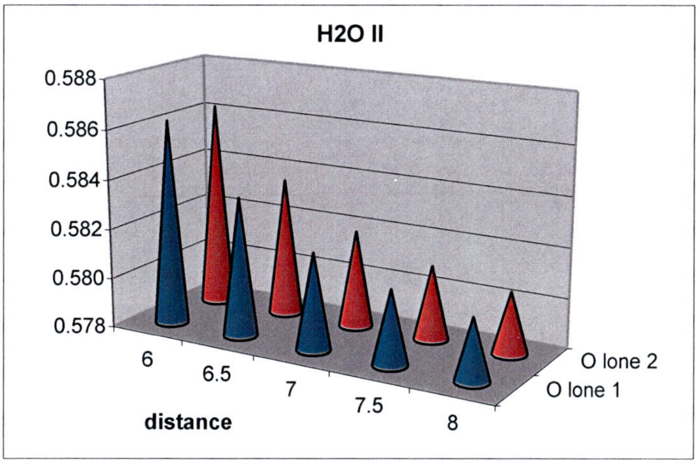

Figure 5.3.

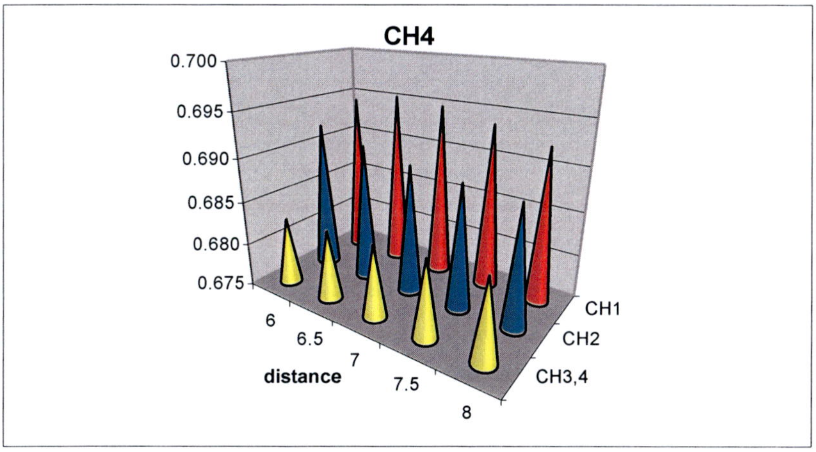

Figure 5.4.

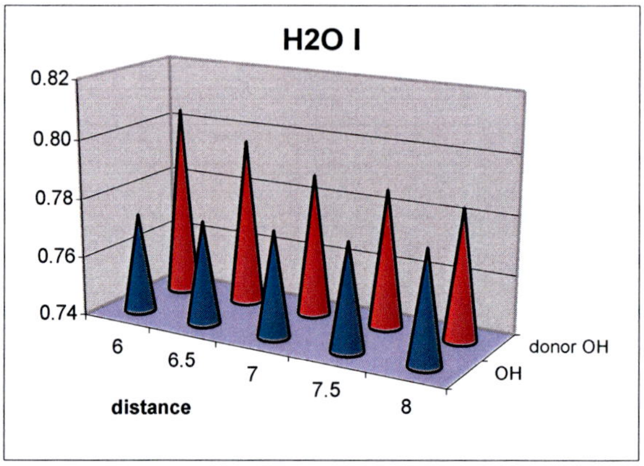

Figure 5.5.

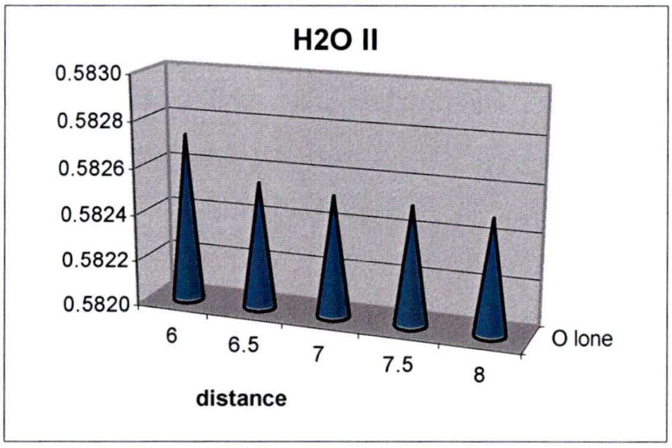

Figure 5.6.

In the CH_4 + H_2O system, depending on the proton-donor molecules, different LMOs deserve special attention. The spatial symmetry reduces further the number of LMOs to be investigated. In the case of methane (as a proton donor) two specific type CH bonds can be identified. One of them is a single CH-bond contributing to the formation of an H-bridge. The other three CH-bonds belong to the other type of bonding, two of them are equivalent. The water molecule behaving as a proton donor has two OH bonds that obviously belong to two different types. The partition of the LMOs of proton accepting molecules containing H can be done on the same footing. An important difference is the presence of lone LMOs in the case of the H_2O molecule. These orbitals might have a major significance especially for proton acceptors.

Two different geometric arrangements were tested. One of them was made up from the results of energy optimization. The optimization geometry has been given by E.S. Kryachko. We also applied model geometry that is defined as a regular tetrahedron for the CH_4 molecule with C-H=2.066 bohr, meanwhile for the H_2O molecules the monomer value of O-H=1.809 bohr, with H-O-H angle =104.52° was used. As a function of the interaction, the position of the centroids of the LMOs has been investigated profoundly. We have systematically analyzed the first order electron moments of the LMOs in the methane + water system as a function of the distance between the molecules. The calculated centroid lengths have been compared to those of the linear water dimers. This procedure has been performed both for the donor and the acceptor molecules for all relevant LMOs. We focused on the following effects:

i) How do the centroids of the LMOs change as a function of the distance depend on the donor/acceptor nature (e.g. do the LMO centroids' variation depend on whether the central atom is an O or a C)?

ii) As a function of the intermolecular distance can one characterize the proton donor or acceptor LMOs in the course of building the H-bond? (Hereby we have already studied the water molecule as a proton donor and acceptor in a linear water dimer, their behavior can be compared to the results obtained there.)

iii) Do the centroids of the LMOs considerably alter under the slight change of geometry such as when introducing the optimized geometry?

In Figure 5.1 we have depicted the electric moments of the LMOs associated with the different CH bonds. In Figures 5.2 and 5.3 we have shown the change of the first order electric moments of the LMOs in the water molecule – behaving as a proton acceptor – as a function of the intermolecular distance. Similar Figures show the change of the centroids in the case when the H_2O acts as a proton donor in the CH_4+H_2O supermolecule. The Figure 5.4 refers to methane and Figures 5.5 and 5.6 refer to H_2O. The values in the Figures have been obtained using the model geometry. Based on these Figures and on the results known for a linear dimer water molecule the answers for the above questions are the following.

According to the results found in recent study, the intermolecular distance dependence of the centroids of the bond LMO participating in the H-bond within a proton donor molecule seems not to be affected by the charge of the central atom. This is a promising statement and is in agreement, in addition, with the results obtained previously for the linear water dimer. Let us compare the curves presented in Figures 5.1 and 5.5. The centroid of the CH (donor CH1 in the figure) and OH (donor OH in the figure) bonds mediating in the proton donor process approach towards the central atom as the intermolecular distance decreases. It holds for both the OH and the CH bonds within the H_2O and the CH_4 molecule that do not participate in the formation of the H-bond; they are changing in a similar fashion. Namely, with the decrease of the distance a smaller effect arises and the displacement of opposite sign, i.e. a divergence is observed away from the C and O atoms, see Figures 5.1 (CH4) and 5.5 (OH). The curves are qualitatively similar for both model and optimized geometries (not shown here, see below next Table).

As far as the proton donor or acceptor nature and its importance are concerned for the H_2O molecule, it is instructive to compare the results obtained for two different supermolecules, CH_4+H_2O and H_2O+H_2O dimers. Let us give an investigation of those OH LMO centroids that do not directly contribute to the formation of the H bond and they are the centroids of OH bonds on the proton acceptor molecule. Values are given in Table 5.1, in atomic units. The lengths are measured from the H atoms. The results for the model and optimized geometries are different since the calculated OH distances are different in these cases, but the values imply an identical dependence on the intermolecular distance.

The results show that within range between 6.0 and 8.0 bohr distance the rate of change of the OH centroid lengths considered for both systems are quite similar. The variation of the centroid length of OH bond in the CH_4+H_2O structure with the intermolecular distance in this interval is as smooth as that of the OH bond LMO of the proton acceptor molecule in the linear water dimer. This suggests that at the equilibrium distance (around 6.7 bohr units of intermolecular separation) the displacement of the OH centroid does not contribute remarkably to the formation of a bonding between the CH_4 and H_2O molecules. It was found, however, that at the total energy minimum of the linear water dimer (distances between oxygen molecules is about 5.5 bohr), the centroid of the OH bond considered moves significantly toward the oxygen atom of proton acceptor (see values in Table 5.1).

Table 5.1.

intermolecular distance	**$CH_4 + H_2O$**		**H_2O+H_2O**
	model geometry	**optimized geometry**	
5,0	-	-	0,8048
5,5	0,7985	0,8076	0,7969
6,0	0,7906	0,8006	0,7914
6,5	0,7868	0,7961	0,7879
7,0	0,7846	0,7933	0,7857
7,5	0,7833	0,7917	0,7839
8,0	0,7827	0,7909	0,7828

We do have to investigate also whether the same conclusion may be drawn in the case of the change of the centroid of the H_2O as a proton acceptor with its oxygen's lone LMO in the systems above. The centroid of lone LMOs is measured from the oxygen atom, as a convenience. The values are given in Table 5.2.

Table 5.2.

intermolecular distance	**$CH_4 + H_2O$**		**H_2O+H_2O**
	model geometry	**optimized geometry**	
5,0	-	-	0,6378
5,5	0,5914	0,5908	0,6211
6,0	0,5876	0,5873	0,6055
6,5	0,5845	0,5844	0,5941
7,0	0,5825	0,5827	0,5807
7,5	0,5813	0,5813	0,5804
8,0	0,5807	0,5807	0,5802

In this case the results for the model and the geometry are remarkably closer to each other; the deviation is less than 0.1%. At the same time for 6.5 bohr distance down to 5.0 bohr the values are remarkably changing for the water dimer. In contrary, no significant deviation occurs for the lone LMO in the CH_4+H_2O system. The study on both types of LMOs of the acceptor molecule (H_2O) in this structure resulted thus in a coherent conclusion. Apparently the water molecule as a proton acceptor does not intend to participate in the formation of the $CH_4 + H_2O$ complex.

Based on the analysis of the centroids of the bond and lone LMOs of the contributing molecules, as a function of the intermolecular distance, significant difference is detected between the two supermolecules. Using the specific example when either the water molecule or the methane molecule provides a proton donor against a water molecule we found the following results. The OH bonds do provide a distinction concerning the proton acceptor nature of the water molecule at the relevant equilibrium geometries. A similar effect is remarkable when the relevant lone pair orbital density is considered. We have to emphasize that lone LMO contributes basically to the formation of the H bonding. One can thus notify, the estimated parameters applied by the Molecular Mechanics (MM) method that are based on the lone LMOs only can be considered very rough ones with large uncertainty. Our results affirm, the given lone LMO value itself does not determine the strength of the H-bond even though the distance from the facing H atom is taken into account (cf. the MM method). On the one hand it is not the position of the H atom but the centroid of the H-bond (e.g. CH or OH) participating in the H-bridge could provide the relevant value for the charge redistribution. On the other hand the charge redistribution of all the orbitals of either the proton donor or the proton acceptor molecules should be taken into account in order to calculate the interaction at an acceptable level. Many authors emphasized the latter fact. We have to add to this statement that the charge redistribution add up as vectors, which means that even small contributions may play a role, i.e. all the charge centers (in our method the centroids of the LMOs) should be calculated. These sums may differ considerably even when the environment is taken into account. Focusing exclusively to a single bond the nature of the H-bridge can sometimes be described approximately only, in many cases leading to an error. In addition the comparison to experimental values enforces the calculation of the total moments by summing up all the centroid values.

According to the values we have found in our study on the $CH_4 + H_2O$ structure: in this case the proton acceptor ability (or more precisely its rejection) of H_2O can be determined. The results are based on the investigation on each LMO centroid of the supermolecule. These centroids can be calculated accurately using relatively small basis even at the Hartree-Fock level. The method is very promising and we hope to extend it to the investigation of other weakly interacting systems as well.

VI. Study of the Bisphenol-A

Information on Bispenol-A one can get from, e.g. Wikipedia. Bisphenol-A was first synthesized in 1891 (as it can be found in Th. Zincke, Mittheilungen aus dem chemischen Laboratorium der Universität Marburg, *Justus Liebigs Annalen der Chemie*, 343 (1905), pp. 75-131.) Bisphenol A was investigated in the 1930s during the search for various synthetic estrogens. The first evidence of BPA's estrogenicity came from experiments in the 1930s. Some compounds turned out to be more powerful as estrogen, so bisphenol-A was not utilized at that time. Nowadays, however, it is currently used in the manufacture of polycarbonate plastic and of epoxy resins. Bisphenol-A is considered for using as an antioxidant and as a polymerization inhibitor. Polycarbonates are widely used in various consumer products, such as in water and food containers and some baby bottles. Some polymers serving as dental fillings may also contain bisphenol-A. The epoxy resins containing bisphenol-A are well-known coatings for the inside of cans while canning food.

Bispenol-A as chemical has been found in human (tested in the USA). Its role may be that it can activate estrogen receptors and this effect leads to similar physiological effects as the body's own estrogens. Bispenol-A presented some hormone disrupting effects in the studies performed both on animals and human cancer cells. It has been seriously claimed that these effects might lead to health problems, especially in men. The plastics industry, on the other hand, has enounced that Bisphenol-A is safe at the current levels of human exposure.

In 2006, however, two authors published a detailed analysis of why some studies failed to replicate findings on low-level effects of Bispenol-A. Independent review by a scientific panel has found through reanalyzing the data that it in fact had found an effect, even though its conclusions stated otherwise. Regarding both the positive controls and the negative controls, there were notified certain contradiction. The Bispenol-A is thus a challenging, at least bi-faced molecule. Its geometry is given as follows:

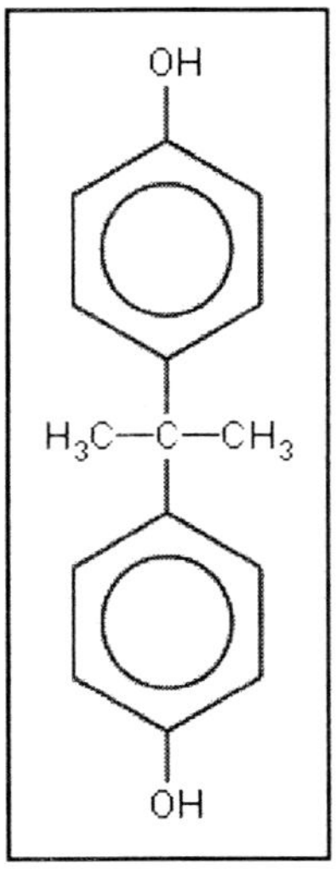

Figure 6.1.

The (energy optimized) structure of the Bisphenol-A (BspA) molecule has been calculated using the well-known 6-31G* basis set. The total energy values have been determined by two widely known quantum chemical methods: Hartree-Fock and density functional. The values are the followings (in atomic units):

Table 6.1.

	Hartree-Fock method	**density functional theory**
electronic energy	-1879.9619858322	-1877.6041539146
nuclear energy	1152.9089079620	1145.9365790257
total energy	-727.0530778702	-731.6675748889

The geometry of the optimized structures by both methods for the Bisphenol-A molecule are given in Tables A1 and A2 (in **Appendix**), in terms of Cartesian coordinates for the atomic nuclei, in bohr (atomic) units. According to these configurations, the electronic structure of the molecule has been subjected to localization. The first order electric moments (see **Chapter IV**) have been computed by the standard Boys localization method. The components of the centroid vectors for each LMO have been determined, providing the Cartesian elements (*x, y* and *z*) of the centroid charge vectors. These are given on Table A3 and Table A4 obtained by the different methods used.

The localized densities are often called core, bond and lone pair charge LMO densities in scientific literature. The differences between some features of localized charge densities, namely, turned out to sufficient large to identify the characteristics of three types of LMO densities to a rather good approximation in parent molecules. The properties of these types of localized distributions are different mainly in their spatial extension (see **Chapter IV**). The core LMO densities are of very small extension and in most of molecules are unambiguously determined by the position of nuclei. The situation is not as simple for the lone and bond LMO densities. The problem for their identification is especially acute when the given molecule has a complex form (has a distorted geometry, with various intra molecular bondings etc). The lone localized charge densities are in the neighborhood of one atom but they have much larger extension than the core ones do. The bond pair densities are more or less situated in the vicinity of two neighboring atoms. The 'neighborhood', however, can not be identified unambiguously. It is a possible procedure to attribute the given LMO to the atom where the endpoint of its centroid is the closest to; that was used in **Chapter V**. It is also a convenience, that in case of bondings where H atoms participate, the endpoints are measured from the 'heavy' atom nucleus. In case of BspA, e.g., these nuclei are the carbon or the oxygen, respectively. The LMO centroids for molecules published in literature are usually given according to this rule. The above assignment tactic is acceptable for both small and large, extended molecules.

Without additional recipe, it is not utilizable, however, for interacting systems. Even in weakly bonded structures, such of van der Waals type, the charge densities of a given contributing molecule are affected by the presence of other atomic or molecular electronic 'cloud'. Under the effects due to intermolecular interaction, certain centroid of charge densities might leave the vicinity of a heavy atom nucleus (to which it has originally been attributed, i.e. when the parent molecule was considered alone). Although the displacement often is not too large, it may be, however, superior to the distance (more precisely the half-distance) between the neighboring heavy atom nuclei. This could thus result in that the given charge density ought to be attributed to another heavy atom, if we have to be consequent in applying the tactic described above. The rigorous application of that rule would thus lead to a non-desired confusion. To avoid this problem, we will use in each case the original attribution. In interacting systems, namely, the numbering of the atoms in the parent molecule will be kept as reference points. (The results obtained for the BspA+H_2O will be presented in the next chapter considering this argument.)

Let us now analyze here the results obtained for the localized charge densities of Bisphenol-A molecule. As a first and important question, whether the values on centroid lengths are comparable to those obtained for small molecules (such as CH_4 or H_2O, see previous chapter). To answer this, the electric moments have been determined by the GAMESS-UK software, using two different methods. Second, we would shed light on the usefulness of the Hartree-Fock method and that of based on the density functional theory: can one notify significant differences in the LMO charge densities calculated by these procedures?

Not to be abundant, we selected certain localized charge densities of the BspA molecule, which are expected to reflect the specificity of the various types of LMOs. The centroid lengths of the following types of orbitals have been chosen and are presented in Figures 6.2 and 6.3:

- core orbital of a C atom, to which a methyl group is linked,
- core orbital of an Oxygen atom,
- CH bond orbital LMO densities in the selected methyl group,
- OH bond orbital attributed to the selected O atom,
- oxygen lone pair orbital densities belonging to the selected O atom.

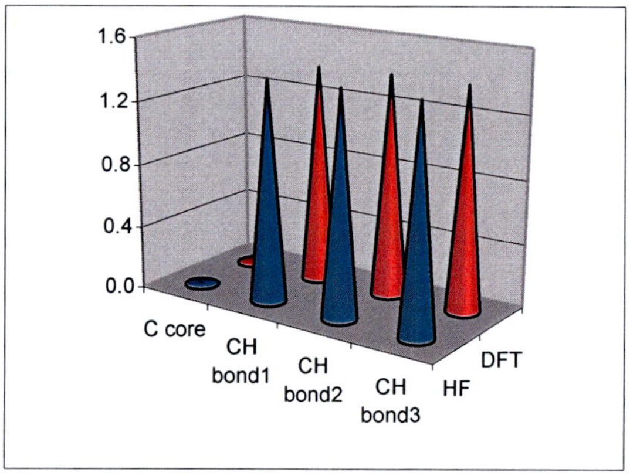

Figure 6.2.

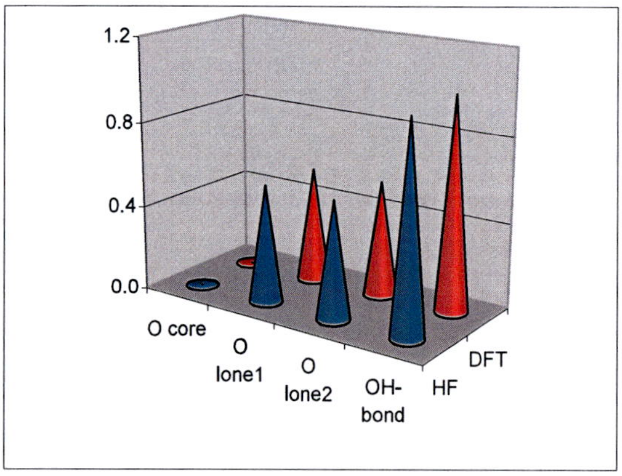

Figure 6.3.

The lengths of the centroid of charge vectors (having been calculated by the HF and the density functional methods) are demonstrated using a standard EXCEL program package. The results are grouped according to the selection described above.

- Figure 6.2 : three CH orbitals of a methyl group
- Figure 6.3: orbitals in the vicinity of an Oxygen

The values allow drawing the conclusions as follows. The core orbitals' extension is very small, as it could have been learned from literature published so far for smaller molecules. The lengths of CH orbitals can be compared to those resulted in calculations performed for other molecules such as methane: the values show a good agreement. This conclusion suggests, the CH localized charge densities reserve their properties in more extended systems such as in a BspA molecule. It should be added, that the CH bond pair LMO densities are not identical for the three CH orbitals, reflecting their different position in the molecule. A similar conclusion holds for the OH bond and lone LMO centroid lengths. They can be compared to the values resulted, e.g. in a water molecule.

As to the potential of the procedures used, it is remarkable: no significant differences could be found between the values obtained by the Hartree-Fock and the density functional methods, respectively.

To sum up, the analysis performed on the localized charge densities in the Bisphenol-A molecule suggests: the centroid lengths in this molecule are well comparable to those obtained previously by various authors and calculations for smaller molecules, which have the same residuals. We thus conclude, the methods are utilizable and the basis set is sufficient large to a further study of the localized charge densities. It is a next question, whether the significant properties of LMO charge densities would be applicable for analyzing weak interactions. The variation of centroid lengths of BspA molecule have been investigated under the effect of an external electric 'cloud'. This latter has been considered in terms of a small molecular entity, namely by the approach of a water molecule toward BspA.

VII. Study of the Bisphenol-A+ H2O Structure

The structure of the Bisphenol-A (BspA) molecule is symmetric, with phenol groups at both sides of the molecule. On one hand, based on chemical intuition the OH groups of the BspA molecule are considered sensitive to participate in/contribute to an intermolecular interaction. It is also interesting on the other hand to which extent are the methyl groups sensitive if such an event occurs. To answer these questions, a series of calculations have been performed, taking into account the structural symmetry of the Bisphenol-A molecule. In the model study the geometry of BspA molecule has been kept frozen according to the coordinates obtained in the geometry optimization procedure (see previous chapter and the appendix).

The intermolecular interaction has been considered in the presence of a water molecule. This latter entity was taken with geometrical parameters (OH = 1.809 bohr, HOH angle = 104.52) known as experimental values, and this geometry has not been altered. The H2O molecule has been 'added' to the BspA molecule, while approaching it with one of its H atom toward the O atom of the phenol group of BspA. The intermolecular separation has been varied: the distance of molecules was taken 6.0 Angstrom, then 5.0, 4.0, 3.0 and 2.0 Angstrom units, respectively. It is to be emphasized: although a model structural arrangement has been considered for the adduct, the contributing molecules have realistic geometry parameters. The symmetry of BspA will be broken in the presence of an additional H2O molecule. To be consistent with our interest expressed above, this leads to the study of the sensitivity of the following groups:

- OH group at the side where the H_2O molecule is approaching,
- OH group at the opposite side,
- CH_3 groups (they are not identical).

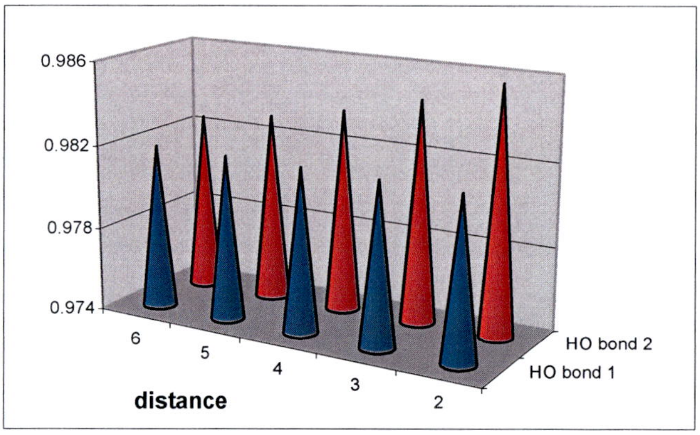

Figure 7.1.

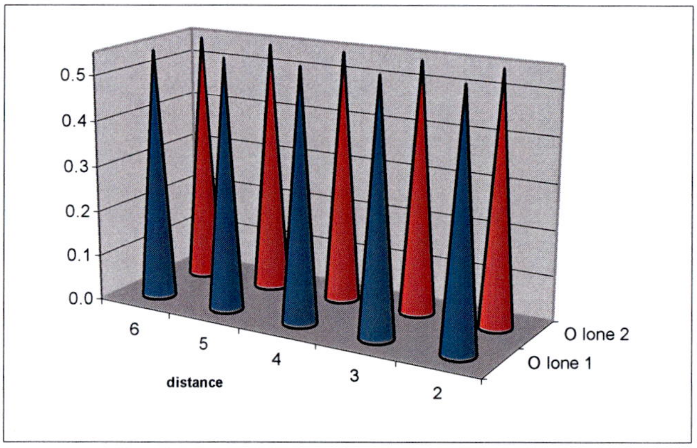

Figure 7.2.

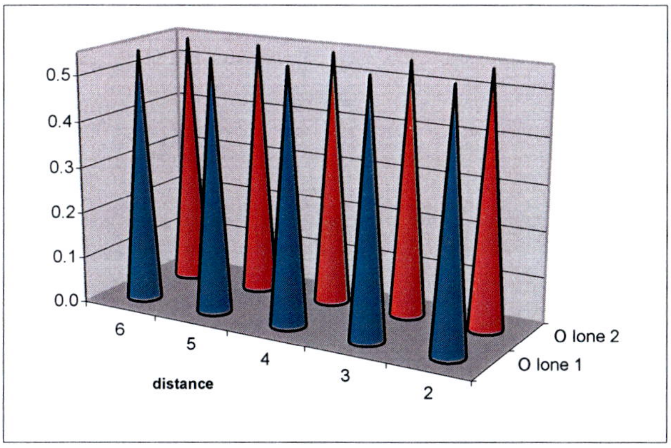

Figure 7.3.

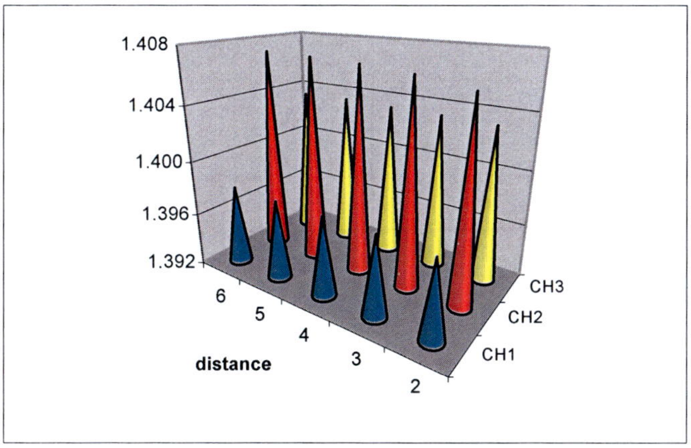

Figure 7.4.

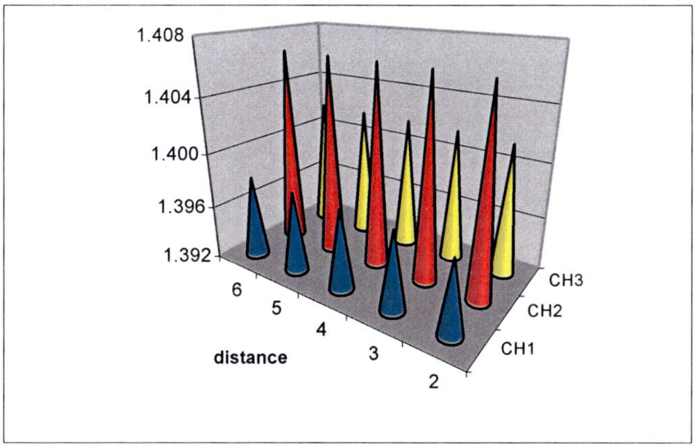

Figure 7.5.

It follows from the previous chapters, that we will analyze the effect of presence/approach of a water molecule on the BspA molecule in terms of the variation of the localized charge densities in the Bisphenol-A molecule.

i) The various distances of separation is expected to shed light, whether the centroid lengths of LMOs of the BspA are sensitive to an external effect provided by a small molecule. It is hoped, that such a detailed analysis is helpful to get insight into the molecular electronic structure, if it is subjected to participate in an interaction.

ii) It may also happen that the LMO electric moments are not changing exclusively in the neighborhood of the site of attack, i.e. only near the OH group, where the water molecule is approaching.

iii) It is also a question, whether the changes could be compared to the corresponding values having been obtained previously on molecules (in interaction) containing similar groups.

Taking into account the impressive similarities of the results given in the previous chapter, only the values based on the Hartree-Fock method are presented here. The results obtained for the Bisphenol-A + water system are exhibited both in Figures 7.1 to 7.5.

To answer the question *i)* established above, we present the results obtained for some LMO densities in the BspA+water supermolecule. The values given on the Figures depict the centroid lengths (bond and lone LMO densities) of both OH groups and the CH bond lengths for the CH_3 groups. A comparison of the extent of changes of various LMO centroid lengths (see Figures7.1-7.5) viz. the variation of the intermolecular separation gives a significant contribution to answer also the question *ii)*. Each quantity is measured from the central, O and C, nuclei.

As the OH groups are the centers of our interest, for setting up a comparison, the corresponding values have been taken from different supermolecules, containing OH groups (and thus OH bond LMOs) of various types. It is natural that the water dimer structures have been chosen for that purpose. Three well-known dimers of water molecules have been considered: the bifurcated, the cyclic and the linear ones. Having been performed a series of calculations (using a basis set as that applied for the study of the BspA molecule), the changes in the centroid length of bond pair LMO densities against the variation of the intermolecular distance have been analyzed. The numerical values reflecting the displacement of the centroid of different bond LMO densities as the distance is varied are given in the Table 7.1:

Table 7.1.

O-O distance (a.u.)	Bifurcated configuration		Cyclic configuration		Linear configuration		
	1/1,2	2/1,2	1,2/1	1,2/2	DON1	DON2	ACC/1,2
8.0	1.0242	1.0237	1.0272	1.0222	1.0208	1.0297	1.0257
7.0	1.0240	1.0234	1.0281	1.0221	1.0152	1.0310	1.0222
6.0	1.0222	1.0212	1.0294	1.0208	1.0002	1.0368	1.0187
5.0	1.0171	1.0164	1.0298	1.0160	0.9546	1.0525	1.0071
4.0	1.0102	1.0107	1.0315	1.0033	0.8359	1.0941	0.9838

Table 7.2.

O-O distance (Å)	in BspA	
	attacked OH_1	opposite OH_2
6.0	0.982556	0.981866
5.0	0.982883	0.981851
4.0	0.983671	0.981824
3.0	0.984599	0.981777
2.0	0.985725	0.981617

It is obvious, the variation of the lengths of the bond pair LMO charge densities of that OH group on BspA is worthwhile considering which is 'directly' attacked by the approaching water molecule. For a comparison those numerical values are given in Table 7.2, which resulted in the calculations performed for both OH groups of the BspA molecule. The results show, that the attacked OH groups' HO bond LMO density in the BspA+water structure has a very similar trend viz. the variation of the intermolecular distance as one of the OH bond LMOs in the cyclic structured water dimer (named **1,2/1**). The trend is clear, as the intermolecular distance decreases, the center of charge density of OH bond LMO moves towards the H atom.

To answer *iii)*: the values are well comparable to those given in **Chapter V**. This finding is supported by values given in Figures 7.1-7.5 obtained for the centroid lengths of BspA+H_2O.

The correspondence between BspA+ H_2O and the cyclic water dimer is highly supported by the numerical values referring to the geometry arrangement of the OH groups considered (in the BspA and in the water molecule, see the coordinates and distances given in Appendix, Table A5). The results thus demonstrate, the use of preliminary established values for localized electric moments (in our case the various lengths of the OH bond LMOs in water dimers) could help to identify a geometry configuration for an 'unknown' structure. Such a procedure could replace certain time-consuming geometry optimization for similar systems. A detailed investigation of this possibility is the subject of one of our next works.

Conclusion

Our work was based on using localized representation particularly applying localized molecular orbitals. A very restricted number of theoretical investigations for so-called biomolecules are known from literature at the same level of approximation. No results have yet been reported on the Bisphenol-A and its complexes in localized representation. In our work two quantum mechanical methods have been used: the Hartree-Fock method and the density functional theory. The Boys localization procedure was then applied. The results obtained by the two procedures are quite similar. The values presented here due to simplicity did not cover the whole regime of the electric charge distribution of the complexes. In this work the following results have been summarized:

- obtained by applying one particular (but widely known) basis set,
- the outcome by using two methods (some results are presented only for one),
- allows characterizing the BspA electronic structure in the presence of a H_2O molecule.

The values of most interest have been presented on various tables and figures.

From theoretical point of view the results affirm, the LMO electric moments are useful to identify the effects of (weak) interactions. This conclusion could be unambiguously drawn on the study of different complexes composed from a methane and a water molecule. The standard Boys localization method was used in the actual calculations. It can be added, that the application of the SMO procedure surely allows drawing more detailed conclusion (see our previous works). We intend thus to incorporate the SMO method into various quantum chemical program packages.

As to the investigation performed on the 'interaction ability' of Bisphenol-A molecule, the results affirm: 'to utilize' LMO electric moments is a useful strategy. The outcome obtained in the actual series of computation suggests, a simple method may provide us with information of equivalent level to that resulted in much more complicated analysis. The study having been carried out demonstrates, the sensitive LMO densities leading to differences in the corresponding values as the intermolecular separation is varied, are certainly located at some well-defined region of the Bispenol-A molecule. In addition, the method enabled us to 'quantify' the sensitivity of the various residuals in BspA viz. the approach of the water molecule, in terms of the displacement of LMO centroid vectors.

The results thus suggest that the method used is both sensitive and selective. This allows establishing two fields of applications. One of them could be the 'identification' of the sensitive regimes of any large (bio) molecule while an interaction occurs. Another useful way to apply the method described above can lead to an establishment of elements of database. The centroids of LMOs are, namely, expected to characterize in a simply manner the van der Waals interactions between a (bio)molecule and its interacting partner (not exclusively one water molecule). One of these possibilities has been pointed out above in the previous chapter.

The steps to build up an LMO moment database would be as follows.

i) Collect a large amount of numerical values of LMO electric moments (i.e. obtained for the lengths of the centroid of charge vectors of LMO densities) while investigating various weak interactions between different molecular structures.

ii) Use the elements of the database of LMO electric moments of (*i*) to predict an estimated value of certain characteristics of a given molecule with a related structure.

iii) Using a carefully established amount of data of LMO electric moments of (*ii*) the sensitive sites of a related molecule against a known (i.e. previously computed) entity could be predicted to a rather good approximation.

In this way an affected part of a molecule by an interaction could be a priori identified without doing time-consuming calculations.

We are intended to describe the steps of construction of an LMO moment database in a next contribution. Researchers doing informatics will presumably be involved into that work. The effective use of this database on certain examples will also be demonstrated there.

Acknowledgement

Thanks are due to Dr. I. Mayer for valuable discussions. We are indebted to Dr. I. Nagy for a careful reading of the draft. The authors are grateful to O. Kapuy, PhD student at the Budapest University of Technology and Economics, for her help in preparation of the Figures.

Appendix

Table A1. Bisphenol-A nuclear coordinates, basis: 6-31 G* (Hartree-Fock method)

atom	*x*	*Y*	*z*
C	0.0008968	2.7295713	0.0214812
C	-2.3849529	1.0623357	0.0177311
C	0.1486754	4.4404770	2.3769980
C	2.3864007	1.0604128	-.0067056
C	-0.1460682	4.4841086	-2.3024209
H	1.8958610	5.5096115	2.3333308
H	-1.4027728	5.7809040	2.4352497
H	0.1298453	3.3561586	4.1153453
H	1.3980997	5.8335541	-2.3268130
H	-0.1096580	3.4347668	-4.0616225
H	-1.8987514	5.5435875	-2.2460496
C	4.3944646	1.4331515	-1.6601408
C	6.5554934	-0.0438165	-1.5747837
C	6.7604609	-1.9534648	0.1952178
C	4.7864708	-2.3694553	1.8683714
C	2.6436604	-0.8807334	1.7522484
H	4.3134765	2.8968789	-3.0609972
H	8.0863546	0.2721878	-2.8703082
O	8.9020714	-3.3516518	0.2134367
H	4.9144068	-3.8576283	3.2521458
H	1.1305597	-1.2618469	3.0531811
C	-2.6648308	-0.8155368	-1.8234154
C	-4.7940541	-2.3016831	-1.9744020
C	-6.7490214	-1.9595484	-0.2525734
C	-6.5231413	-0.1296353	1.5886253
C	-4.3586565	1.3604063	1.7080807
H	-1.1664773	-1.1336279	-3.1579473
H	-4.9739932	-3.7379647	-3.3978691
O	-8.8002696	-3.4710489	-0.4713322
H	-8.0178389	0.1536767	2.9419923
H	-4.2647235	2.7624981	3.1698918
H	8.8282119	-4.5987694	1.4942363
H	-10.0118019	-3.0719204	0.7835246

Table A2. Bisphenol-A nuclear coordinates, basis: 6-31 G* (DFT method)

atom	*x*	*y*	*z*
C	0.0010726	2.7457758	0.0195576
C	-2.3822293	1.0693974	0.0179182
C	0.1480745	4.4534148	2.3896988
C	2.3873566	1.0727062	-0.0108534
C	-0.1476727	4.5026681	-2.3143623
H	0.9019508	5.5498830	2.3439729
H	-1.4316778	5.7917117	2.4619364
H	0.1458671	3.3470786	4.1381320
H	1.4265527	5.8483202	-2.3533086
H	-0.1362944	3.4355399	-4.0866111
H	-1.9053214	5.5921900	-2.2479609
C	4.4096730	1.4478676	-1.6792637
C	6.5840475	-0.0402795	-0.0402795
C	6.7885062	-1.9639442	0.2007361
C	4.7954411	-2.3793651	1.8881657
C	2.6377711	-0.8793566	1.7673937
H	4.3174159	2.9257682	-3.0968180
H	8.1313457	0.2704123	-2.9002578
O	8.9565795	-3.3773699	0.2169379
H	4.9263757	-3.8827631	3.2875154
H	1.1004225	-1.2562539	3.0756082
C	-2.6562766	-0.8180701	-1.8342710
C	-4.8045831	-2.3181086	-1.9898934
C	-6.7752028	-1.9729683	-0.2599866
C	-6.5495383	-0.1215753	1.6024253
C	-4.3752822	1.3754553	1.7263080
H	-1.1334732	-1.1354008	-3.1750025
H	-4.9858950	-3.7694497	-3.4284455
O	-8.8515539	-3.5013593	-0.4858997
H	-8.0651180	0.1608556	2.9656383
H	-4.2714180	2.7963027	3.2002005
H	8.8299755	-4.6339815	1.5447076
H	-10.0578601	-3.0520843	0.8185152

Table A3. Boys localization: Contributions to the electric dipole moment of the LMOs in a.u. (HF method).

atom	*x*	*y*	*z*
1	-8.80017	-3.47066	-0.47098
2	8.90164	-3.35169	0.21375
3	-6.74938	-1.95981	-0.25260
4	6.76082	-1.95373	0.19525
5	0.00090	2.72954	0.02148
6	-4.35869	1.36039	1.70809
7	2.64368	-0.88075	1.75225
8	4.39447	1.43314	-1.66014
9	-2.66484	-0.81553	-1.82340
10	-2.38486	1.06238	0.01772
11	2.38631	1.06047	-0.00670
12	4.78651	-2.36947	1.86836
13	-6.52316	-0.12966	1.58862
14	-4.79409	-2.30166	-1.97436
15	6.55547	-0.04385	-1.57474
16	0.14865	4.44041	2.37691
17	-0.14605	4.48404	-2.30233
18	-9.44243	-3.22610	0.22994
19	8.82805	-4.02362	0.92548
20	-3.15862	0.65157	1.03884
21	3.16393	0.67298	-0.99331
22	-.92309	5.35215	2.40830
23	-7.56220	0.05602	2.51668
24	2.48483	0.09861	0.89275
25	-5.80345	-2.16032	-1.12945
26	6.61621	-0.44415	-0.36680
27	-.11623	3.75735	-3.50084
28	7.59471	0.17693	-2.46050
29	-1.64022	-1.02388	-2.73608
30	5.80932	-2.20617	1.05864
31	-5.45581	0.61706	1.65794
32	.06314	3.56699	1.21399
33	-8.04257	-2.85867	-0.33314
34	8.05455	-2.85370	0.26038
35	-2.49436	0.13311	-0.92144
36	3.70411	-1.94737	1.28144
37	-8.57412	-3.96861	-0.37745
38	-1.34260	5.20617	-2.25432
39	-6.64776	-0.48351	0.37291
40	4.33901	2.43610	-2.61975
41	-8.98326	-3.32967	-0.97591
42	9.28019	-2.96306	0.33435
43	4.89666	-3.40082	2.81476

Table A3. Continued.

atom	*x*	*y*	*z*
44	-4.29717	2.32030	2.71052
45	-.06119	3.58864	-1.15445
46	-4.14402	-1.14585	-2.07408
47	.92019	5.40159	-2.31007
48	8.88250	-3.58238	-0.29171
49	5.48502	0.69791	-1.62440
50	1.34093	5.16930	2.33772
51	-4.91895	-3.27909	-2.94222
52	.12982	3.68965	3.56243
53	1.61017	-1.13415	2.64414
54	1.19137	1.88034	.04664
55	-3.63947	1.41195	.35904
56	3.62808	1.40022	-.28861
57	1.19152	1.87986	-.01957
58	-3.67413	-1.88564	-1.38552
59	4.15913	-1.23942	2.00270
60	-6.16529	-1.22748	1.06006
61	6.15609	-1.14338	-1.07646

Table A4. Boys localization: Contributions to the electric dipole moment of the LMOs in a.u. (DFT method).

atom	*x*	*y*	*z*
1	-8.85146	-3.50104	-0.48563
2	8.95623	3.37739	0.21718
3	-6.77552	-1.97319	-0.26001
4	6.78881	-1.96417	0.20076
5	0.00107	2.74577	0.01956
6	-4.37530	1.37542	1.72629
7	4.79547	-2.37935	1.88814
8	-6.54954	-0.12161	1.60239
9	-2.38216	1.06944	0.01791
10	2.63782	-0.87937	1.76737
11	2.38728	1.07275	-0.01085
12	4.40968	1.44784	-1.67924
13	-2.65631	-0.81807	-1.83424
14	-4.80461	-2.31807	-1.98983
15	6.58402	-0.04032	-1.58953
16	0.14806	4.45334	2.38960
17	-0.14766	4.50259	-2.31427
18	-9.49159	-3.23277	0.23893
19	8.85803	-4.05398	0.95139
20	-5.47936	0.62656	1.67355
21	2.48494	0.10598	0.89623
22	-0.93138	5.34702	2.42481

Table A4. Continued.

atom	*x*	*y*	*z*
23	-2.49061	0.13666	-0.92398
24	4.34748	2.44883	-2.63897
25	-3.65523	1.41088	0.37547
26	6.64090	-0.46897	-0.35285
27	0.92760	5.40063	-2.32657
28	-6.18057	-1.23253	1.04326
29	-1.62749	-1.02355	-2.74118
30	5.82893	-2.21592	1.07173
31	7.62313	0.17249	-2.47342
32	-1.18339	1.89523	0.04548
33	-8.07850	-2.87791	-0.34401
34	8.09247	-2.86981	0.26471
35	-4.30735	2.33700	2.72513
36	4.13207	-1.2263	2.01755
37	-9.00272	-3.33341	-0.97872
38	5.50843	0.70520	-1.63958
39	-4.93029	-3.29453	-2.95543
40	1.60003	-1.12661	2.65405
41	-7.59391	0.05710	2.52395
42	4.91084	-3.41053	2.83243
43	0.13934	3.69225	3.56301
44	-8.59689	-3.96843	-0.38402
45	1.33007	5.18863	2.34525
46	8.90074	-3.58758	-0.28106
47	-0.13347	3.76813	-3.50265
48	-1.33280	5.23293	-2.25572
49	9.29544	-2.97111	0.34036
50	-3.66549	-1.87588	-1.39264
51	0.07058	3.57736	1.21381
52	-0.06951	3.60139	-1.15601
53	3.19311	0.69340	-1.01907
54	-5.82302	-2.17489	-1.14001
54	1.18537	1.89703	-0.02197
56	3.65209	1.41639	-0.31982
57	-3.18203	0.66014	1.04680
58	-4.13012	-1.14326	-2.07648
59	3.67949	-1.93170	1.29863
60	-6.65705	-0.49718	0.36331
61	6.18058	-1.16473	-1.05644

Table A5. Cartesian vector elements of some localized charge densities in the Bisphenol-A molecule + H_2O molecule approaching at one side of BspA. Distance is measured between the Oxygen of BspA and one of the Hydrogen atom 1. of water. Calculated using Hartree-Fock method, 6-31 G* basis set.

Additional H_2O molecule at a distance of 6.0 Angstrom from the BspA molecule (side II) with the Cartesian coordinates of some significant atoms in a.u.

Atom	*x*	*y*	*z*
Central carbon in BspA	-0.2537296	-2.0472798	-0.0048753
Phenol group at side I. of BspA			
Oxygen atom	8.9151142	5.7997670	-0.5026450
Hydrogen atom	10.1751002	5.8304745	0.7675014
Phenol group at side II. of BspA			
Oxygen atom	-7.8366649	0.0646425	0.0824070
Hydrogen atom	-8.1791779	1.2940301	1.3363991
Coordinates of the approaching water molecule			
Hydrogen atom 1.	-18.8093430	2.9189432	-0.0253811
Oxygen atom	-20.5414351	3.3695088	-0.0423960
Hydrogen atom 2.	-20.7324875	4.5888042	1.2538523
Central carbon in BspA	-1.0396529	-1.7870904	-0.0111900
Phenol group at side I. of BspA			
Oxygen atom	9.0392440	4.8512138	-0.5014661
Hydrogen atom	10.2923013	4.7239283	0.7695278
Phenol group at side II. of BspA			
Oxygen atom	-8.2988385	1.2569442	0.0713896
Hydrogen atom	-8.4857086	2.5194132	1.3253661
Coordinates of the approaching water molecule			
Hydrogen atom 1.	-11.8085983	2.6585087	0.0331803
Oxygen atom	-13.4706977	3.3222407	0.0150857
Hydrogen atom 2.	-13.5085889	4.5557435	1.3114178

References

[1] Moskowitz, J.W.; Harrison, M.C. *J. Chem. Phys.* 1965, 43, 3550-3555.

[2] Csizmadia, I.G.; Harrison, M.C.; Moskowitz, J.W.; Sutcliffe, B.T. *Theoret. Chim. Acta (Berl.)* 1966, 6, 191-216.

[3] Salez, C.; Veillard, A. *Theoret. Chim. Acta (Berl.)* 1968, 11, 441-451.

[4] Veillard, A. *Theoret. Chim. Acta (Berl.)* 1968, 12, 405-411.

[5] Mehler, E.L.; Paul, C.H. *Chem. Phys. Lett.* 1979, 63, 145.

[6] Raffenetti, R.C.; Ruedenberg, K. *J. Chem. Phys.* 1973, 59, 5950.

[7] Adamowich, L.; Bartlett, R.J. *Chem, Phys, Lett.* 1984, 110, 365.

[8] Flores, J.R. *J. Chem. Phys.* 1993, 98, 5642.

[9] Zhang, R.Q.; Chu, T.S.; Lee, S.T. *J. Chem. Phys.* 2001, 114, 5531.

[10] Jensen F. *J. Chem. Phys.* 2001, 115, 9113.

[11] Petersson, G.A.; Zhong S.; Montgomery, J.A. Jr.; Frisch, M.J. *J. Chem. Phys.* 2003, 118, 1101

[12] Weigend, F.; Furche, F.; Ahlrichs, R. *J. Chem. Phys.* 2003, 119, 12753.

[13] Ozaki, T.; Kino, H. *J. Chem. Phys.* 2004, 121, 10879-10888.

[14] Huzinaga, S. *J. Chem. Phys.* 1965, 42, 1293.

[15] Tavouktsoglou, A.N.; Huzinaga, S. *J. Chem. Phys.* 1980, 72, 1385.

[16] Tatewaki, H.; Huzinaga, S. *J. Comp. Chem.* 1980, 1, 205-228.

[17] Tatewaki, H.; Sakai, Y.; Huzinaga, S. *J. Comp. Chem.* 1981, 2, 96-99.

[18] Krishnan, R.; Binkley, J.S.; Seeger, R.; Pople, J.A. *J. Chem. Phys.* 1980, 72, 650.

[19] Frisch, M.J.; Pople, J.A.; Binkley, J.S. *J. Chem. Phys.* 1984, 80, 3265.

[20] Curtiss, L.A.; Raghavachari, K,; Pople, J.A. *J. Chem. Phys.* 1993, 98, 1293.

[21] Smith, B.J.; Radom, L. *Chem. Phys. Lett.* 1994, 231, 345-351.

[22] Curtiss, L.A.; Redfern, P.C.; Raghavachari, K.; Rassolov, V.; Pople, J.A. *J. Chem. Phys.* 1999, 110, 4703.

[23] Dunning, T.H. Jr. *J. Chem. Phys.* 1971, 55, 716.

[24] Dunning, T.H. Jr. *J. Chem. Phys.* 1989. 90, 1007.

[25] Silver, D.M.; Nieuwpoort, W.C. *Chem. Phys. Lett.* 1978, 57, 421.

[26] Silver, D.M.; Wilson, S.; Nieuwpoort, W.C. *Int. J. Quantum. Chem.* 2004, 14, 635-639.

[27] Mezey, P.G. *Theoret. Chim. Acta (Berl.)* 1979, 53, 183-192.

[28] Moncrieff, D.; Kobus, J.; Wilson, S. *J. Phys. B* 1995, 28, 4555-4557.

[29] Pedersen, L.; Morokuma, K. *J. Chem. Phys.* 1967, 46, 3941-3947.

[30] Poppinger, D. *Chem. Phys.* 1976, 12, 131-137.

[31] Thomas, J.R.; DeLeeuw, B.J.; Vacek, G.; Crawford, T.D.; Yamaguchi, Y.; Schaefer III, H.F. *J. Chem. Phys.* 1993, 99, 403-410.

[32] Bär, M.R.; Sauer, *J. Chem. Phys. Lett.* 1994, 226, 405-412.

[33] Rosenberg, B.J.; Shavitt, I. *J. Chem. Phys.* 1975, 63, 2162.

[34] Rosenberg, B.J.; Ermler, W.C.; Shavitt, I. *J. Chem. Phys.* 1976, 65, 4072.

[35] Feller, D. *J. Chem. Phys.* 1993, 98, 7059.

[36] McDowell, K. *J. Chem. Phys.* 1978, 68, 4151.

[37] Klahn, B.; Bingel, W.A. *Theoret. Chim. Acta (Berl.)* 1977, 44, 27-43.

[38] Klahn, B.; Bingel, W.A. *Theoret. Chim. Acta (Berl.)* 1977, 44, 9-26,

[39] Klahn, B.; Bingel, W.A. *Intern, J, of Quantum Chem.* 1977, XI, 943-957.

[40] Schmidt, M.W.; Ruedenberg, K. *J. Chem. Phys.* 1979, 71, 3951.

[41] Wilson, S.; Guest, M.F. *Chem. Phys. Lett.* 1980, 73, 607.

[42] Klahn, B.; Morgan III, J.D. *J. Chem. Phys.* 1984, 81, 410,

[43] Klahn, B. *J. Chem. Phys.* 1985, 83, 5754.

[44] Szász, O. *Math Ann.* 1916, 77, 482,

[45] Rosas-Garcia, V.M.; Crawford, T.D. *J. Chem. Phys.* 2003, 118, xxx-

[46] Burton, P.G. *J. Chem. Phys.* 1979, 70, 3112

[47] Hariharan, P.C.; Pople, J.A. *Theoret. Chim. Acta (Berl.)* 1975, 28, 213-222.

[48] Hobza, P.; Sauer, J. *Theoret. Chim. Acta (Berl.)* 1984, 65, 279-290.
[49] Dingle, T.W.; Huzinaga, S.; Klobukowski, M. *J. Comp. Chem.* 1989, 10, 753-769.
[50] Latajka, Z.; Scheiner, S. *J. Comp. Chem.* 1987, 8, 663-673.
[51] Latajka, Z.; Scheiner, S. *J. Comp. Chem.* 1987, 8, 674-682.
[52] Huh, S.B.; Lee, J.S. *J. Chem. Phys.* 2003, 118, 3035.
[53] Boys, S.F.; Bernardi F. *Mol. Phys.* 1970, 19, 553-566.
[54] Cullen, J.M. *Intern. J. of Quantum Chem.* 1991, 25, 193-207.
[55] Gutowski, M.; Chalasinski, G. *J. Chem. Phys.* 1993, 98, 5540.
[56] Davidson, E.R.; Chakravorty, S.*J. Chem. Phys. Lett.* 1994, 217, 48.
[57] van Duijneveldt, F.B.; van Duijneveldt-van de Rijdt, J.G.C.M.; van Lenthe, J.H. *Chem. Rev.* 1994, 94, 1873-1885.
[58] Van Duijneveldt, F.B. Basis set superposition error. Molecular Interactions 1997, 81-104.
[59] Van Duijneveldt, F.B.; van Duijneveldt-van de Rijdt, J.G.C.M.; van Lenthe, J.H. *Chemical Reviews* 1994, 94, 1873-1885.
[60] Gutowski, M.; Van Duijneveldt-Van de Rijdt, J.G.C.M.; Van Lenthe, J.H.; Van Duijneveldt, F.B. *J. Chem. Phys.* 1993, 98, 4728-4737.
[61] Vos, R.J.; Van Duijneveldt, F.B.; Van Lenthe, J.H. *Mol. Phys.* 1989, 67, 1011-1020.
[62] Van Lenthe, J.H.; Vos, R.J.; Van Duijneveldt-Van de Rijdt, J.C.G.M.; Van Duijneveldt, F.B. *Chem. Phys. Lett.* 1988, 143, 435-438.
[63] Van Lenthe, J.H.; Van Duijneveldt-Van de Rijdt, J.G.C.M.; Van Duijneveldt, F.B. Adv. *Chem. Phys.* 1987, 69, 521-566.
[64] Gutowski, M.; Van Duijneveldt, F.B.; Chalasinski, G.; Piela, L. *Mol. Phys.* 1987, 61, 233-247.
[65] Gutowski, M.; Van Duijneveldt, F.B.; Chalasinski, G.; Piela, L. Comment. *Chem. Phys. Lett.* 1986, 129, 325-328.
[66] Gutowski, M.; Van Lenthe, J.H.; Verbeek, J.; Van Duijneveldt, F.B.; Chalasinski, G. *Chem. Phys. Lett.* 1986, 124, 370-375.
[67] Van Lenthe, J.H.; Dijkstra, F. *Mol. Phys.* 1999, 96, 555-558.
[68] Karlström, G.; Sadlej, A.J. *Theoret. Chim. Acta (Berl.)* 1982, 61, 1-9.
[69] Leclercq, J.M.; Allavena, M.; Bouteiller, Y. *J. Chem. Phys.* 1983, 78, 4606.
[70] Schwenke, D.W.; Truhlar, D.G. *J. Chem. Phys.* 1985, 82, 2418.
[71] Latajka, Z.; Scheiner, S. *J. Chem. Phys.* 1984, 81, 407.
[72] Pullman, A.; Claverie, P.; Cluzan, M-C. *Chem. Phys. Lett.* 1985, 117, 419.
[73] Del Bene, J.E. *J. Comp. Chem.* 1987, 8, 810-815.
[74] Alagona, G.; Ghio, C.; Cammi, R.; Tomasi, J. *Intern. J. Quantum. Chem.* 1987, XXXII, 207-226.
[75] Alagona, G.; Ghio C.; Cammi, T.; Roberto, J. *Intern, J, Quantum Chem*, 1987, XXXII, 227-248.
[76] Tsuzuki, S.; Tanabe, K. *J. Phys. Chem.* 1991, 95, 2272-2278.
[77] Feller, D. *J. Chem. Phys.* 1992, 96, 6104.
[78] Wind, P.; Heully, J-L. *Chem. Phys. Lett.* 1994, 230, 35-40.
[79] Simon, S.; Duran, M.; Dannenberg, J.J. *J. Chem. Phys.* 1996, 105, 11024.
[80] Simon, S.; Duran, M.; Dannenberg, J.J. *J. Phys. Chem.* A 1999, 103, 1640-1643.
[81] Tschumper, G.S.; Kelty, M.D.; Schaefer III, H.F. *Mol. Phys.* 1999, 96, 493-504.
[82] Hamza, A.; Mayer, I. *J. of Quantum Chem.* 2001, 82, 53-59.

[83] Sordo, J.A.; Sordo, T.L.; Fernández, G.M.; Gomperts, R.; Chin, S.; Clementi, E. *J. Chem. Phys.* 1989, 90, 6361.
[84] Siu, F.M.; Ma, N.L.; Tsang, C.W. *J. Chem. Phys.* 2001, 114, 7045.
[85] Kaczmarek, A.; Sadlej, A.J.; Leszczynski, J. *J. Chem. Phys.* 2004, 120, 7837.
[86] Clementi, E. *Proc. Nat. Acad. Sci.* 1972, 69, 2942-2944.
[87] Clementi, E. In Computational aspects for large chemical systems. Clementi, E.; Ed. Springer, Berlin, Germany. 1980. *Lecture notes in Chemistry* **19**, pp. 178-184.
[88] Tang, K.C.; Edmiston, C. *J. Chem. Phys.* 1970, 52, 997.
[89] Bender, C.F. *J. Comp. Phys.* 1972, 9, 547-554.
[90] Dyczmons, V. *Theoret. Chim. Acta (Berl.)* 1973, 28, 307-310.
[91] Diercksen, G.H.F. *Theoret. Chim. Acta (Berl.)* 1974, 33, 1-6.
[92] Pendergast, P.; Fink, W. *J. Comp. Phys.* 1974, 14, 286-300.
[93] Karlström, G. *J. Comp. Chem.* 1981, 2, 83-86.
[94] Strout, D.L.; Scuseria, G.E. *J. Chem. Phys.* 1995, 102, 8448.
[95] Almlöf, J.; Faegri Jr, K.; Korsell, K. *J. Comp. Chem.* 1982, 3, 385-399.
[96] Almlöf, J.; Taylor, P.R. In *Advanced Theories and Computational Approaches to the Electronic Structure of Molecule.* Dykstra, C.E.; Ed. Riedel, Dordrecht, Holland. 1984, 107-125.
[97] Cremer, D.; Gauss, J. *J. Comp. Chem.* 1986, 7, 274-282.
[98] Häser, M.; Ahlrichs, R. *J. Comp. Chem.* 1989, 10, 104-111.
[99] Saebo, S.; Almlöf, *J. Chem. Phys. Lett.* 1989, 154, 83.
[100] Ahlrichs, R.; Bär, M.; Häser, M.; Kölmel, Ch.; Sauer, *J. Chem. Phys. Lett.* 1989, 164, 199.
[101] Rossi, M. *Theoret. Chim. Acta (Berl.)* (1966) 4, 30-36.
[102] Dacre, P.D. *Chem. Phys. Lett.* 1970, 7, 47.
[103] Winter, N.W.; Ermler, W.C.; Pitzer, R.M. *Chem. Phys. Lett.* 1973, 19, 179.
[104] Takada, T.; Sasaki, F. *Intern. J. of Quantum Chem.* 1980, XVIII, 1157-1163.
[105] Carsky, P.; Hess Jr, B.A.; Schaad, L.J. *J. of Comp. Chem.* 1984, 5, 280-287.
[106] Pulay, P. *Theoret. Chim. Acta (Berl.)* 1979, 50, 299-312.
[107] Pulay, P.; Saebo, S. *Theor. Chim. Acta* 1986, 69, 356.
[108] Saebo S. *Intern. J. of Quantum Chem.* 1992, 42, 217-239.
[109] Paizs, B.; Baker, J.; Suhai, S.; Pulay, P. *J. Chem. Phys.* 2000, 113, 6566.
[110] Bandyopadhyay, P.; Gordon M.S. *J. Chem. Phys.* 2002, 116, 5023-5032.
[111] Li, H.; Gordon, M.S.; Jensen, J.H. *J. Chem. Phys.* 2006, 124, 214108.
[112] Guest, M.F.; Bush, Van Dam, H.J.J.; Sherwood, P.; Thomas, J.M.H.; Van Lenthe, J.H.; Havenith, R.W.A.; Kendrick, J. *Mol. Phys.* 2005, 103, 719-747.
[113] Frisch, M. J.; Trucks, G. W.; Schlegel, H. B.; Scuseria, G. E.; Robb, M. A.; Cheeseman, J. R.; Montgomery, J. A., Jr.; Vreven, T.; Kudin, K. N.; Burant, J. C.; Millam, J. M.; Iyengar, S. S.; Tomasi, J.; Barone, V.; Mennucci, B.; Cossi, M.; Scalmani, G.; Rega, N.; Petersson, G. A.; Nakatsuji, H.; Hada, M.; Ehara, M.; Toyota, K.; Fukuda, R.; Hasegawa, J.; Ishida, M.; Nakajima, T.; Honda, Y.; Kitao, O.; Nakai, H.; Klene, M.; Li, X.; Knox, J. E.; Hratchian, H. P.; Cross, J. B.; Bakken, V.; Adamo, C.; Jaramillo, J.; Gomperts, R.; Stratmann, R. E.; Yazyev, O.; Austin, A. J.; Cammi, R.; Pomelli, C.; Ochterski, J. W.; Ayala, P. Y.; Morokuma, K.; Voth, G. A.; Salvador, P.; Dannenberg, J. J.; Zakrzewski, V. G.; Dapprich, S.; Daniels, A. D.; Strain, M. C.; Farkas, O.; Malick, D. K.; Rabuck, A. D.; Raghavachari, K.; Foresman, J. B.; Ortiz, J.

V.; Cui, Q.; Baboul, A. G.; Clifford, S.; Cioslowski, J.; Stefanov, B. B.; Liu, G.; Liashenko, A.; Piskorz, P.; Komaromi, I.; Martin, R. L.; Fox, D. J.; Keith, T.; Al-Laham, M. A.; Peng, C. Y.; Nanayakkara, A.; Challacombe, M.; Gill, P. M. W.; Johnson, B.; Chen, W.; Wong, M. W.; Gonzalez, C.; Pople, J. A. *Gaussian* **03**, Revision C.02; Gaussian, Inc.: Wallingford, CT, 2004

[114] http://www.wavefun.com/

[115] Wulfov, A.L. *Chem. Phys. Lett.* 1996, 255, 300-308.

[116] Rudberg, E.; Salek, P.; *J. Chem. Phys.* 2006, 125, 084106.

[117] Edmiston, C.; Ruedenberg, K. Rev. Mod. Phys. 1963, 35, 457.

[118] Edmiston, C,; Ruedenberg, K. In *Quantum Theory of Atoms, Molecules and Solid State*, Löwdin, P.O.; Ed.; Academic Press: NewYork, NY, 1966, pp. 263-280.

[119] England, W.; Salmon, L.S.; Ruedenberg, K. In *Topics in Current Chemistry* 1971, 23, 31-123.

[120] England, W. *Intern. J. of Quantum Chem.* 1971, V, 683-697.

[121] Ruedenberg, K. Sundberg, K.R. In *Quantum Science.* Calais, J.L.; Goscinski, O.; Linderberg, J.; Öhm, Y.; Eds. Plenum, New York, NY. 1976, pp.505-515.

[122] England, W.; Gordon, M.S.; Ruedenberg, K. *Theoret. Chim. Acta (Berl.)* 1975, 37, 177-216.

[123] Boys, S.F. *Rev. Mod. Phys.* 1998, 32, 296-298.

[124] Boys S.F. In Quantum Theory of Atoms, Molecules and Solid State, Löwdin, P.O.; Ed.; Academic Press: NewYork, NY, 1966, pp.253-262.

[125] Kaldor, U. *J. Chem. Phys.* 1967, 46, 1981-1987.

[126] Daudel, R.; Kapuy, E,; Kozmutza, C.; Stephens, M.E. *Theoret. Chim. Acta (Berl.)* 1979, 53, 147-157.

[127] Daudel, R.; Kapuy, E.; Kozmutza, C. *Theoret. Chim. Acta (Berl.)* 1980, 56, 259-267.

[128] Kapuy, E.; Kozmutza, C. *J. Chem. Phys.* 1991, 94, 55-65.

[129] Kozmutza, C.; Kapuy, E. *Int. J. Quant. Chem.* 1990, 38, 665-673.

[130] Kozmutza, C.; Varga, I.; Udvardi, L. *J. Mol. Structure (Theochem)* 2003, 666-667, 95-97.

[131] Magnasco, V.; Perico, A. *J. Chem. Phys.* 1967, 47, 971-981.

[132] Von Niessen, W. *J. Chem. Phys.* 1971, 55, 1948.

[133] Von Niessen, W. *Theoret. Chim. Acta (Berl.)* 1973, 31, 111-135.

[134] Peters, D. *J. Am. Chem. Soc.* 1972, 94, 707.

[135] Peeters, D.; Leroy, G. *Theoret. Chim. Acta (Berl.)* 1978, 50, 135-144.

[136] Ryback, W.; Poirier, R.; Kari, R. *Intern. J. Quantum Chem.* 1978, XIII, 1-16.

[137] Sundberg, K.R.; Erano, J.; Lipscomb, W.N. *J. Chem. Phys.* 1979, 71, 1515.

[138] Luken, W.L. *J. Chem. Phys.* 1983, 78, 5729.

[139] Cizek J.; Förner W.; Ladik J. *Theoret. Chim. Acta (Berl.)* 1983, 64, 107-116.

[140] Leonard, J.M.; Luken, W.L. *Intern. J. of Quantum Chem.* 1984, XXV, 355-365.

[141] Aslangul, C.; Constanciel, R.; Daudel, R.; Esnault, L. *Intern. J. of Quantum Chem.* 1974, VIII, 499-522.

[142] Bader, R.F.W.; Stephens, M.E. *Chem. Phys. Lett.* 1974, 26, 445-449.

[143] Bader, R.F.W.; Stephens, M.E. *J. Am. Chem. Soc.* 1975, 97, 7391.

[144] Diner, S.; Malrieu, J.P.; Claverie, P. *Theoret. Chim. Acta (Berl.)* 1969, 13, 1-17.

[145] Malrieu, J.P.; Claverie, P.; Diner, S. *Theoret. Chim. Acta (Berl.)* 1969, 13, 18-45.

[146] Diner, S.; Malrieu, J.P.; Jordan, F.; Gilbert, M. *Theoret. Chim. Acta (Berl.)* 1969, 15, 100-110.
[147] Jordan, F.; Gilbert, M.; Malrieu, J.P.; Pincelli, U. *Theoret. Chim. Acta (Berl.)* 1969, 15, 211-224.
[148] Cullen, J.M.; Zerner, M.C. *Intern. J. of Quantum Chem.* 1982, XXII, 497-535.
[149] Saebo, S.; Pulay, P. *Chem. Phys. Lett.* 1985, 113, 13.
[150] Saebo, S.; Pulay, P. *J. Chem. Phys.* 1987, 86, 914.
[151] Saebo, S.; Pulay P. *Annu. Rev. Phys. Chem.* 1993, 44, 213-236.
[152] Werner, H-J.; Manby F.R.; Knowles, P.J. *J. Chem. Phys.* 2003, 118, 8149.
[153] Pipek, J. *Intern. J. of Quantum Chem.* 1985, XXVII, 527-546.
[154] Pipek, J.; Mezey, P.G. *J. Chem. Phys.* 1989, 90, 4916.
[155] Pipek, J.; Varga, I.; Nagy, T. *Intern. J. of Quantum Chem.* 1990, XXXVII, 529-537.
[156] Coffey, P. *Intern. J. of Quantum Chem.* 1974, VIII, 777-782.
[157] Bauschlicher, C.W, Jr. *J. Chem. Phys.* 1980, 72, 880.
[158] Böhm, M.C. *Theoret. Chim. Acta (Berl.)* 1981, 59, 609-624.
[159] Reed, A.E.; Weinhold, F. *J. Chem. Phys.* 1985, 83, 1736.
[160] Pipek, J.; Ladik, *J. Chem. Phys.* 1986, 102, 445-458.
[161] Rajzmann, M.; Brenier, B.; Purcell, K.F. *Theor. Chim Acta* 1987, 72, 13-26.
[162] Bartha, F.; Bogár, F.; Kapuy, E. *Intern. J. of Quantum Chem.* 1990, XXXVIII, 215-219.
[163] Nemukhin, A.V.; Weinhold F. *J. Chem. Phys.* 1992, 97, 1095.
[164] Crawford, T.D.; Schaefer III, H.F. *J. Chem. Phys.* 1996, 105, 1060.
[165] Flood, E.; Pulay, P.; Boggs, J.E. *J. Am. Chem. Soc.* 1977, 99, 5570.
[166] Surján, P.R.; Mayer, I.; Kertész, M. *J. Chem. Phys.* 1982, 77, 2454-2459.
[167] Silvi, B.; Kryachko, E.S.; Tishchenko, O.; Fuster, F.; Nguyen, M.T. *Mol. Phys.* 2002, 100, 1659-1675.
[168] Aufderheide, K.H. *J. Chem. Phys.* 1980, 73, 1777.
[169] Aufderheide, K.H.; Chung-Phillips A. *J. Chem. Phys.* 1980, 73, 1789.
[170] Ayers, P.W. *J. Chem. Phys.* 2000, 113, 10886.
[171] Pendás A.M. *J. Chem. Phys.* 2002, 117, 965.
[172] Weaire, D.; Williams, A.R. *J. Phys. C: Solid State Phys.* 1977, 10, 1239.
[173] Iftimlie, R.; Thomas, J.W,; Tuckerman, M.E. *J. Chem. Phys.* 2004, 120, 2169.
[174] Kolos, W.; Wolniewicz, L. *J. Chem. Phys.* 1965, 43, 2429-2441.
[175] Kolos, W. *Intern. J. of Quantum Chem.* 1967, 1, 169-186.
[176] Collins, J.B.; Schleyer, P.R.; Binkley, J.S.; Pople, J.A.; Radom, L. *J. Am. Chem. Soc.* 1976, 98, 3436.
[177] Burton, P.G. *J. Chem. Phys.* 1977, 67, 4696.
[178] Meyer, W.; Hariharan, P.C.; Kutzeinigg, W. *J. Chem. Phys.* 1980, 73, 1880.
[179] Stärck, J.; Meyer, W. *Chem. Phys. Lett.* 1994, 225, 229-232.
[180] Van Mourik, T.; Dunning, Jr T.H. *J. Chem. Phys.* 1999, 111, 9248.
[181] Chalasinski, G.; van Lenthe, J.H.; Groen, Th.P. *Chem. Phys. Lett.* 1984, 110, 369
[182] Tao F-M. *J. Chem. Phys.* 1993, 98, 3049.
[183] Cybulski, S.M.; Toczylowski, R.R. *J. Chem. Phys.* 1999, 111, 10520.
[184] Blomberg, M.R.A.; Siegbahn, P.E.M. *Intern. J. of Quantum Chem.* 1978, XIV, 583-592.
[185] Jones, R.O. *J. Chem. Phys.* 1979, 71, 1300.

[186] Harrison, R.J.; Handy, N.C. *Chem. Phys. Lett.* 1983, 98, 97.
[187] Diercksen, G.H.F.; Kell, V.; Sadlej A.*J. Chem. Phys.* 1986, 103, 55-74.
[188] Wright, J.S.; Barclay, V.J. *J. Chem. Phys.* 1991, 12, 697-704.
[189] Berthod, H.; Pullman, A. *Chem. Phys. Lett.* 1980, 70, 434.
[190] Vaden, T.D.; Lisy, J.M. *J. Chem. Phys.* 2006, 124, 214315
[191] Morokuma, K.; Pedersen, L. *J. Chem. Phys.* 1967, 48, 3275.
[192] Morokuma, K.; Winick, J.R. *J. Chem. Phys.* 1970, 52, 1301.
[193] Matsuoka, O.; Clementi, E.; Yoshimine, M. *J. Chem. Phys.* 1976, 64, 1351.
[194] Szczesniak, M.M.; Scheiner, S. *J. Chem. Phys.* 1986, 84, 6328.
[195] Szalewicz, K.; Cole, S.J.; Kolos, W.; Bartlett, R.J. *J. Chem. Phys.* 1988, 89, 1662.
[196] Vos, R.J.; Hendriks, R.; and van Duijneveldt, F.B. *J. Comp. Chem.* 1990, 11, 1-18.
[197] Smith, B.J.; Swanton, D.J.; Pople, J.A.; Schaefer III, H.F. *J. Chem. Phys.* 1990, 92, 1240.
[198] Feller, D. *J. Chem. Phys.* 1992, 96, 6104.
[199] Specchio, R.; Famulari, A.; Sironi, M.; Raimondi, M. *J. Chem. Phys.* 1999, 111, 6204.
[200] Kim, K.S.; Mhin, B.J.; Choi, U.S.; Lee, K. *J. Chem. Phys.* 1992, 97, 6649.
[201] Chakravorty, S.J.; Davidson, E.R. *J. Phys. Chem.* 1993, 97, 6373.
[202] Chen, H.Y.; Sheu WS. *J. Chem. Phys.* 1999, 110, 9032.
[203] Tschumper, G.S.; Leininger, M.L.; Hoffman, B.C.; Valeev, E.F.; Schaefer, H.F.; Quack, M. *J. Chem. Phys.* 2002, 116, 690.
[204] Leforestier, C.; Gatti, F.; Raymond S.F.; Saykally, R.J. *J. Chem. Phys.* 2002, 117, 8710.
[205] Watanabe, Y.; Taketsugufoot, T.; Wales, D.J. *J. Chem. Phys.* 2004, 120, 5993.
[206] Misquitta, A.J.; Szalewicz K. *J. Chem. Phys.* 2005, 122, 214109.
[207] Kistenmacher, H.; Lie G.C.; Popkie H.; Clementi E. *J. Chem. Phys.* 1974, 61, 546.
[208] Belford, D.; Campbell, E.S. *J. Chem. Phys.* 1987, 86, 7015.
[209] Choi J.Y.; Davidson E.R. *J. Comp. Chem.* 1988, 10, 163-175.
[210] Pastor N.; Ortega-Blake I. *J. Chem. Phys.* 1993, 99, 7899- 8035.
[211] Xantheas, S.S.; Dunning, Jr Th.H. *J. Chem. Phys.* 1993, 99, 6774.
[212] Mhin, B.J.; Kim, J.S.; Lee, S.; Lee, J.Y.; Kim, K.S. *J. Chem. Phys.* 1994, 100, 4484.
[213] Xantheas, S.S. *J. Chem. Phys.* 1995, 102, 4505.
[214] Klopper, W.; Almlöf, J. *J. Chem. Phys.* 1993, 99, 5167.
[215] Jensen, J.O.; Krishnan, P.N.; Burke L.A. *Chem. Phys. Lett.* 1995, 246, 13-19.
[216] Khan, A. *J. Chem. Phys.* 2002, 116, 6628.
[217] Gregory, J.K.; Clary, D.C. *J. Chem. Phys.* 1995, 103, 8924.
[218] Wales, D.J.; Walsh, T.R. *J. Chem. Phys.* 1996, 105, 6957.
[219] Kryachko, E.S. *Chem. Phys. Lett.* 1997, 272, 132-138.
[220] Schütz, M.; Rauhut, G.; Werner, H-J. *J. Phys. Chem.* A 1998, 102, 5997-6003.
[221] Rodriguez, J.; Laria, D.; Marceca, E.J.; Estrin, D.A. *J. Chem. Phys.* 1999, 110, 9039.
[222] Grigorenko, B.L.; Nemukhin, A.V.; Topol, I.A.; Burt, S.K. *J. Chem. Phys.* 2000, 113, 2638.
[223] Kozmutza, C.; Kryachko, E.; Tfirst, E. *J. Mol. Structure (Theocem)* 2000, 501-502, 435.
[224] Losada, M.; Leutwyler, S. *J. Chem. Phys.* 2002, 117, 2003.
[225] Keutsch, F.N.; Saykally, R.J.; Wales, D.J.*J. Chem. Phys.* 2002, 117, 8823.
[226] Lischka, H. *J. Am. Chem. Soc.* 1974, 96, 4761.

[227] Frisch, M.J.; Pople, J.A.; Del Bene, J.E. *J. Phys. Chem.* 1985, 89, 3664-3669.
[228] Seel, M.; Del Re, G. *Intern. J. of Quantum Chem.* 1986, XXX, 563-566.
[229] Frisch, M.J.; Del Bene, J.E.; Binkley, J.S.; Schaefer III, H.F. *J. Chem. Phys.* 1986, 84, 2279.
[230] Del Bene, J.E. *J. Comp. Chem.* 1989, 10, 603-615.
[231] Szczesniak, M.M.; Brenstein, R.J.; Cybulski, S.M.; Scheiner, S. *J. Phys. Chem.* 1990, 94, 1781-1788.
[232] Racine, S.C.; Davidson, E.R.; Racine S.C.; Davidson, E.R. *J. Phys. Chem.* 1993, 97, 6367.
[233] Peterson, K.A.; Dunning T.H, Jr. *J. Chem. Phys.* 1995, 102, 2032.
[234] Langlet, J.; Caillet, J.; Caffarel, M. *J. Chem. Phys.* 1995, 103, 8043.
[235] Nelson, Jr., D.D.; Fraser, G.T.; Klemperer, W. *Science* 1987, 238, 1670.
[236] Latajka, Z.; Scheiner, S. *J. Comp. Chem.* 1987, 8, 674-682.
[237] Hassett, D.M.; Marsden, C.J.; Smith, B.*J. Chem. Phys. Lett.* 1991, 183, 449.
[238] Muguet, F.F.; Robinson, G.W. *J. Chem. Phys.* 1995, 102, 3648.
[239] Boese, A.D.; Amalendu, C.; Martin, J.M.L.; Marx, D. *J. Chem. Phys.* 2003, 119, 5965.
[240] Kolos, W.; Ranghino, G.; Clementi, E.; Novaro, O. *Intern. J. of Quantum Chem.* 1980, XVII, 429-448.
[241] Turi, L.; Dannenkey, J.J. *J. Phys. Chem.* 1993, 97, 7899-7909.
[242] Gu, Y.; Kar, T.; Scheiner, S. *J. Am. Chem. Soc.* 1999, 121, 9411-9422.
[243] Kryachko, E.S.; Zeegers-Huyskens, Th. *J. Phys. Chem.* 2001, 105, 7118-7125.
[244] Kryachko, E.S.; Zeegers-Huyskens, Th. *J. Phys Chem. A* 2002, 106, 6832.
[245] Scheiner, S.; Kar, T. *J. Phys. Chem.* A 2002, 106, 1784-1789.
[246] Ahlström, M.; Jönsson, B.; Karlström, G. *Mol. Phys.* 1979, 38, 1051-1059.
[247] Sadlej, J.; Roos, B.O. *Theor. Chim. Acta* 1989, 76, 173-185.
[248] Stibor, I.; Holy, P.; Hobza, P.; Cársky, P. *J. Comp. Chem.* 1990, 11, 837-847.
[249] Jaffe, R.L.; Smith, G.D. *J. Chem. Phys.* 1996, 105, 2780.
[250] Kryachko, E.S.; Nguyen, M.T. *J. Phys. Chem.* A 2002, 106, 4267-4271.
[251] Gao, S.; Hahn, J.R.; Ho, W. *J. Chem. Phys.* 2003, 119, 6232.
[252] Tsuzuki, S.; Honda, K.; Uchimaru, T.; Mikami, M. *J. Chem. Phys.* 2004, 120, 647.
[253] Sahu, P.K.; Lee, S-L. *J. Chem. Phys.* 2005, 123, 044308.
[254] Berzelius, J.J. *J. Chem. Phys.* 1813, 6, 119.
[255] Kekule, A. *Z. Chem.* 1867, 3, 217.
[256] Mulliken, L.S. Lewis acids and bases and molecular complexes. *J. Chem. Phys.* 1951, 19, 514.
[257] Hollister, C.; Sinanoglu, O. *J. Am. Chem. Soc.* 1966, 88, 13
[258] Ruedenberg K. *Rev. Mod. Phys.* 1962, 34, 326-376.
[259] Feinberg, M.J.; Ruedenberg, K. *J. Chem. Phys.* 1971, 55, 5804.
[260] Ruedenberg, K. *Localization and Delocalization in Quantum Chemistry* 1975, 1, 223-245.
[261] Paolini, J.P. *J. Comp. Chem.* 1990, 11, 1160-1163.
[262] Schneider, W.G. *J. Chem. Phys.* 1955, 23, 26.
[263] Daudey, J.P.; Claverie, P.; Malrieu, J.P. *Intern. J. of Quantum Chem.* 1974, VIII, 1-15.
[264] Allinger, N.L.; Kok, R.A.; Imam, M.R. *J. Comp. Chem.* 1988, 9, 591-595.

[265] Van Lenthe, J.H.; van Duijneveldt-van de Rijdt, J.G.C.M.; van Duijneveldt, F.B. In Ab Initio Methods in Quantum Chemistry - II. Lawley, K.P.; Ed. Wiley, New York, NY, 1987, pp. 521.
[266] Gutowski, M.; Piela, L. *Mol. Phys.* 1988, 64, 337-355.
[267] Magnasco, V.; Figari, G.; Costa, C. *J. Mol. Struct. (Theochem)* 1992, 261, 237-253.
[268] Mitchell, J.B.O.; Price, S.L. *J. Comp. Chem.* 1990, 11, 1217-1233.
[269] Hobza, P.; Zahradnik, R. *Intern. J. of Quantum Chem.* 1992, 42, 581-590.
[270] Amovilli, C.; McWeeny, R. *Chem. Phys.* 1995, 198, 71-78.
[271] Cybulski, S.M.; Chalasinski, G.; Moszynski, R. *J. Chem Phys.* 1990, 92, 4357.
[272] Kapuy, E.; Kozmutza, C. *Chem. Phys. Lett.* 1994, 226, 484.
[273] Roeggen, I.; Wind, P.A. Theory. *J. Chem. Phys.* 1996, 105, 2751.
[274] Mo, Y.; Gao, J.; Peyerimhoff, S.D. *J. Chem. Phys.* 2000, 112, 5530.
[275] Langlet, J.; Caillet, J.; Berges, J.; Reinhard, P. *J. Chem. Phys.* 2003, 118, 6157.
[276] Salvador, P.; Mayer, I. *J. Chem. Phys.* 2004, 120, 5882.
[277] Kristyán, S.; Pulay, P. *Chem. Phys. Lett.* 1994, 229, 175-180.
[278] Novoa, J.J.; Sosa, C. *J. Phys. Chem.* 1995, 99, 15837-15845.
[279] Wu, X.; Vargas, M.C.; Nayak, S.; Lotrich, V.; Scoles, G. *J. Chem. Phys.* 2001, 115, 8748.
[280] Wu, Q.; Yang, W. *J. Chem. Phys.* 2002, 116, 515.
[281] Cybulski, S.M.; Seversen, C.E. *J. Chem. Phys.* 2005, 122, 014117.
[282] Johnson, E.R.; Becke, A.D. *J. Chem. Phys.* 2005, 123, 024101.
[283] Scrocco, E.; Tomasi, J. *Adv. in Quant. Chem.* 1978, II, 116-191.
[284] Burnham, C.J.; Xantheas, S.S. *J. Chem. Phys.* 2002, 116, 5115.
[285] Cabaleiro-Lago, E.M.; Rios, M.A. *J. Chem. Phys.* 1999, 110, 6782.
[286] Chalasinski, G.; Szczesniak, M.M.; Cybulski, S.M. *J. Chem. Phys.* 1989, 92, 2481.
[287] Bhattacharya, A.; Anderson J.B. *J. Chem. Phys.* 1994, 100, 8999.
[288] Gonzalez-Lezana. A; Rubayo-Soneira, J.; Miret-Artés, S.; Gianturco, F.A.; Delgado-Barrio, G.; Villarreal, P. *J. Chem. Phys.* 1999, 110 (18), 9000.
[289] Umezawa, N.; Tsuneyuki, S.; Ohno, T.; Shiraishi, K.; Chikyow, T. *J. Chem. Phys.* 2005, 122, 224101.
[290] Sudhakar, P.V.; Lammertsma, K. *J. Chem. Phys.* 1993, 99, 7929.
[291] Kuo, J.L. *J. Phys.:Conf. Ser.* 28 2006, 87-90.
[292] Muguruma, Ch.; Koga, N.; Kitaura, K.; Morokuma, K. *Chem. Phys. Lett.* 1994, 224, 139-144.
[293] Muguruma, Ch.; Koga, N.; Kitaura, K.; Morokuma, K. *J. Chem. Phys.* 1995,103, 9274.
[294] Masella, M.; Flament, J-P. *J. Chem. Phys.* 1999, 110, 7245.
[295] Cabaleiro-Lago, E.M.; Rios MA. *J. Chem. Phys.* 2000, 113, 9523.
[296] Rincon, L.; Almeida, R.; Garcia-Aldea, D.; Diez y Riega, H. *J. Chem. Phys.* 2001, 114, 5552.
[297] Patwari, G.N.; Ebata, T.; Mikami, N. *J. Chem. Phys.* 2002, 116, 6056.
[298] Van der Vaart, A.; Merz, Jr, K.M. *J. Chem. Phys.* 2002, 116, 7380.
[299] Chalasinski, G.; Cybulski, S.M.; Szczesniak, M.M.; Scheiner, S. *J. Chem. Phys.* 1989, 91, 7048.
[300] Cabaleiro-Lago, E.M.; Otero, J.R. *J. Chem. Phys.* 2002, 117, 1621.
[301] Gálvez, O.; Gómez, P.C.; Pacios, L.F. *J. Chem. Phys.* 2003, 118, 4878.
[302] Vaden, T.D.; Lisy, J.M. *J. Chem. Phys.* 2005, 123, 074302.

[303] Gregory, J.K.; Clary, D.L. *J. Chem. Phys.* 1996, 105, 6626.
[304] Tarakeshwar, P.; Kim, K.S.; Brutschy, B. *J. Chem. Phys.* 2001, 114, 1295.
[305] Kuo, J-L.; Clobanu, C.V.; Ojamäe, L.; Shavitt, I.; Singer, S.J. *J. Chem. Phys.* 2003, 118, 3583.
[306] Girardi, M.; Figueiredo, W. *J. Chem. Phys.* 2004, 120, 5285.
[307] Kulkarni, A.D.; Ganesh, V.; Gadre, S.R. *J. Chem. Phys.* 2004, 121, 5043.
[308] Furuhama, A.; Dupuis, M.; Hirao, K. *J. Chem. Phys.* 2006, 124, 164310.
[309] Yoon, B.J.; Morokuma, K.; Davidson, E.R. *J. Chem. Phys.* 1985, 83, 1223.
[310] Hermansson, K.; Ojamae, L. *Intern. J. of Quantum Chem.* 1992, 42, 1251-1270.
[311] Lee, C.T.; Chen, H.; Fitzgerald, G. *J. Chem. Phys.* 1995, 102, 266.
[312] Hamann, D.R. *Phys. Rev. B* 1997, 55, R10 157.
[313] Kuo, J-L.; Coe, J.V.; Singer, S.J.; Band, Y.B.; Ojamäe, L. *J. Chem. Phys.* 2001, 114, 2527.
[314] Saitta, A.M.; Strassle, T.; Rousse, G.; Hamel, G.; Klotz, S.; Nelmes, R.J.; Loveday, J.S. *J. Chem. Phys.* 2004, 121, 8430.
[315] Monson, A.; Rigby, M.; Steele, W.A. *Mol. Phys.* 1983, 49, 893-898.
[316] Tanaka, H. *J. Chem. Phys.* 2000, 112, 799.
[317] Da Silva, F.L.B.; Olivares-Rivas, W.; Degréve, L.; Ákesson, T. *J. Chem. Phys.* 2001, 114, 907.
[318] Tsunekawa, N.; Milyagawa, H.; Kitamura, K.; Hiwatan Y. *J. Chem. Phys.* 2002, 116, 6725.
[319] Yu, H.; Hansson, T.; van Gunsteren, W.F. *J. Chem. Phys.* 2003, 118, 221.
[320] Jensen, L.; Van Duijnen, P.T.; Snijders, J.G. *J. Chem. Phys.* 2003, 118, 514.
[321] Senn, H.M.; Margl, P.M.; Schmid, R.; Ziegler, T.; Blöchl, P.E. *J. Chem. Phys.* 2003, 3, 1089.
[322] Prendergast, D.; Grossman, J.C.; Galli, G. *J. Chem. Phys.* 2005, 123, 014501.
[323] Cabral do Couto, P.; Estácio, S.G.; Costa, C.J. *J. Chem. Phys.* 2005, 123, 054510.
[324] Yoshizawa, K.; Hoffmann, R. *J. Chem. Phys.* 1995, 103, 2126-2131.
[325] Fink, K.; Staemmler, V. *J. Chem. Phys.* 1995, 103, 2603,
[326] Parra, R,D,; Bulusu, S,; Zeng, X.C. *J. Chem. Phys.* 2003, 118, 3499.
[327] Otto, P.; Ladik, J.; Laki, K.; Szent-Györgyi, *Proc. Natl. Acad. Sci. USA* 1978, 75, 3548-3550.
[328] Hobza, P.; Sandorfy, C. *J. Am. Chem. Soc.* 1987, 109, 1302.
[329] Elstner, M.; Hobza, P.; Frauenheim, Th.; Suhai, S.; Kaxiras, E. *J. Chem. Phys.* 2001, 114, 5149.
[330] Magyar, A.; Szendi, Z.; Kiss, J.T.; Pálinkó, I. *J. Mol. Struct. (Theochem)* 2003, 666-667, 163-168.
[331] Khaliullin, R.Z.; Head-Gordon, M.; Bell, AT. *J. Chem. Phys.* 2006, 124, 204105
[332] Haley, T.P.; Graybill, E.R.; Cybulski, S.M. *J. Chem. Phys.* 2006, 124, 204301.
[333] Boutonnet, N.; Van Belle, D.; Wodak, S.J. *Theor. Chem. Accounts* 2001, 106, 10-21.
[334] Van Lenthe, JH.; den Boer, D.H.W.; Havenith, R.W.A.; Schauer, R.; Siebert, H-C. *J. Mol. Struct. (Theochem)* 2004, 677 , 29-37.
[335] Lu, D.; Liu, Z.; Liu, Z.; Zhang, M.; Ouyang, P. *J. Chem. Phys.* 2005, 122, 134902.
[336] Movahed, P.; Jönsson, B.A.G.; Birnir, B.; Wingstrand, J.A.; Jorgensen, T.D.; Ermund, A.; Sterner, O.; Zygmunt, P.M.; Högestätt, E.D. *J. Biol. Chem.* 2005, 46, 38496-38504.
[337] Güner, U.; Arkun, Y.; Erman, B. *J. Chem. Phys.* 2006, 124, 134911.

[338] Clementi, E.; Popkie, H. *J. Chem. Phys.* 1972, 57, 1077-1094.
[339] Kistenmacher, H.; Popkie, H.; Clementi, E. *J. Chem. Phys.* 1974, 61, 799.
[340] Corongiu, G.; Clementi, E. *J. Chem. Phys.* 1978, 69, 4885.
[341] Newsome, J.R.; Neilson, G.W.; Enderby, J.E. J. Phys. C. Solid State Phys. 1980, 13, L923.
[342] Floris, F.M.; Martinez, J.M.; Tomasi, J. *J. Chem. Phys.* 2002, 116, 5460.
[343] Kassab, E.; Evleth, E.M.; Hamon-Tahra, Z.D. *J. Am. Chem. Soc.* 1990, 112, 103.
[344] Wang, X-B.; Nicholas, J.B.; Wang, L-S. *J. Chem. Phys.* 2000, 113, 10837.
[345] Raugel, S.; Klein, M.I. *J. Chem. Phys.* 2001, 116, 196.
[346] Klein, M.L. *Science* 2001, 291, 2106.
[347] Kallen, R.G.; Jencks, W.P. *J. Biol. Chem.* 1966, 241, 5864- 5878.
[348] Williams, I.H.; Spangler, D.; Femec, D.A.; Maggiora, G.M.; Schowen, R.L. *J. Am. Chem. Soc.* 1983, 105, 31-40.
[349] Gould, I.R,; Cornell, W.D,; Hillier, I.H. *J. Am. Chem. Soc.* 1994, 116, 9250-9256.
[350] Kryachko, E.S.; Nakatsuji, H. *J. Phys. Chem. A* 2002, 106, 731-742.
[351] Wei, D.Q.; Truchon, J-F.; Sirois, S.; Salahub, D. *J. Chem. Phys.* 2002, 116, 6028-6038.
[352] Dudowicz, J.; Freed, K.F.; Shen, M.Y. *J. Chem. Phys.* 2003, 118, 1989.
[353] Van Erp T.S.; Melijer, E.J. *J. Chem. Phys.* 2003, 118, 8831.
[354] Faraone, A.; Liu, L.; Chen, S.H. *J. Chem. Phys.* 2003, 119, 6302.
[355] Estácio, S.G.; Cabral do Couto, P.; Guedes, R.C.; Costa Cabral, B.J.; Martinho Simoes, J.A. *Theor Chem. Acc.* 2004, 112, 282-289.
[356] Chaudhari, A.; Sahu, P.K.; Lee, S-L. *J. Chem. Phys.* 2004, 120, 170.
[357] Herrera, B.; Toro-Labbe, A. *J. Chem. Phys.* 2004, 121, 7096.
[358] Kim, S.; Wheeler, S.E.; Schaefer III, H.F. *J. Chem. Phys.* 2006, 124, 204310.
[359] Jensen, J. H.; Li, Hui.; Robertson, A. D.; Molina, P. A. *J. Phys. Chem. A* 2005, 109, 6634-6643.
[360] Van Thiel, M.; Becker, E.D.; Pomentel, G.C. *J. Chem. Phys.* 1957, 27, 486.
[361] Peterson, K.I.; Fraser, G.T.; Nelson, Jr., D.D.; Klemperer, W. In. Ab Initio Quantum Chemistry with Experiment for small Molecules. Bartlett, R.J. Ed. Riedel, Dodrecht, Holland. 1985, pp.217-244
[362] Silvi, B.; Wieczorek, R.; Latajka, Z.; Alikhani, M.E.; Dkhissi, E.; Bouteiller, Y. *J. Chem. Phys.* 1999, 111, 6671.
[363] Fajardo, M.E.; Tam, S. Observation *J. Chem. Phys.* 2001, 115, 6807.
[364] Dorbez-Sridi, R.; Cortés, R.; Mayer, E.; Pin, S. *J. Chem. Phys.* 2003, 116, 7269.
[365] Souda, R.; Kawanowa, H.; Kondo, M.; Gotoh, Y. *J. Chem. Phys.* 2003, 119, 6194.
[366] Van Wijngaarden, J.; Jäger, W. *J. Chem. Phys.* 2002, 116, 2379.
[367] Björneholm, O.; Federmann, F.; Kakar, S.; Möller, T. *J. Chem. Phys.* 1999, 111, 546.
[368] Isaacs, E.D.; Shukla, A.; Platzman, P.M.; Hamann, D.R.; Barbiellini, B.; Tulk, C.A. *Phys. Rev. Lett.* 1998, 82, 600.
[369] Isaacs, E.D.; Shukla, A.; Platzman, P.M.; Hamann, D.R.; Barbiellini, B.; Tulk, C.A. *J. Phys. Chem.* Sol. 2000, 61, 403-406.
[370] Romero, A.H.; Silvestrelli, P.L.; Parrinello, M. *J. Chem. Phys.* 2001, 115, 115.
[371] Kropman, M.F.; Bakker, H.J. *Science* 2001, 291, 2118.
[372] Wade, E.A.; Cline, J.I.; Lorenz, K.T.; Hayden, C.; Chandler, D.W. *J. Chem. Phys.* 2003, 116, 4755.

[373] Arunan, E.; Emilsson, T.; Gutowsky, H.S.; Fraser. G.T.; de Oliveira, G.; Dykstra, C.E. *J. Chem. Phys.* 2002, 117, 9766.

[374] Daigoku, K.; Ishiuchi, S.; Sakai, M.; Fujii, M.; Hashimoto, K. *J. Chem. Phys.* 2003, 119, 5149.

[375] Lavrich, R.J.; Plusquellic, D.F.; Suenram, R.D.; Fraser, G.T.; Walker, A.R.H.; Tubergen, M.J. *J. Chem. Phys.* 2003, 118, 1253.

[376] Hu, Y.; Yang, S. *J. Chem. Phys.* 2003, 119, 8321.

[377] Tubergen, M.J.; Torok, C.R.; Lavrich, R.J. *J. Chem. Phys.* 2003, 119, 8397.

[378] Alavi, S.; Ripmeester, J.A.; Klug, D.D. *J. Chem. Phys.* 2005, 123, 051107.

[379] Itatani, J.; Levesque, J.; Zeidler, D.; Niikura, H.; Päpin, Kieffer, H.J.C.; Corkum, P.B.; Villeneuve, D.M. *Nature* 2004, 432, 867.

[380] Szczesniak, M.M.; Chalasinski, G.; Cybulski, S.M.; Cieplak, P. *J. Chem. Phys.* 1993, 98, 3078.

[381] Cao, Z.T.; Tester, J.W.; Trout, B.L. *J. Chem. Phys.* 2001, 115, 2550.

[382] Omolulu, A-O.; Szalewicz, K. *J. Chem. Phys.* 2005, 123, 134311.

In: Quantum Chemistry Research Trends
Editor: Mikas P. Kaisas, pp. 57-125
ISBN: 978-1-60021-620-6

Chapter 2

ELECTRONIC STRUCTURE AND REACTIVITY OF INORGANIC, ORGANIC, ORGANOELEMENT AND COORDINATION COMPOUNDS: AN EXPERIENCE IN THE AREA OF APPLIED QUANTUM CHEMISTRY

Alexei N. Pankratov*

Division of Analytical Chemistry and Chemical Ecology, Department of Chemistry, N. G. Chernyshevskii Saratov State University, 83 Astrakhanskaya Street, Saratov 410012, Russia

Dedicated to my mother Elena M. (Trepak) Pankratova
and to the bright memory of my father Nickolai D. Pankratov
and my grandmother Vasilisa K. (Miron'ko) Trepak

Abstract

A brief overview of the author's line of scientific inquiry has been presented. Discussed are results of the establishment of predictive structure - property relationships, the study of interrelation between the substances reactivity and the molecules electronic structure, the elucidation of mechanisms of chemical reactions and the allowance for medium effects on their proceeding, the solving of other interpretational tasks of chemistry. The methods of different hierarchy and theory level, adequate to the problems to be solved, are used, from semiempirical to modern sophisticated *ab initio* and DFT methods, along with comprehensive approaches and techniques including the NBO, energetic deletion, AIM, orbital interactions, electrostatic potentials analyses, calculation of vibrational, electronic absorption and NMR spectra, scaling of vibrational frequencies, separation and analysis of normal vibrations, ZPVE, thermal and BSSE corrections, localization of transition states, IRC, SCRF, *et al.* The capabilities and potentialities of the current quantum chemistry have been considered. The

* E-mail address: PankratovAN@chem.sgu.ru; PankratovAN@info.sgu.ru,
Web: chem..saratov.ru/pankratov; www.chem.saratov.ru/pankratov; sgu.ru/faculties/chemical/pankratov; www.sgu.ru/faculties/chemical/pankratov

problems of systematization and generalization of the information resources on natural sciences, on ecology are also involved.

The major problem facing the chemistry today and in the future is to predict the time course of chemical process and its ultimate result starting from the information on structure and properties of the molecular systems of reactants, intermediates, products, as well as on the transition states, taking into account the solvation shells, *i.e.* to gain information on the chemical reaction mechanism.

The reaction mechanism is the reaction description in detail, taking into account all the intermediate stages and the substances emerging here (intermediates), the nature of reactive species interaction, the character of bonds cleavage, the change in energy of the chemical system over the whole pathway of its transfer from initial to final state [1].

Reaction mechanism is a sum of information on the nature of reactive species of reactants, intermediates, yield products, including the solvation shells, on the change in concentrations of the above species with time, as well as on the transition states and other specific points on the potential energy surface, in the perfect case - a comprehensive information on the potential energy surface of chemical reaction [1, 2].

The term "reaction mechanism" is at least by the order of magnitude more complicated compared to the term "structure". The mechanism definition requires obligatorily the description in detail of such items as how the structure and chemical bonds change in reactants during the every separate step of chemical transformation [3].

The reactions proceeding in one step are called simple (elementary) reactions, and the reactions that involve several steps are complex [1].

The term of reaction mechanism is used mainly in two senses. For complex reactions involving several stages, the mechanism is a sum of stages resulting in the transformation of initial substances into products. For the simple reaction (elementary reaction, elementary stage), which could not be subdivided into more simple chemical acts, the elucidation of reaction mechanism means the identification of physical processes underlying the essence of chemical transformation. For single particle (molecule in ground or excited state, ion, radical, ion radical, diffuse pair, singlet or triplet radical pair, complex) or two (rarely three) particles (molecules, ions, radicals, ion radicals, *etc.*) in the definite quantum states, the changes in atomic nuclei positions and electronic states consist the essence of their transformations into another species with the intrinsic quantum states. In the physical processes under consideration, the acts of energy transfer from one species to another in the explicit form are frequently involved. For the elementary reactions in solution, the mechanism comprises the changes in adjacent solvation shell of transformed species [1].

In the case of complex reactions, the unstable intermediate species - ions, free radicals, ion radicals, *etc.* - are formed at some stages. Their relative stability and, consequently, a probability for their formation increases with the more actual possibility of delocalization (dissipation) of charge in ion, or of uncoupled electron in radical [1].

For the comprehensive description of complex reaction mechanism, the following data are usually required:

1) the reaction subdivision into the separate steps and equilibrium stages;
2) the characteristics of intermediate products and evaluation of their lifetime;

3) the description of transition state for every step of reaction on the basis of: i) composition; ii) geometry; iii) solvation viewpoints; iiii) energetics;
4) overall description of the processes, both preceding the every transition state and following that, starting from the information on energy levels (electronic, vibrational, rotational, *etc.*) in the ground and excited states [3].

The aim of studying the reaction mechanism is to gain a possibility of managing the reaction course, its direction and efficiency. A knowledge of reaction mechanisms is by far necessary for managing the yields of any products, optimizing the laboratorial and industrial processes, systematizing the reactions, revealing the resemblance of different reactions, predicting the consequences of altered reaction conditions (the substituent nature and position, solvent, temperature, *etc.*) on the reaction course, predicting new reactions [4]. Finally, the sequential elucidation of chemical reactions at a molecular level provides a considerable intelligence satisfaction. Many chemists look upon this problem as being the essence of chemistry discipline and think that it deserves especially extensive studies regardless to the practical importance of the investigation [4]. Let us refer in this relation to the belief by L.P. Hammett, a leading scientist in physical organic chemistry, one of the founders of correlation analysis: "The successful decoding of a complicated mechanism provides an esthetic satisfaction, and that always attracted chemists, even if they would have none of that sometimes".

The special role in substantiating the mechanisms of reactions is assigned to quantum chemical modeling of their pathways (routes) with the localization of transition states and the construction of energy profiles.

The prognostic function of chemistry determines a situation, when the necessity for elucidating a relation of physicochemical, analytical and other properties, reactivity, biological activity of substances on the structure of their molecules and ambient factors, first of all the medium (solvent), is brought in the forefront. In the opinion of Professor, Academician of the Russian Academy of Sciences, Nickolai S. Zefirov, the problem of structure - property relationship is the most important in chemistry and, possibly, in the whole natural sciences. The above problem is especially urgent under the conditions when, as Professor, Academician of the Russian Academy of Sciences Vladimir I. Minkin tells, a mutation of chemistry has occurred.

An increasing importance in chemistry for approaching uniformly to two topical chemical problems - constructing the reaction barriers and molecular design, for explaining regioselectivity and ambivalence (nucleophilic-electrophilic dichotomy) ([5, 6] *et al.*), for describing and modeling chemical bond, properties of molecules and crystals [6], intermolecular interactions [7], extraction [8], for treating the chemical subjects phenomena from the viewpoint of general theory of systems [7], gains the concept of electronegativity [9, 10]. The analysis of electronic effects of atomic groups in molecules is extremely important to explore the physicochemical, analytical properties of substances and molecules, reactivity, catalysis, solvation.

In the division "4. Chemical sciences on materials" in the List "The principal lines of fundamental investigations", ratified by the Presidium of the Russian Academy of Sciences [11], the aforesaid aspects are reflected in the first paragraph: "4.1. Theory of chemical structure and chemical bond, kinetics and mechanisms of chemical reactions, reactivity of chemical compounds, stereochemistry, chemistry of crystalline compounds". The second paragraph 4.2 includes a point "Synthesis and study of novel substances, development of materials and nanomaterials with the given properties and functions". Realization of lines of fundamental investigations in chemistry presented in other paragraphs of "The List", requires the use of methods, approaches, the solvation of definite scientific problems, all of which obeying the kinds of 4.1 and 4.2. The latter is totally concerned with the realization of "Fundamentals of Russian Federation Policy in development of science and technologies for up to 2010 and further prospects" [12].

The aforesaid items occupy one of the central positions within the rating arrangement of the general problems of chemistry, that could be presented [13] as follows: the art of chemical synthesis, chemical structure and function, chemical processes control, chemical materials science, chemical engineering science, chemical energetic science, chemical analytics and diagnostics, chemistry of life.

In the newspaper of Russian educational and scientific community *Search* (in Russian: *Poisk*) [14], the popularity rating of scientific journals included in the database of Scientific Electronic Library of the Russian Foundation for Basic Research, for March, 2000, was published. Of 46 editions concerned the first place by amount of appeals to papers over the above period is occupied by the journal *Chemical Physics Letters*. The third place is for the *Journal of Molecular Structure. Theochem*. For the substantiated conclusions on the structure and dynamics of the scientific interest, the rating analysis over the more prolonged period is required. Nevertheless, it is obvious that the problems of studying the structure, reactivity and mechanisms are the matter of priority for the Russian scientists.

It is likely that the citation index is insufficient and not completely objective criterion of the scientific works evaluation. However, if its value is much greater than the mean quantity in the corresponding branch of knowledge, this fact confirms, to a definite degree, the availability of scientific works. The site [15] contains the information on 1000 famous chemists who possesses the highest ratio of citation number to total number of publications over the period since 1981 to 1997. In the list of these scientists the 1st and 4th places are occupied by A. Bax and R.R. Ernst, respectively, the specialists in radiospectroscopy - an assembly of powerful methods of studying the structure, reactivity and mechanisms of the reactions, the 2nd, 3rd, 6th, 14th, 15th, 22nd, 23rd, 24th, 28th, 35th, 40th - by the theoretically-oriented chemists contributing to the quantum chemistry and its applications: J.A. Pople (Nobel Prize Winner), P.v.R. Schleyer, H.F. Schaefer, D.G. Truhlar, J.J.P. Stewart, P.A. Kollman, M.J.S. Dewar, W.L. Jorgensen, C.W. Bauschlicher, K.N. Houk, R. Hoffmann (Nobel Prize Winner), 5th by G.M. Whitesides (molecular drug design), 41th by A. Zewail, the inviter of femtosecond spectroscopy allowing the reactions (at least, relatively simple ones by now) to be studied in the real-time mode.

In the communication [16], the Russian scientists, whose works were cited until 2000 inclusively not less than 1000 times, are mentioned. Let us list those of them, who has made the outstanding contributions to the development of namely the theoretical chemistry, of the knowledge of structure and reactivity of substances (pointed out in the brackets is number of the position in the list [16]): N.S. Zefirov (3), V.I. Minkin (5), A.A. Ovchinnikov (6), V.I. Nefedov (13), A.E. Shilov (14), I.P. Beletskaya (16), Yu.A. Zhdanov (17), Yu.A. Zolotov (18), A.L. Buchachenko (23), A.F. Pozharskii (24), L.V. Vilkov (26), G.M. Zhidomirov (27), V.V. Boldyrev (28), L.M. Epshtein (33), R.A. Evarestov (39), Yu.N. Molin (45), L.A. Gribov (47), V.L. Tal'roze (48), I.I. Moiseev (52), O.M. Nefedov (54).

The work [17] deals with the citation indices (1991-2004) of chemists-analysts of Russia, some of whom are engaged in solving the problems of theoretical chemistry and apply the methods of quantum chemistry in their scientific research.

All the aforesaid testifies to the doubtless actuality of studying the problems of structure, reactivity, reaction mechanisms in chemistry, including the studies assisted by theoretical and first of all quantum chemical research methods.

One of the avenues of scientific inquiries developed at the Division of Analytical Chemistry and Chemical Ecology, Department of Chemistry of N. G. Chernyshevskii Saratov State University, inexorably associated with the problems of analytical science, could be formulated in the following manner: the establishment of quantitative structure - property relationships in the series of inorganic, organic, organoelement, coordination compounds; the statement of the reactivity interrelation to the molecules electronic structure in the ground and excited states on the basis of refining the understanding of electronic effects, electronegativities of atomic groups, hydrogen bond, generalization of the views on reaction mechanisms (including oxidation and reduction, nitrosation, nitration, azo coupling, alkoxylation, condensations, other electrophilic, nucleophilic and radical processes, complexation, ligand exchange, molecular and ionic association, dissociation, tautomerism and ambivalent reactivity, isomerization, proton, hydrogen atom and "hydride ion" transfer), regioselectivity of homolytic (oxidative and reductive) coupling reactions for the substances of different classes; study of the medium influence on chemical processes occurrence; systematization and generalization of information about Web resources on natural sciences, on ecology.

In 1997 the information on range of scientific interests and line of scientific inquiry was included to *Science Contacts Database* - the edition of Science International (Cambridge). The author refereed manuscripts presented for publication in *Journal of the American Chemical Society, The Journal of Organic Chemistry, Journal of Coordination Chemistry, Journal of Analytical Chemistry (Russia)*. On November 1, 2005 the author has been elected a Regular Member of the American Chemical Society.

There are over 530 publications, among which are articles in *Helvetica Chimica Acta, Canadian Journal of Chemistry, The Journal of Physical Chemistry A, Journal of Molecular Structure. Theochem, International Journal of Quantum Chemistry, Monatshefte für Chemie, Journal of Coordination Chemistry, Heteroatom Chemistry, Phosphorus, Sulfur, and Silicon and the Related Elements, International Journal of Molecular Sciences, Journal of Biochemistry and Molecular Biology, Food Technology and Biotechnology, Online Information Review, Internet Journal of Chemistry, Journal of the Serbian Chemical Society, Croatica Chemica Acta, Revue Roumaine de Chimie, Afinidad, The Journal of the Argentine Chemical Society, Indian Journal of Chemistry, et al.*, in the academic journals of Russia and Former Soviet Union countries. The author is the scientific editor of two Collections of articles.

The scientific area under discussion is contributed by the Ph. Ds in Chemistry Andrew E. Shchavlev, Inna M. Uchaeva, Alexei V. Shalabay, Assistant Professor of the Saratov State Medical University Olga A. Chaplygina, student Nickolai A. Bychkov and other Colleagues.

Presented below are the selected scientific results.

The quantitative relationships providing a possibility of *a priori* evaluation of heats and free energies of formation, entropies, ionization potentials, electron affinities, potentials of chemical and electrochemical oxidation and reduction, pK_a values, stability constants, dipole moments, vibrational frequencies, electronegativities, inductive and mesomeric parameters of atomic groups of inorganic, organic, organoelement, coordination compounds, products yields and reaction rates of electrophilic, nucleophilic and radical aromatic substitution, rates of enzymatic oxidation of amines and phenolic substrates, transport rate of surfactants ions through nano exclusion membranes (molecular sieves) ([18-43] *et al.*).

A priori evaluation of the above parameters by quantum chemical methods is important for molecular design of compounds with given properties, may serve as a base for expert decision on reasonability of synthesis that is sometimes labour- and time-consuming, requiring complicated device, rigid conditions, expensive and toxic chemicals.

The predictive power of the correlations stated (hundreds of dependences for hundreds of molecular systems) is determined by the fact that those are not of the formal character, but have an explicit physical meaning, are characterized by great values of correlation coefficient, are deduced for a considerable amount of substances in different media, therewith the compounds are various in nature and structure, the substituents in molecules essentially differ by their nature and positions.

For rather simplified obtaining predictive quantitative relations, often the semiempirical quantum chemical methods would be reasonable to use.

Let us present a few examples of predictive interrelationships.

By the semiempirical MNDO, AM1 and PM3 methods, standard values of entropy, heats and free energies of formation, first ionization potentials and dipole moments of the molecules have been computed for the compounds CH_3X (X = H, CH_3, SiH_3, GeH_3, SnH_3, $CH{=}CH_2$, C_6H_5, CCH, CHO, $COCH_3$, COOH, CN, NH_2, PH_2, AsH_2, OH, SH, $SOCH_3$, SO_2CH_3, OCN, NCO, SCN, NCS, NC, NO_2, F, Cl, Br, I); n-$C_5H_{11}X$ (X = H, CH_3, OH, NO_2, F, Cl, Br, I); C_6H_5X (X = H, $CH{=}CH_2$, C_6H_5, CCH, CHO, $COCH_3$, COOH, CN, NH_2, PH_2, OH, SH, SeH, SO_2H, SO_2CH_3, SO_3H, NCO, SCN, SeCN, NCS, NC, NO, NO_2, F, Cl, Br, I), 90 mono-, bi-, tri-, poly- and spirocyclic non-aromatic hydrocarbons, 148 three-, five-, six-, seven-membered N-, O-, S-, Se-, B-containing unsaturated heterocycles, 91 quinones, 62 tropones and tropolones, 63 anilines, 232 pyridines, inorganic, coordination and organic arsenic, cadmium, antimony, thallium, bismuth compounds [18, 20, 25, 27, 29, 31-35, 41, 43]. Correlation equations allowing one to evaluate *a priori* thermodynamic and molecular properties of various substances have been obtained. Deviations from the above dependences have been noted.

A facilitated theoretical methodology for the prediction of solid-phase (and condensed state in general) heats of formation (ΔH_f) of chemical species has been developed using quinones as the examples [41, 43]. Such a prognosis may be performed on the base of simple semiempirical gaseous-phase calculations. Below given are predictive dependences linking experimental solid-phase ΔH_f values with the computed ones (Table 1; Figures 1-3).

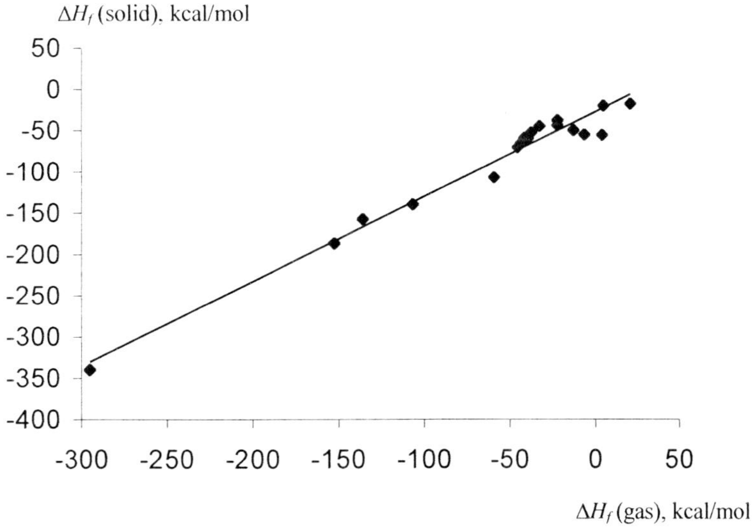

Figure 1. Dependence of standard heats of formation of quinones in solid [44] and gaseous (MNDO) states.

Table 1. Dependences between experimental standard heats of formation of solid quinones [44] and computed standard heats of formation of gaseous quinones*

Method	ΔH_f (solid) = $a + b\Delta H_f$ (gas)			ΔH_f (solid) = $c\Delta H_f$ (gas)		r
	a, kcal/mol	b	S_0^2	c	S_0^2	
MNDO	−27.23 ± 8.14	1.026 ± 0.091	183.9	1.217 ± 0.134	652.3	0.9842
AM1	−35.90 ± 8.26	1.060 ± 0.102	212.4	1.308 ± 0.194	1133	0.9817
PM3	−26.32 ± 8.20	1.057 ± 0.094	184.3	1.247 ± 0.133	615.6	0.9842

*S_0^2 is adequacy dispersion, r is correlation coefficient

The values corresponding to abscissae and ordinates start from the same origin. That is why linear regression models both with and without the intercept may be adequate. Confidence limits for the regression coefficients allow any of the two types of relationships mentioned.

In accordance with the criterion of adequacy dispersion, preferable are linear dependences of common type, *i.e.* with non-zero intercept (Table 1). Then the slope is close to unity.

Thus, the crystal field effect over the series of compounds under study makes, with respect to the heat of formation, an essentially constant contribution expressed as a scale shift in the Y direction.

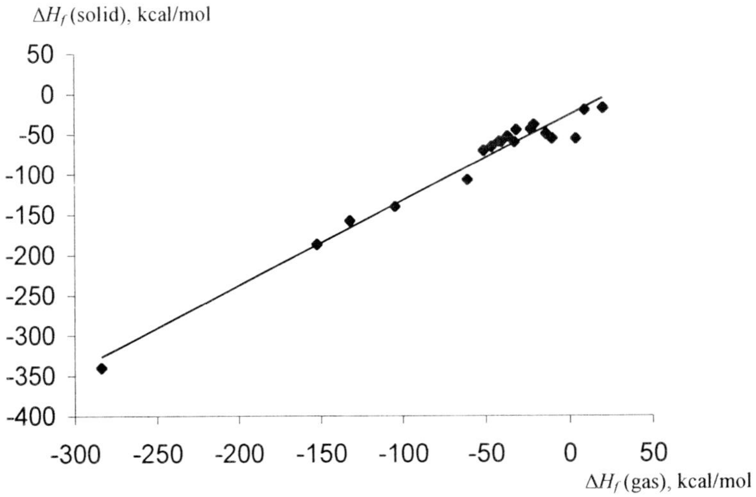

Figure 2. Dependence of standard heats of formation of quinones in solid [44] and gaseous (AM1) states.

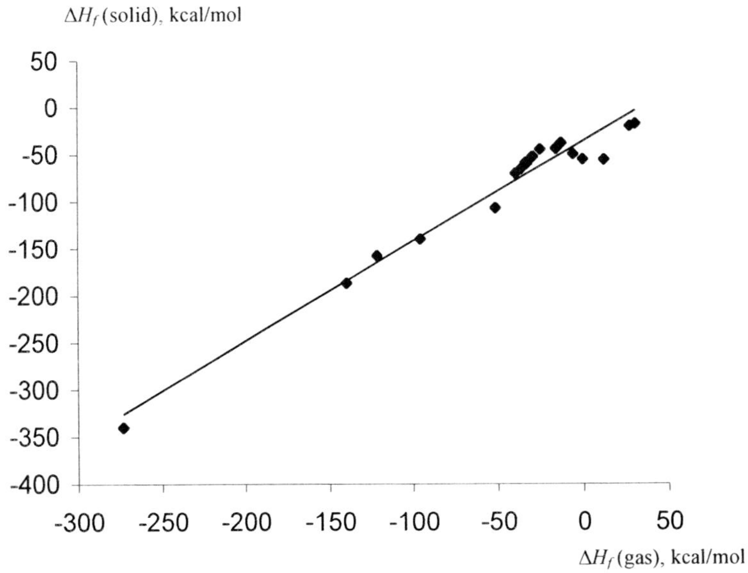

Figure 3. Dependence of standard heats of formation of quinones in solid [44] and gaseous (PM3) states.

Let us emphasize that the sample of compounds we have investigated, for which the linear relationships are valid, includes both *ortho* quinones and *para* quinones. Consequently, the contribution of intermolecular interactions in the solid state has appeared to be constant over the series of substances involving both 1,2-quinones and 1,4-quinones.

The quinone molecules contain the strongly polar C=O bonds. That is why the principal contribution to intermolecular interactions in crystalline structure should be made by electrostatic interactions, especially with the C=O group participation.

Such assumption is favoured by X-ray data [45]: oxygen atoms of the 1,2-benzoquinone molecule participate in shortened intermolecular contacts.

An unchanged electrostatic contribution is possible provided that the electron density distributions in the *ortho* and *para* quinone molecules (above all in the carbonyl group) are analogous.

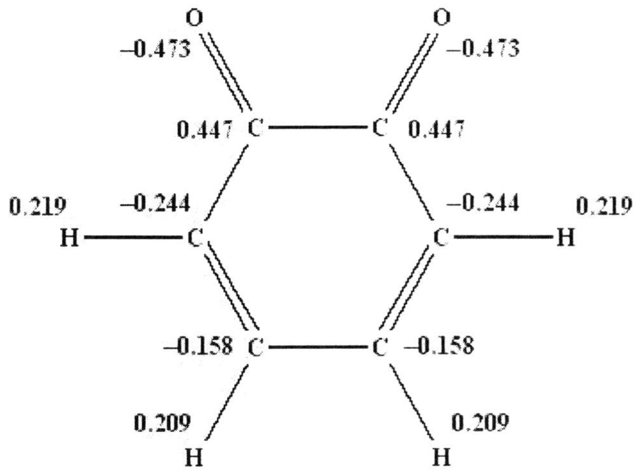

Figure 4. Natural atomic charges in the 1,2-benzoquinone molecule.

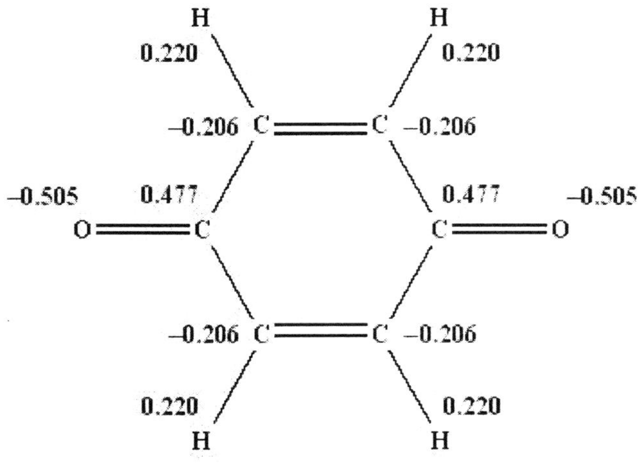

Figure 5. Natural atomic charges in the 1,4-benzoquinone molecule.

For justifying the assumption, we have studied by the hybrid B3LYP/6-311++G(3df,3pd) method within the DFT approach, the electronic structure of the 1,2-benzoquinone and 1,4-benzoquinone molecules. Indeed, the electron distribution in polar carbonyl groups and in adjacent CH groups show similar characteristics for isomeric quinones. Figures 4 and 5 depict natural charges on atoms (according to the NBO-analysis data), which characterize the molecules' electronic structures more objectively as compared to Mulliken charges, and Table 2 - Wiberg bond orders.

Table 2. Wiberg bond indexes in the natural atomic orbital basis

Bond	Wiberg index in the molecule of	
	1,2-benzoquinone	1,4-benzoquinone
C=O	1.800	1.741
C=C	1.771	1.821
C(O)-C(H)	1.071	1.040
C(CO)-H	0.917	0.918

Furthermore, the data of X-ray structure analysis [45, 46] show that elementary cells in the crystal structure of 1,2-benzoquinone and 1,4-benzoquinone contain the same number of molecules, equal to two, and the compounds have close values of density in the solid state (Table 3).

Table 3. Values of solid state density (ρ_{calc}) deduced from X-ray structure analysis data

Compound	ρ_{calc}, g·cm^{-3}	Reference
1,2-Benzoquinone	1.424	45
1,4-Benzoquinone	1.4	46

The character of electron density distribution in molecules and the similarity of structure parameters are a prerequisite for a constant contribution of the crystal field effect to the heat of formation of both *ortho* and *para* quinones.

Therefore, the fact is clarified that the unified linear dependences are valid for series of substances involving such structurally dissimilar compounds as 1,2-quinones and 1,4-quinones are.

The above methodology may be extended to compounds of other classes provided that the intermolecular interactions are near-constant over the series under study. The latter would be fulfilled *e.g.*, at the dominating contribution of the molecule frame (determining the compound's assignment to one or other class or series) or separate atomic groups (in particular, polar ones) in the composition of the frame fragment to the intermolecular interactions. Although the condition of constant energies of intermolecular interactions is not of universal character in general, nevertheless such condition should be naturally followed for the numerous series of the compounds similar in their structures.

The dependence of electronegativity (determined by various experimental and theoretical methods and expressed against different scales), inductive parameter *iota* of substituents X, mesomeric dipole moments upon the molecules CH_3X, XCH_2COOH, *n*-$C_5H_{11}X$, C_6H_5X, 4-$XC_6H_4NH_2$, 4-$XC_6H_4NO_2$, 2,6-disubstituted pyridines and N-methylpyridinium cations electronic structure indices (according to different quantum chemical methods data) has been stated. The possibility of a facilitated estimation of electronegativity, inductive and mesomeric parameters of atomic groups using 90 correlations has been shown [19, 22].

Commonly used constants following from the linear free energies relationships principle possess a number of limitations (multiplicity of sets, ambiguous interpretation, *etc.*). The quantities evaluated in the works [19, 22] overcome the aforesaid drawbacks to the significant extent, because they are expressed *via* the values having clear physical meaning: effective charge of valence shell, effective principal quantum number, group dipole moments.

It has been shown [38] that aryldiazonium cations present a series of molecular systems, on the basis of which the predictive interrelationships (alternative with respect to those proposed in the works [19, 22]) could be stated for evaluating the mesomeric dipole moments (μ_m) of atomic groups. If this problem is solved using the quantitative relations μ_m *vs* electron affinity, thus the substituted diazonium cations can be involved whether these exist or not really. An important point is solely that phenyldiazonium is appropriate "matrix", on the basis of which the mesomeric effect of substituents is simulated.

Table 4. Logarithms of stepwise stability constants for palladium (II) chloride complexes, *d* and *e* coefficients in equations $\lg K_{stab} = d + e\Delta H^0$, and values of correlation coefficient *r*

By the ZINDO/1 data

$\lg K_1$	$\lg K_2$	$\lg K_3$	$\lg K_4$	Ref.	*d*	*e*, mol/kcal	*r*
4.34	3.54	2.68	1.68	47*	2.519 ± 0.144	$-0.008597 \pm 0.001097^{+}$	-0.9991^{+}
					2.531 ± 0.249	$-0.008403 \pm 0.001867^{\wedge}$	$-0.9973^{\wedge}$
3.98	3.24	2.30	2.00	48*	2.460 ± 0.534	$-0.006672 \pm 0.004075^{+}$	-0.9804^{+}
					2.467 ± 0.445	$-0.006572 \pm 0.003337^{\wedge}$	$-0.9864^{\wedge}$
3.88	3.06	2.14	1.34	49-51**	2.083 ± 0.123	$-0.008299 \pm 0.000938^{+}$	-0.9993^{+}
					2.094 ± 0.093	$-0.008128 \pm 0.000699^{\wedge}$	$-0.9996^{\wedge}$

By the PM3 data

$\lg K_1$	$\lg K_2$	$\lg K_3$	$\lg K_4$	Ref.	*d*	*e*, mol/kcal	*r*
4.34	3.54	2.68	1.68	47*	2.295 ± 0.370	$-0.009484 \pm 0.002814^{+}$	-0.9953^{+}
					2.299 ± 0.371	$-0.009431 \pm 0.002809^{\wedge}$	$-0.9952^{\wedge}$
3.98	3.24	2.30	2.00	48*	2.280 ± 0.494	$-0.007435 \pm 0.003758^{+}$	-0.9865^{+}
					2.282 ± 0.453	$-0.007409 \pm 0.003434^{\wedge}$	$-0.9886^{\wedge}$
3.88	3.06	2.14	1.34	49-51**	1.864 ± 0.211	$-0.009182 \pm 0.001603^{+}$	-0.9984^{+}
					1.868 ± 0.172	$-0.009136 \pm 0.001304^{\wedge}$	$-0.9989^{\wedge}$

*$\lg K_{stab}$ values are determined by means of spectrophotometric method
**$\lg K_{stab}$ values are determined by means of solubility, potentiometric and spectrophotometric methods
+Correlations are found assuming *cis*-$[Pd(H_2O)_2Cl_2]$ participation in reactions by second and third steps
^Correlations are found assuming *trans*-$[Pd(H_2O)_2Cl_2]$ participation in reactions by second and third steps

On the base of quantum chemical investigations the character of ligands H_2O, H_3O^+, Cl^- mutual influence in palladium (II) complexes has been revealed. By the first three steps, the substitution of hydronium ion in the palladium atom internal sphere is thermodynamically more favourable as compared to the displacing of water molecule from the coordination sphere. Stepwise stability constants (K_{stab}) of the palladium (II) chloride complexes correlate with the computed standard enthalpy values (ΔH^0) of the substitution of water molecules by chloride ion (Table 4) [40].

The existence of the above quantitative $\lg K_{stab}$ *vs* ΔH^0 relationships confirms the fact of being constant for aqueous medium contribution to ΔH^0 at all steps of substitution.

Let us mention that the correlation coefficients are sufficiently high, in spite of the fact that non-uniform quantities are compared: the experimental values of stability constants for aqueous solutions and standard enthalpies of the reactions estimated on the basis of gaseous-phase quantum chemical computations. That underlines reliability and predicting significance of the correlations.

Use of the like correlation equations for another central atoms and ligands enables one to evaluate *a priori* the stepwise constants of complexation and to put into practice the directional synthesis of complexes with the given interrelationship of stability constants by different steps.

The results obtained are of importance when studying Pd(II) complexation with organic ligands, including that in analytical chemistry. As shown in the present work, the rate of exchange of hydronium ion by chloride ion in palladium (II) complexes much exceeds the rate of substitution of water molecule. Provided that a certain organic reagent is involved instead of chloride ion, the analytical reaction would be expected to proceed more readily in acidic medium. The reaction should be conducted at such optimal acidity, at which water molecule in the internal coordination sphere of metal atom is protonated, but even though the reagent molecule is protonated, so not to the reaction centre of complexation.

An approach has been developed to the selectivity prediction and molecular design of analytical redox reagents [23, 28]. Selectivity of such reagents is determined by their standard (E^0) or close to them formal (E^f) redox potentials. We have obtained quantitative expressions linking the E^f values with quantum chemically computed Koopmans ionization potentials. Professor Svetlana P. Mushtakova *et al.* (Division of General and Inorganic Chemistry, Department of Chemistry, N. G. Chernyshevskii Saratov State University) have synthesized 5 new reagents and measured their E^f quantities [52]. Using our relationships, we have evaluated the above E^fs theoretically (Table 5).

Table 5. Formal redox potentials (E^f) of sulphodiphenylamines and sulphotriphenylamines: experimental ones determined in H_2SO_4 aqueous solutions [52] and those evaluated from first Koopmans ionization potentials of amines and their conjugate acids

Compound	E^f_{theor} (V) from ionization potential of		E^f_{exper}, V
	Molecular form	Protonated form	
2-Nitro-4'-sulphodiphenylamine	1.07	0.96	0.91
3-Nitro-4'-sulphodiphenylamine	1.15	0.98	0.93
4-Nitro-4'-sulphodiphenylamine	1.16	0.98	1.1
4-Sulphotriphenylamine	0.94	0.91	0.89
4,4'-Disulphotriphenylamine	1.00	0.92	0.94

As it can bee seen, theoretical E^f values coincide well with the experimental ones. The most perfect agreement with the experiment is observed as a rule under the theoretical evaluations based on the ionization potentials of conjugate acids of amines. That confirms indirectly our experimental results on studying the mechanisms of diarylamines oxidation in acidic media, in accordance with which the protonated molecules of substrates undergo oxidation. Commonly the oxidation reactions are treated as electrophilic processes [53]. However, in the given case the cation of amine conjugate acid is attacked by the oxidizing agent anion, and the initial act of reaction is nucleophilic.

Based on the study [54] concerned with the rates (v) of amines and phenolic compounds oxidation catalyzed by laccase of basidiomycete *Lentinus edodes* (Berk.) Sing, as well as on the results of PM3 computations, the linear correlations of v and ln$v$ values with first ionization potentials of the substrates molecules and radicals derived from them, spin densities on N and O atoms of the above radicals, and with the radicals reorganization energies have been obtained [24]. The interrelations found could be used for express facilitated (at semiempirical level) predicting v values, *i.e.* for the directive *a priori* selection of substrates (within the series of compounds similar in structure), for molecular design of substrates with the given behaviour on enzymatic oxidation (maximal or optimal oxidation rate, exclusive ability to be oxidized mediated by certain enzyme, *etc.*), along with the expert decisions working out in respect to promising substances synthesis. Thus sensitivity and selectivity of oxidation reactions would be ruled by the choice of substrates with the necessary properties.

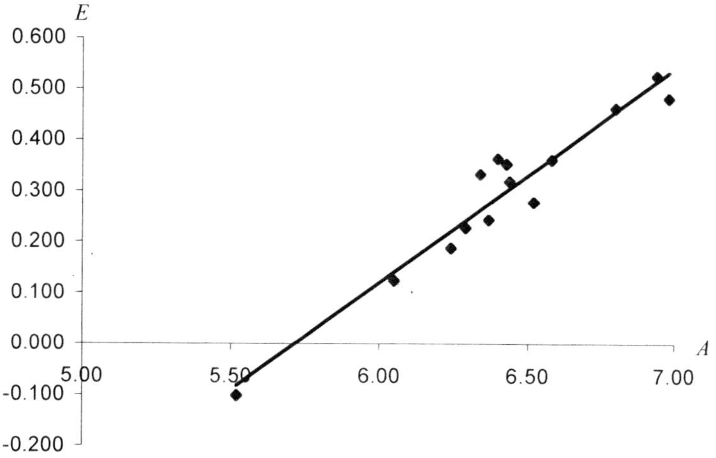

Figure 6. Dependence of half-wave potentials (E, V) of polarographic reduction of aryldiazonium $XC_6H_4N^+{\equiv}N$ cations (measured in nitromethane *vs* saturated Hg_2Cl_2 electrode) upon computed proton affinities (A, eV). X = H, 3-CH_3, 4-CH_3, 4-$N(CH_3)_2$, 3-NO_2, 4-NO_2, 4-OCH_3, 3-F, 4-F, 4-SO_2CF_3, 3-Cl, 4-Cl, 3-Br, 4-Br. Correlation coefficient r = 0.9662.

By the fact of finding 43 relationships, we have shown that the reduction potentials, dimerization potentials and potentials in half-equivalent point on titration of aryldiazonium cations $XC_6H_4N^+{\equiv}N$ (chemical reduction with $K_4[Fe(CN)_6]$ and $TiCl_3$ in water, $(C_2H_5)_3N$, $(i\text{-}C_4H_9)_4N^{+-}OH$, CH_3OK and $C_{10}H_8^{\bullet-}Na^+$ in acetone; polarographic reduction in nitromethane, sulfolane, and N,N-dimethylformamide) are related linearly to the quantum chemically evaluated electron affinities (A) and to the stabilization energies of radicals formed on diazonium cations reduction. Sixty six linear correlations of frequencies (ν) characterizing a collection of bonds stretching vibrations of the $C\text{-}N^+{\equiv}N$ fragment in the $XC_6H_4N^+{\equiv}NY^-$ salts with different anions in vaseline oil, N,N-dimethylformamide, acetone, ethylacetate, methanol, water, with the bonds orders of N≡N and C-N, with the charges on carbon atoms in *para* positions of the C_6H_5X molecules aromatic rings, with the mesomeric dipole moments (μ_m) of X substituents have been found. Twelve quantitative relationships combining the μ_m

and ν quantities with the *A* values have been established. The interrelations obtained have an explicitly expressed physical meaning, are featured by rather high correlation coefficients and have a predictive power in respect to redox properties, electron affinities, vibrational frequencies of aryldiazonia, as well as to mesomeric dipole moments of atomic groups in organic molecules [38]. Examples of the above correlations are shown on Figures 6 and 7.

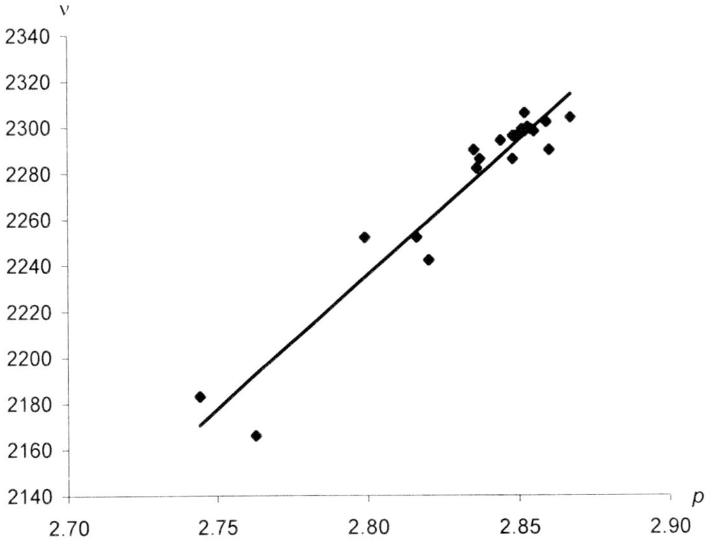

Figure 7. Dependence of stretching vibrational frequencies (ν, cm^{-1}) (in vaseline oil) for fragment $C-N^+{\equiv}N$ bonds in aryldiazonium $XC_6H_4N^+{\equiv}N$ cations upon computed N≡N bond orders (*p*). X = H, 3-CH_3, 4-CH_3, 4-$COCH_3$, 3-$COOCH_3$, 4-$COOCH_3$, 4-$COOC_2H_5$, 4-NH_2, 4-$N(CH_3)_2$, 4-$NHCOCH_3$, 3-NO_2, 4-NO_2, 4-OH, 3-OCH_3, 4-OCH_3, 4-F, 4-SO_2NH_2, 3-Cl, 4-Cl, 4-Br, 4-I. Correlation coefficient r = 0.9633.

The cationic localization energies (Λ^+) for the reactions of monosubstituted benzenes with nitroyl (nitronium) ion have been computed by the PM3 method. Yields of nitration isomeric products correlate linearly with the Λ^+ values. The latter lets one predict nitration yields of monosubstituted benzenes isomers (*ortho*, *meta* and *para*) [26].

The substituent rate factors for the interaction of 1-chloro-2-nitro-4-substituted benzenes with methoxide ion in methanol depend linearily upon the anionic localization energies, which lets one forecast kinetic characteristics of the above reactions [30].

The approach developed could be extended over the analogous reactions of polysubstituted benzenes, different classes of compounds, as well as the reactions of homolytic aromatic substitution involving the step of radical σ-complexes formation.

The quantitative interrelations of rates of surfactant ions transport through nanofiltration membranes (molecular sieves) (measured by the "applied potential" method) with ions linear sizes and hydration enthalpies [21] have been established. Using the correlations found, one can predict the actual values of ionic transport rate, which are not always available from an experimental determination owing to a rather complicated synthesis of surfactants with a large hydrocarbon radical or to poor solubility of short-chain surfactants in water.

Such an important type of intramolecular and intermolecular interaction as hydrogen bonding is actually under our consideration.

Boiling and melting points, vapour pressure, solubility, molar volume, density, mixing heat, heat conductivity and heat expansion, viscosity, surface tension, dielectric constant and dielectric losses, dipole moment, electroconductivity, segnetoelectric properties, diamagnetic susceptibility, ionization, molar refraction, refraction coefficient, another optical properties, spectra, acid-base, tautomeric, analytical properties, reactivity, biological activity of chemical compounds depend on the presence of intramolecular hydrogen bond (IHB) and its thermodynamic parameters ([55, 56] *et al.*). For the directive synthesis of compounds, the quantum chemical evaluation of the IHB effect on the electronic structure of molecules is of great interest.

The appropriate model objects for studying IHB are organic molecules with a planar quasicycle. They are widely distributed and play a very important role in chemistry, being promising building blocks for constructing the supramolecular structures with the specific receptor properties. For the planar quasicyclic systems, the effect of IHB on the electronic structure of molecules can be isolated in the explicit form. Usually, the reference objects on studying the IHB are the isomers without IHB. But such the compounds differ substantially by their properties from the isomers with IHB. In the cases of many molecules with a planar quasicycle, more adequate reference compound could be used - a conformer without IHB, with all other specific interactions being the same as those in the molecule with IHB [57, 58].

We have shown that, in relation to the compounds with planar quasicycle, the specifically parameterized semiempirical methods of quantum chemistry (MNDO/H, I-MNDO, MNDO/HB, MNDO/M) are applicable for the preliminary *a priori* semiquantitative estimation of the heat of IHB formation before proceeding to more sophisticated (*ab initio* or DFT) quantum chemical methods and forthcoming aimed synthesis [59].

We have studied the effect of IHB on the electronic structure of the planar quasiring molecules (Figure 8) (*ab initio* and DFT). IHB has been established to result in the local electron redistribution embedded at the quasicycle, as compared to the conformer without IHB and to the corresponding *para* isomer (in the case of substituted benzenes). Apparently, IHB does not influence the mechanism of chelation of the organic compounds (including the analytical reagents) with the metal ions. But its cleavage is involved in the reaction energetics, that is of importance for the directive design of analytical reagents [60, 61]. Below, some examples of IHB-involving systems will be considered in more detail in view of the medium effects role description.

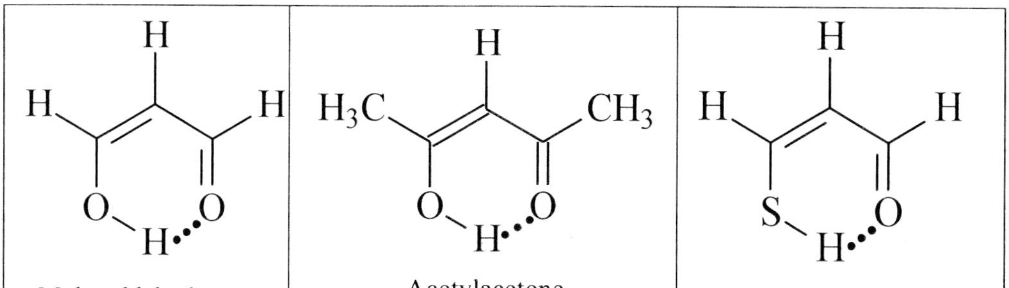

Malonaldehyde | Acetylacetone | Thiomalonaldehyde

	X	Compound
Derivatives of aniline $2\text{-}XC_6H_4NH_2$	CHO	2-Aminobenzaldehyde
	COOH	2-Aminobenzoic (anthranilic) acid
	COO^-	2-Aminobenzoate anion
	NO	2-Nitrosoaniline
	NO_2	2-Nitroaniline
	OH	2-Aminophenol
	OCH_3	2-Methoxyaniline (*ortho*-anisidine)
	SH	2-Aminobenzenethiol
	SCH_3	2-Methylmercaptoaniline
	F	2-Fluoroaniline
	Cl	2-Chloroaniline
	Br	2-Bromoaniline
Derivatives of phenol $2\text{-}XC_6H_4OH$	CHO	2-Hydroxybenzoic (salicylic) aldehyde
	COOH	2-Hydroxybenzoic (salicylic) acid
	COO^-	2-Hydroxybenzoate (salicylate) anion
	NO	2-Nitrosophenol
	NO_2	2-Nitrophenol
	OH	1,2-Dihydroxybenzene (pyrocatechol)
	OCH_3	2-Methoxyphenol (guaiacol)
	SH	2-Hydroxybenzenethiol
	SCH_3	2-Methylmercaptophenol
	F	2-Fluorophenol
	Cl	2-Chlorophenol
	Br	2-Bromophenol
Derivatives of benzenethiol $2\text{-}XC_6H_4SH$	CHO	2-Mercaprobenzoic (thiosalicylic) aldehyde
	COOH	2-Mercaprobenzoic (thiosalicylic) acid
	COO^-	2-Mercaprobenzoate (thiosalicylate) anion
	NO	2-Nitrosobenzenethiol
	NO_2	2-Nitrobenzenethiol
	OCH_3	2-Methoxybenzenethiol
	SH	1,2-Dimercaptobenzene (dithiocatechol)
	SCH_3	2-Methylmercaptobenzenethiol
	F	2-Fluorobenzenethiol
	Cl	2-Chlorobenzenethiol
	Br	2-Bromobenzenethiol

Figure 8. Molecular systems for which the effect of intramolecular hydrogen bond on electronic structure has been studied (continued on next page).

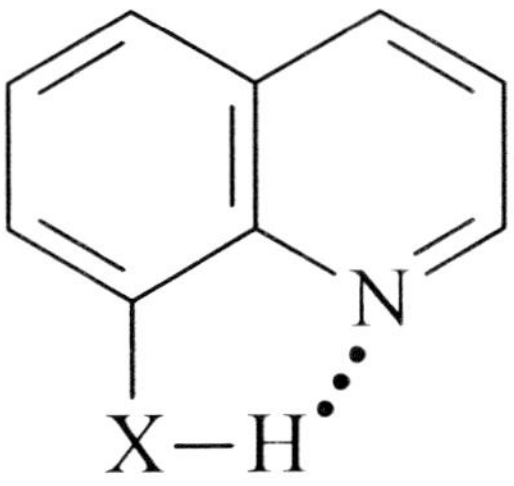	X = O X = S	8-Hydroxyquinoline 8-Mercaptoquinoline

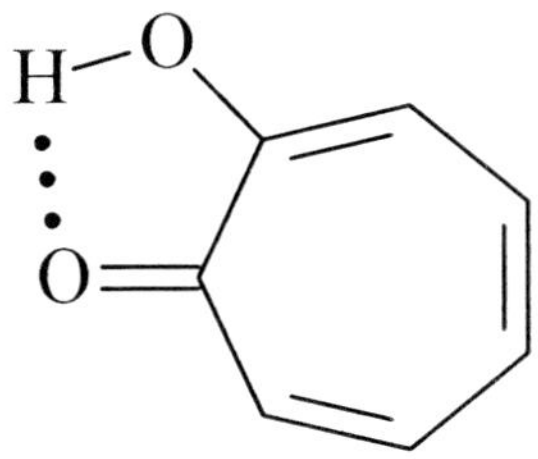

2-Hydroxy-2,4,6-cycloheptatriene-1-one
(tropolone, purpurocatechol)

Figure 8. Molecular systems for which the effect of intramolecular hydrogen bond on electronic structure has been studied (completed).

A local character of the IHB effect is reflected also by the IR and Raman spectra formation in the following planar-quasicycled systems studied at the B3LYP/6-311++G(d,p) level of theory (Figures 9 and 10).

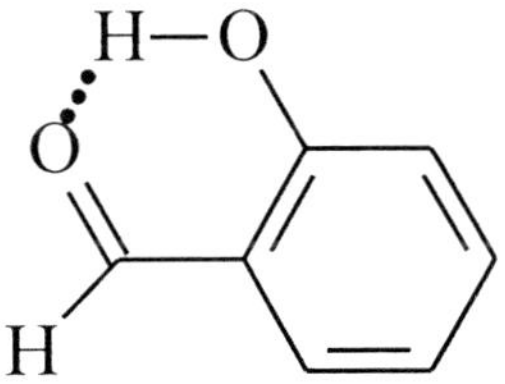

2-Hydroxybenzoic
(salicylic) aldehyde

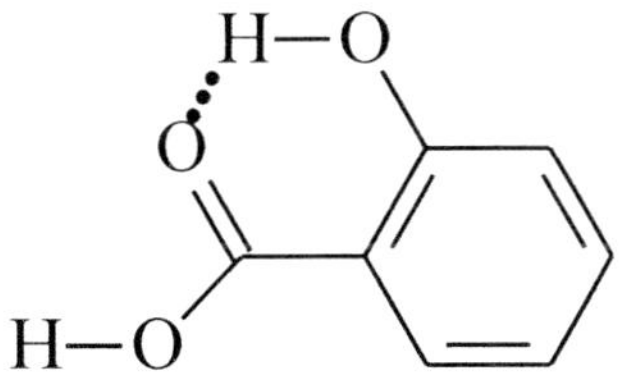

2-Hydroxybenzoic (salicylic)
acid

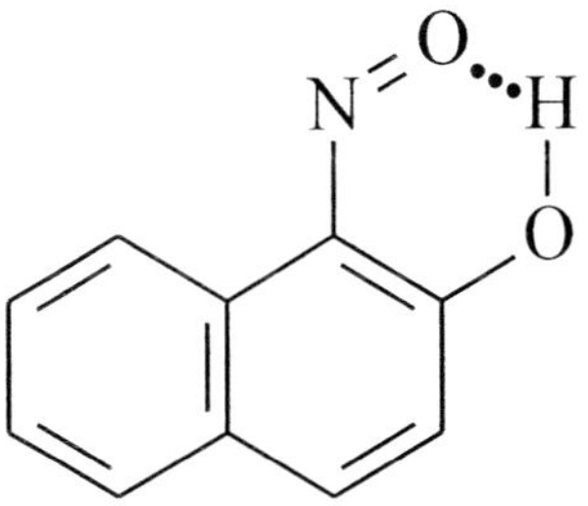

1-Nitroso-2-naphthol

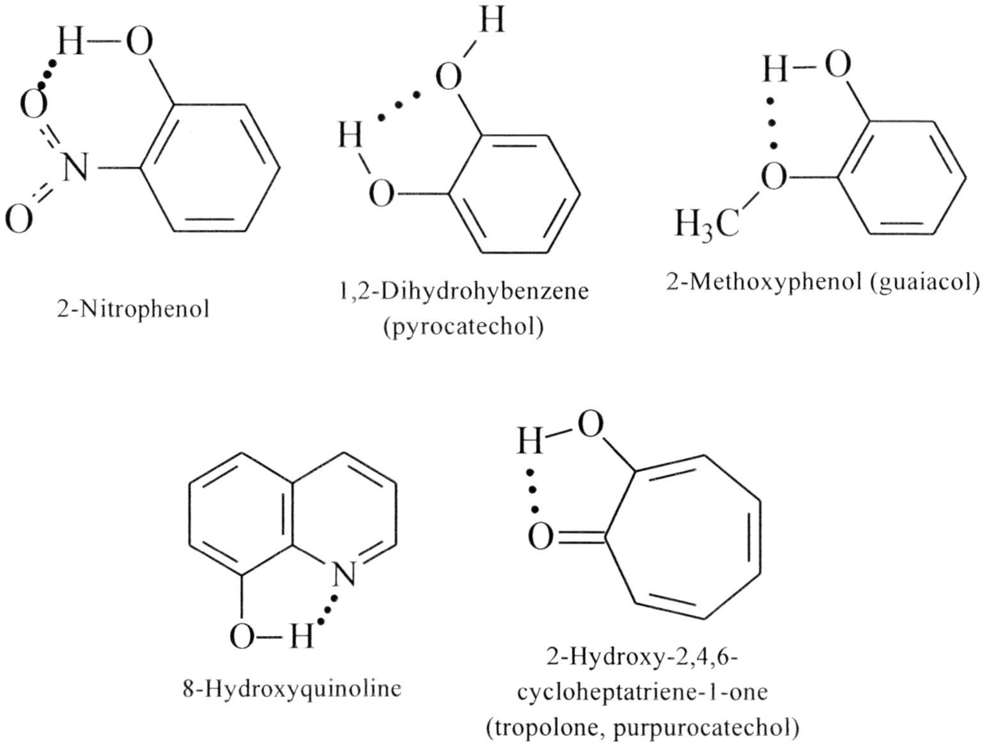

Figure 9. Planar-quasicycled molecules for which the vibrational spectra has been analyzed.

The most profound shifts in frequencies on the IHB formation is exhibited by three vibrations related to the mobility of hydroxyl proton: stretching *q*(OH) (frequency decreases), planar angular β(OH) (frequency increases slightly), non-planar motion of the OH bond as a kind of torsion (ρ(OH)) (frequency increases) [62]. The effect of IHB on the reactivity of the compounds under study should be expected just in the reactions mediated by the OH group, that is particularly confirmed by the dependence of acid-base properties on the IHB [63]. Minor (~40 cm^{-1}) and, as a rule, low-frequency shifts are observed with the vibrations localized on proton-acceptor group. For the majority of aromatic cycle vibrations, the frequencies enhance ($\Delta\nu \leq$ 10-15 cm^{-1}). The above peculiarities occur for the whole series studied. The decrease in stretching vibration frequency *q*(OH) and the increase in non-planar vibration frequency ρ(OH) correlate closely to the value of hydrogen bond enthalpy (ΔH). *E.g.*, for 2-nitrophenol ΔH = –10.26 kcal/mol, $\Delta\nu(q(OH))$ = –341 cm^{-1}, $\Delta\nu(\rho(OH))$ = +361 cm^{-1}; and for guaiacol ΔH = –4.43 kcal/mol, $\Delta\nu(q(OH))$ = –55 cm^{-1}, $\Delta\nu(\rho(OH))$ = +203 cm^{-1} [62].

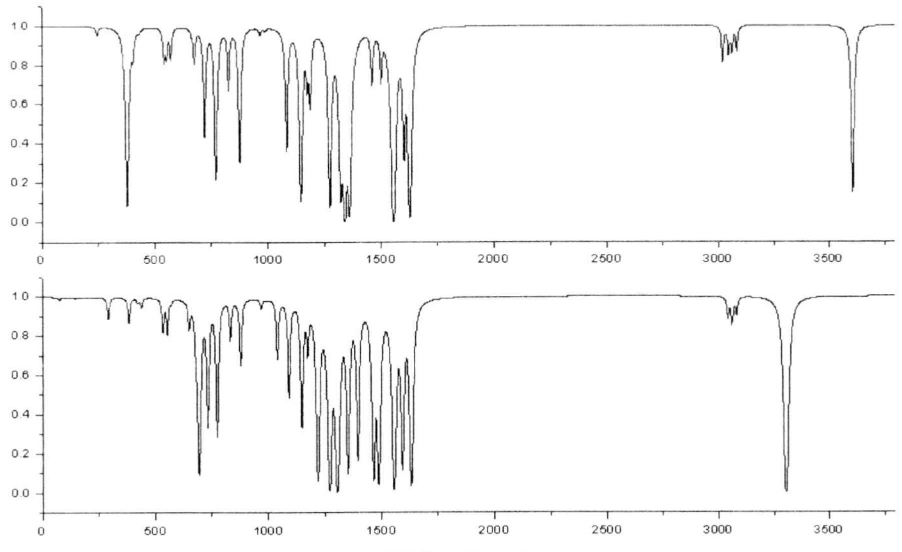

2-Nitrophenol

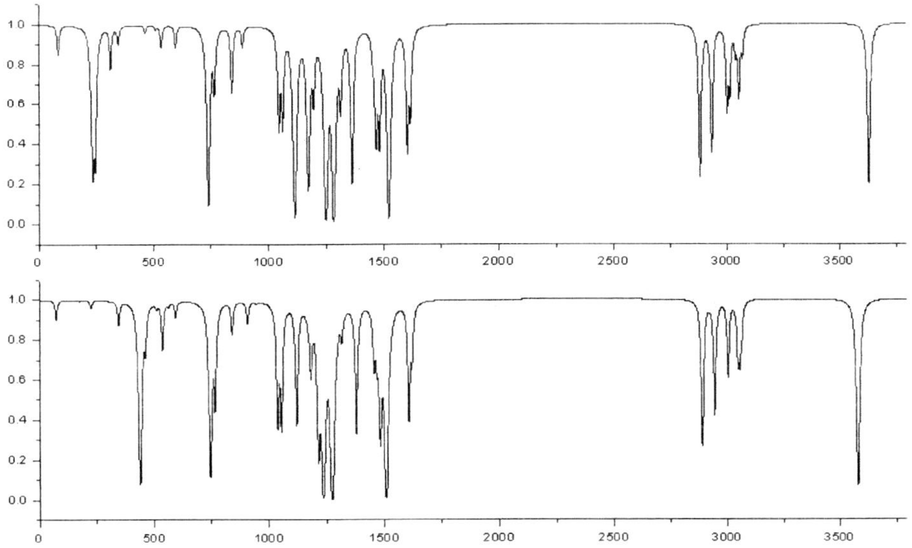

2-Methoxyphenol (guaiacol)

Figure 10. IR spectra calculated at the B3LYP/6-311++G(d,p) level of theory. The vertical axis is for the transmittance values and the horizontal axis is for frequencies (cm^{-1}). For each compound, the upper spectrum refers to the conformer without IHB, and the lower one - to the rotamer with IHB. Spectral curves were approximated by Lorenz contours with the half-width of 5 cm^{-1}.

The comparison between the experimental and theoretical IR spectra for pyrocatechol and guaiacol has showed that within the range of the OH stretching vibrations display, there are two bands ~50 cm^{-1} distant from each other. This shift agrees well with the values obtained from the quantum chemical computations: $\Delta\nu = -41$ cm^{-1} (pyrocatechol), -49 cm^{-1} (guaiacol), and testifies to the simultaneous existence of two rotamers of the aforesaid substances in the gaseous phase. In the spectrum of pyrocatechol (conformer with IHB), there is a band of stretching vibration $q(\mathrm{OH})$ of the second (free) OH group, therewith $\Delta\nu = +56$ $\mathrm{cm^{-1}}$ compared to the band of associated hydroxyl group. For 2-nitrophenol and all other compounds with $\Delta H < -5$ kcal/mol studied, the only intensive band corresponding to the conformer with IHB [62] is observed in the gaseous-phase IR spectra.

Let us consider now the works connected with the amino species oxidation mechanisms and homolytic coupling regioselectivity.

Kinetic, electrophoretic, spectroscopic (electronic absorption spectra, EPR), potentiometric, voltammetric, and quantum chemical investigations have been carried out for aromatic and heterocyclic amino compounds (diphenylamine and its derivatives, triphenylamine, carbazole, acridane, acridone, phenoxazine, phenothiazine and its derivatives), including analytical redox reagents, oxidation mechanisms ([64-66] *et al.*). It has been established that the amino compounds oxidation reactions proceed in acidic, neutral, alkaline aqueous solutions, as well as in acidic organo-aqueous media to form radical intermediates or final products and are of first order in both substrates and oxidants. A limiting stage is bimolecular. A second-order rate constant depends on the medium acidity, that is related to the participation of oxidants and/or substrates in protolytic equilibria.

In particular, on the diarylamines oxidation with the different sufficiently strong oxidants ($Ce(SO_4)_2$, $KMnO_4$, $K_2Cr_2O_7$, NH_4VO_3, $(NH_4)_2S_2O_8$, H_2O_2, KIO_3, KIO_4, *et. al.*), as well as on electrooxidation, the intermediate - cation radical of amine is initially formed (in acidic media, the protonated amine is oxidized, which is commonly attacked by the anion: MnO_4^-, $Cr_2O_7^{2-}$, $HCr_2O_7^-$, VO_3^-, $S_2O_8^{2-}$, $SO_4^{\bullet-}$, IO_3^-, IO_4^-, *et al.* or by the electroneutral species of oxidative agent) ($Ph = C_6H_5$):

$$Ph_2NH \rightleftarrows Ph_2NH^{\bullet+} + \bar{e},$$

$$Ph_2NH_2^+ \rightleftarrows Ph_2NH^{\bullet+} + H^+ + \bar{e}.$$

The cation radicals of amino compounds are sufficiently strong NH-acids, and in the essentially basic media are capable of deprotonating to yield the diarylaminyl radicals:

$$Ph_2NH^{\bullet+} \rightleftarrows Ph_2N^{\bullet} + H^+.$$

In the cation radicals and diarylaminyl radicals, the spin density of uncoupled electron is delocalized over the aromatic rings:

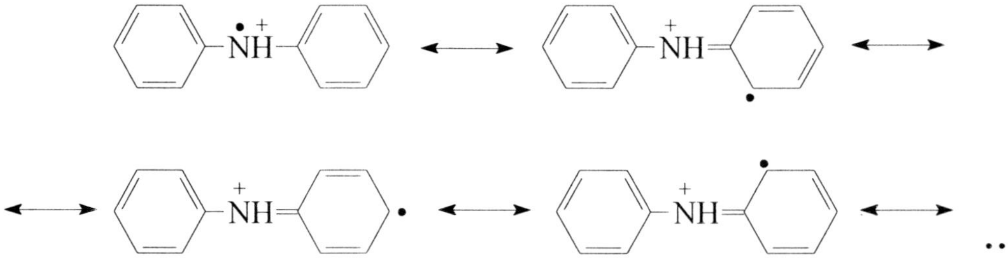

so that their dimerization is possible involving both the nitrogen atom and the rings carbon atoms. For instance, in the cation radical of diphenylamine, according to our data from the quantum chemical computation at the UB3LYP/6-311++G(d,p) level of theory, the values of spin density on the nitrogen atom, carbon atoms in the *ortho* and *para* positions of aromatic ring are equal respectively to 0.360, 0.119 (0.147) and 0.237, the total spin value (S^2) after annihilation being equal to 0.7713 therewith.

Since the highest spin density in the radical intermediates is focused on the nitrogen atom, thus in the weakly-acidic, neutral and weakly-basic media, the dimerization of the N,N-type occurs in a number of cases:

$$2Ph_2NH^{\bullet+} \rightleftarrows Ph_2N\text{-}NPh_2 + 2H^+.$$

However, in the acidic media the N-N bond is labile, and the C,C-coupling of aromatic rings is realized. Therewith the dimerization *via* the carbon atoms in *ortho* positions of rings does not occur. Thus, we have established that if both *para* positions of the diarylamine molecule are occupied, the amine does not undergo oxidation to form N,N'-diaryl-*p*-diphenoquinonediimine dication.

The absence of dimerization has been shown also for diarylaminyl radicals - intermediates of diarylamines oxidation in alkaline media. In particular, 4,4'-dicarboxydiphenylamine can not be oxidized with potassium hexacyanoferrate (III) $K_3[Fe(CN)_6]$ in the medium of 3-10 M KOH.

Consequently, the radical intermediates undergo dimerization by the type of *para,para* junction:

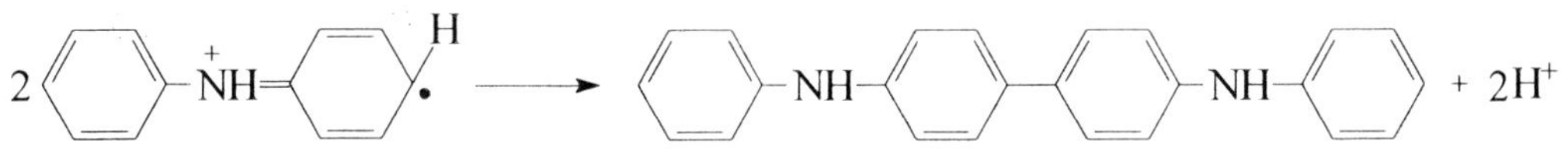

Para,para junction of the aromatic rings is more preferable, as compared to the dimerization involving *ortho* carbon atoms, also from the viewpoint of spin density.

The appearing N,N'-diarylbenzidine is capable of being further oxidized (the example for reaction in acidic medium is shown):

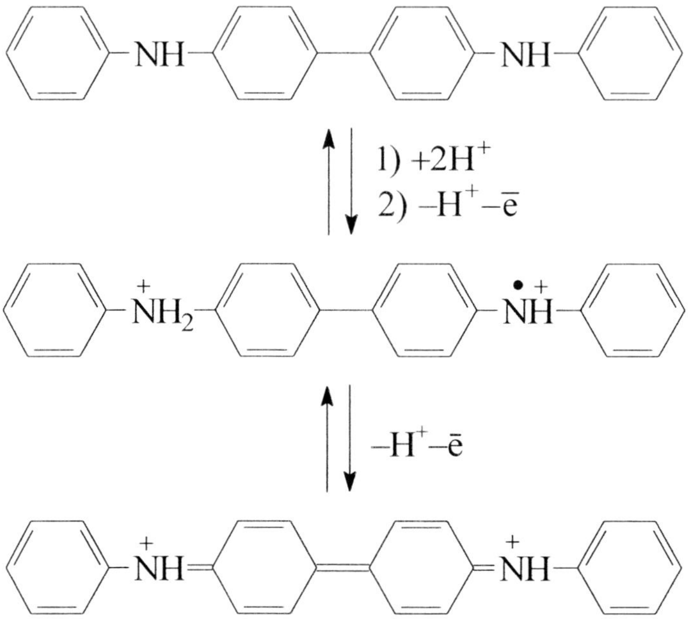

So, the oxidation product of diphenylamine in acidic media is N,N'-diphenyl-*p*-diphenoquinonediimine dication [66]:

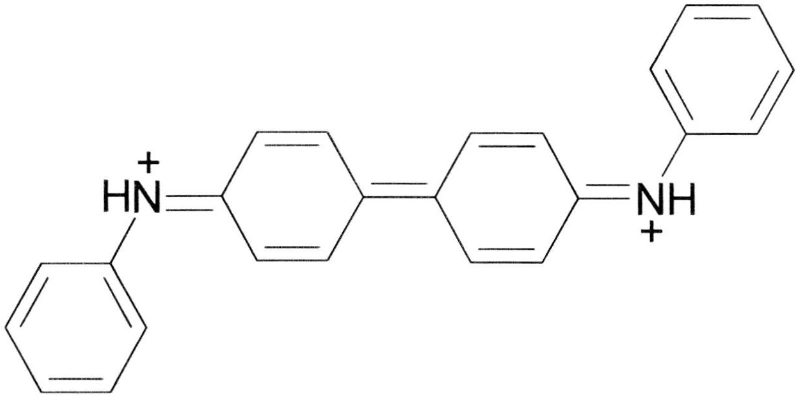

The electronic absorption spectrum of this species computed by means of the PPP method with adaptation of parameters (see below) coincides well with the experimental spectrum: λ_{max} = 581 nm (theor.), 580 nm (exper.) [66].

Diarylamines the molecules of which contain NH_2, NHR, NR_2 or OH groups in *ortho* or *para* positions, are oxidized to yield monomeric, not dimeric, products. Below presented is the oxidation scheme of 4-amino-4'-methoxydiphenylamine (Variamine Blue):

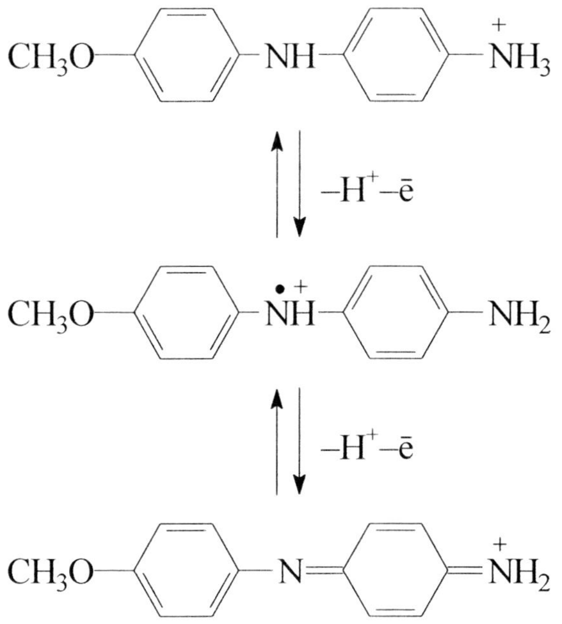

The Variamine Blue electrooxidation on glassy carbon rotating disk electrode in aqueous and organo-aqueous media proceeds in two single-electron steps, the second being rate-determining one.

The equation of polarization curve is the following therewith:

$$E = E^0 - [RT/(1 + \alpha)F]\ln(K\delta/Da) - [RT/(1 + \alpha)F]\ln[(i_d - i)/i],$$

where δ is the diffusion layer thickness; D is the diffusion coefficient of depolarizer; a is the hydronium ion activity; i_d is the limiting current; K is the product of rate constants for the first and second steps; α is the transfer coefficient for the second step.

As discussed for aqueous-dimethylformamide solutions, the pH decrease from 2.45 to 4.52 is accompanied by the limiting stage alteration [65].

Using 52 substances of different types and their derivatives as examples, the general trends of spin density distribution in radical intermediates, as well as the regularities of reactivity and regioselectivity of homolytic (oxidative and reductive) coupling for organic compounds, have been formulated. For explaining and predicting the said reactions pathways, a quantum chemical consideration has been examined in terms of kinetic (including spin density), thermodynamic, and steric factors [67].

The principal regularities of spin density distribution in radical intermediates of homolytic coupling consist in the following: for aromatic hydrocarbons, pyrrole, furan, evident is similar pattern of delocalization of the above quantity in cation radicals and anion radicals - intermediates of oxidative and reductive homolytic processes, respectively; in the diethylmalonate radical, spin density is localized on the dehydrogenized carbon atom, in the benzoate radical - on the oxygen atoms, in the intermediates of the SH-compounds oxidative dimerization - on the sulphur atom; in the cation radicals and radicals with *n*-electron donor substituents of the NH_2 и OH types, benzyl and methylbenzyl radicals, the excess of spin density takes place on the heteroatom, as well as on the carbon atoms in *ortho* and *para* positions of the aromatic rings; in the anion radicals of benzene derivatives with electron acceptor substituents, pyridine and its 1-oxide - on the carbon atom in the position 4 of the cycle [67].

The regioselectivity of homolytic dimerization of organic substances occurs mainly because of definite spin density distribution in the radical intermediates [67]. Such a reactivity index as spin density appears valid in spite of spin contamination (impurity of the spin state due to the admixing of states with higher multiplicities to the wavefunction).

Table 6 represents an example of reactivity predictions based on the spin density distribution in the cation radicals occurring on heterocyclic amino compounds oxidation.

Table 6. Spin densities on atoms of cation radicals - oxidation intermediates or products computed at the UB3LYP/6-311++G(d,p) level of theory

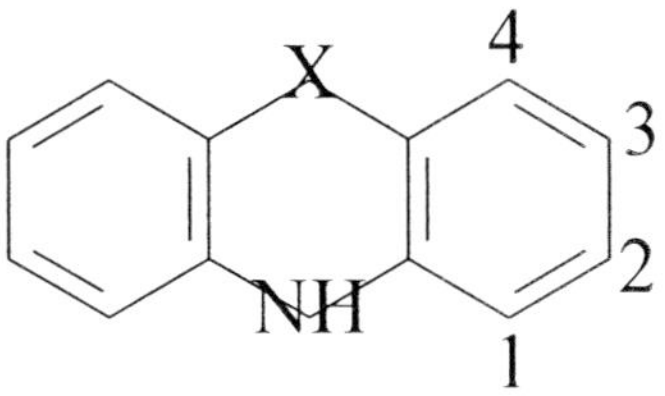

Compound	X	Position of the molecular system		
		1	3	N
Phenothiazine	S	0.012	0.109	0.256
Phenoxazine	O	0.007	0.137	0.300
Carbazole	–	0.197	0.227	0.322

The S^2 values for the above cation radicals are 0.7576, 0.7591 and 0.7689, respectively, which is rather close to the $S^2 = 0.75$ quantity for pure doublet state, the latter being characteristic for radical species with one unshared electron. Thus, spin contamination practically does not distort the information concerning spin density distribution on atoms in the ground states of the cation radicals under study.

The above results let one predict the order of the species capability to undergo oxidative coupling: phenothiazine < phenoxazine < carbazole, which has been confirmed experimentally.

Thus, the main oxidation product for phenothiazine in acidic media is the cation radical, whether the oxidation is chemical ($Ce(SO_4)_2$, $KMnO_4$, $K_2Cr_2O_7$, NH_4VO_3, $(NH_4)_2S_2O_8$, H_2O_2, KIO_3, KIO_4, *et. al.*) or electrochemical.

For phenoxazine, the cation radical occurrence within the initial period of the reaction development, is replaced by a two-electron oxidation product formation.

As for carbazole, its interaction with ammonium vanadate in the medium of 4.25-10.3 M H_2SO_4 is featured by the stoichiometry 1 : 2, *i.e.* the cation radical appeared at first turns into the dimer rapidly. Anodic oxidation of carbazole on glassy carbon rotating disk electrode in the sulphuric acid solutions proceeds under the mixed control conditions and involves one electron transfer due to electrode passivation caused by dimers and/or polymers precipitation onto its surface [64].

The similarity in directions (on Figures 11 and 12, marked by asterisk) for electrophilic (nucleophilic) processes and oxidative (reductive) coupling reactions of thiourea, N-phenylthiourea, many C_6H_5X compounds, naphthalene, anthracene, phenanthrene, carbazole, acridane, acridone, 5,10-dihydrophenazine, phenoxazine, phenothiazine, acridine, N-methylpyridinium, pyrilium, thiopyrilium cations originates from a level of qualitative similarity of electron density distribution in the substrates molecules and spin density distribution in the radical intermediates [67].

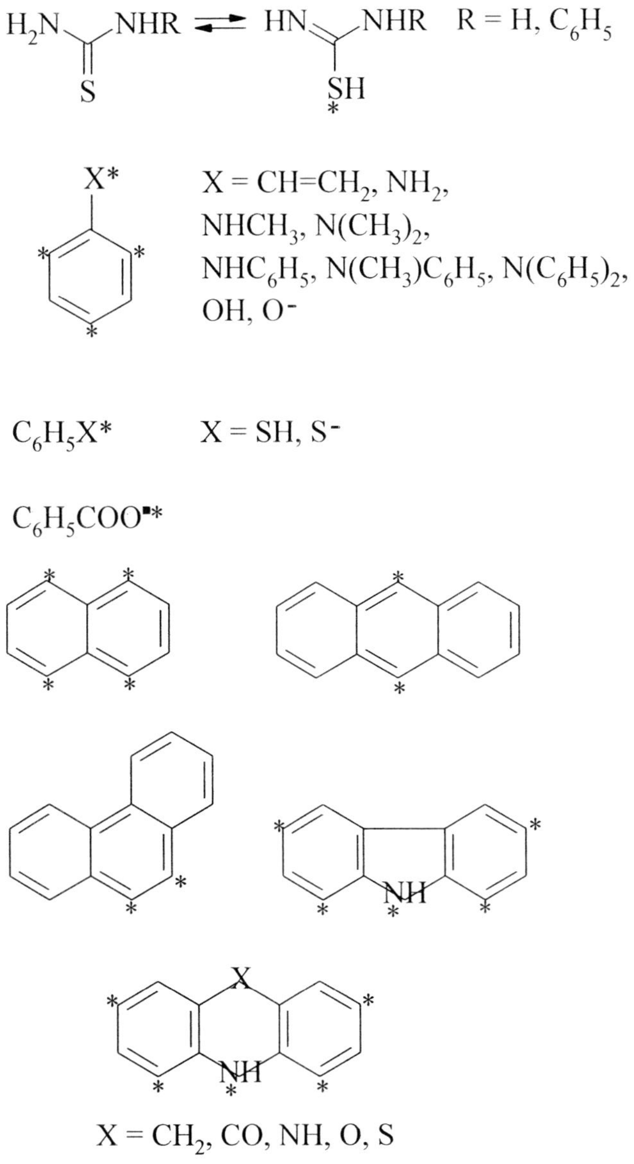

Figure 11. Compounds featured by similar regioselectivity of electrophilic and oxidative coupling reactions.

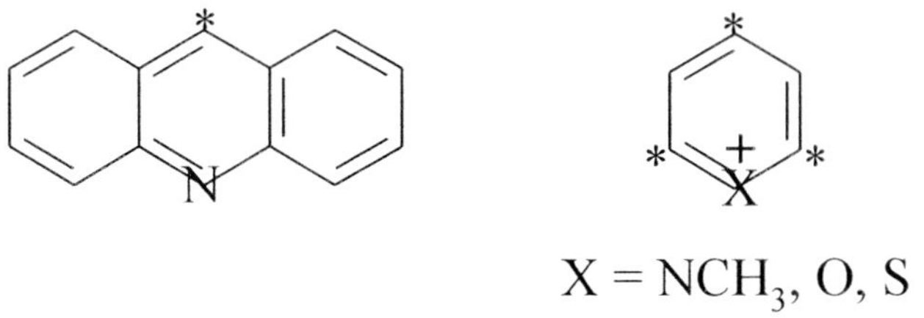

Figure 12. Compounds featured by similar regioselectivity of nucleophilic and reductive coupling reactions.

By means of the ^{14}N NQR method, the electronic structure of nitrogen atom (a reaction centre in redox processes) of diphenylamine and its 4-methoxy, 2,4,4'-trimethoxy and 4-benzoyl derivatives has been explored. Unshared electronic pair donates 0.18-0.27 electrons to π-systems of aromatic rings, that allows this pair to remain reactive with respect to electrophilic attack (oxidation, protonation, *etc.*). A change in both population of the unshared pair orbital and reactivity in electrophilic processes in the aforesaid series of compounds takes place according to signs of mesomeric effect of methoxy and benzoyl groups [68].

Chemical shifts in the NMR ^{13}C spectra of 1-aminohexane, diphenyl, N-methylaniline, diphenylamine and its 4-methyl, 4-methoxy, 2-, 3-, 4-nitro derivatives, phenol, diphenyloxide, are related linearly to charges on carbon atoms of aromatic rings.

It has been shown that substituents in molecules influence both the reactivity of bridge diphenyls:

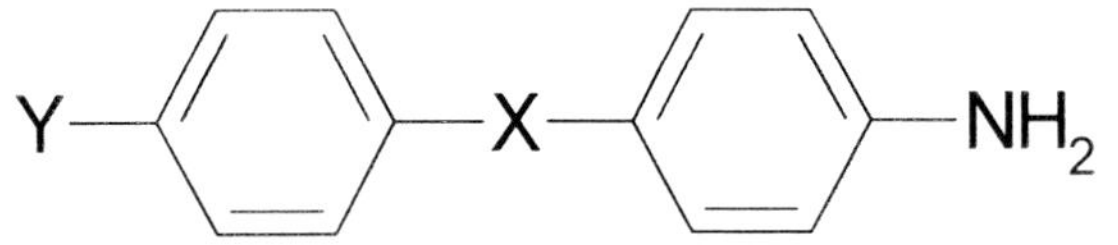

X = CH_2, CH=CH, NH, O, S; Y = H, NO_2

in electrophilic reactions (protonation, acylation) and the diarylamines oxidation rate predominantly *via* the field effect with no regard to the nature of bridge group (CH_2, CH=CH, NH, O, S).

According to semiempirical data [69], the series of electron-acceptor and electron-donor ability of substituents in organic molecules have been obtained.

An oxidation mechanism ([64-66] *et al.*) commonness for various amino compounds is due to the vicinity of the corresponding amino nitrogen-containing groups in the series of changes of electronic effects and electronegativities of substituents.

A variation of electronic influence is the inductive effect [70]. According to the Glossary of Terms Used in Theoretical Organic Chemistry [71], inductive effect is a polarization of chemical bonds due to a shift of involved electron pairs toward a more electronegative group. Inside a molecule, the inductive effect is transferred through space and determined by the electrostatic forces between the interacting centres. Another original model of inductive effect (G.N. Lewis) is a model with transfer of this effect over bonds by means of consequent polarization of the bonds between dipolar or charged substituent and reactive centre. A quantitative evaluation of the inductive effect of groups (substituents) is given, in particular, by R.W. Taft σ^*-constants for aliphatic compounds and by the inductive σ_I-constants for aromatic substances.

In works of 1990-s, the inductive effect is considered as a sum of electronegativity effect and field effect. Electronegativity effect is a shift of electron density under the action of substituent over the σ-bonds chain. Some chemists identify the electronegativity effect with the inductive effect.

There is another terminology. It introduces the term "polar effect" presenting a sum of inductive effect (it is essentially the electronegativity effect) and the field effect.

We will discuss just the electron displacement along the σ-bonds chain and use the term "electronegativity effect".

In the pure state, the above effect could be considered using a saturated hydrocarbon chain as an example. In this case, the influence of substituent in aliphatic molecule is not complicated by mesomeric shift of electron density with a participation of π-bonds.

Elucidated (*ab initio* and DFT) are distance and specificity of action of the effect of atomic groups electronegativity in the saturated systems (Figures 13-25), the sequence of weakening the electron-acceptor and electron-donor properties of substituents in aromatic molecules, electronic structure of the molecules of aniline series compounds.

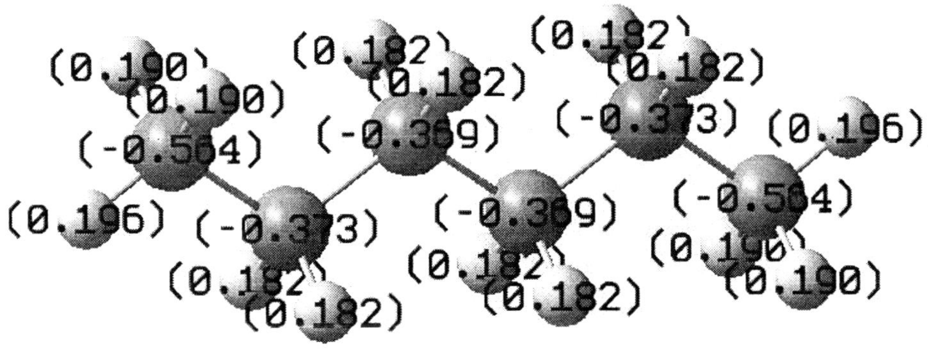

Figure 13. Natural atomic charges in the *n*-hexane molecule at B3LYP/6-311++G(3d,3p).

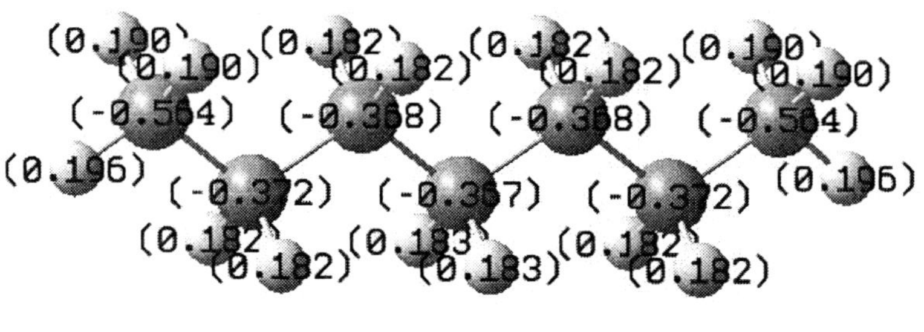

Figure 14. Natural atomic charges in the *n*-heptane molecule at B3LYP/6-311++G(3d,3p).

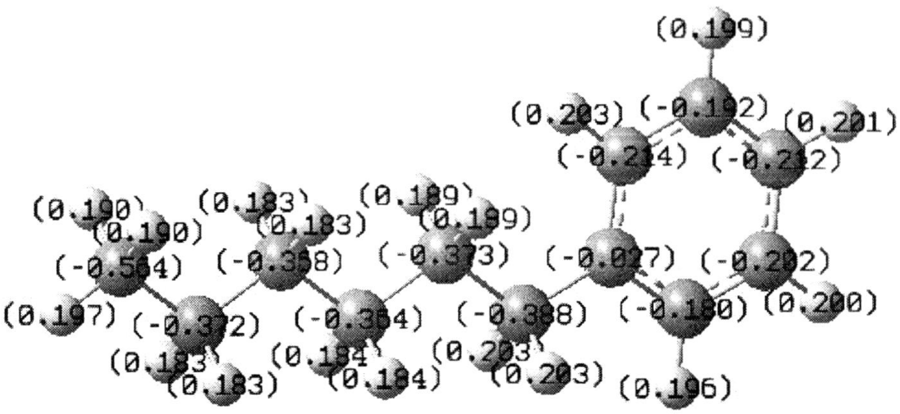

Figure 15. Natural atomic charges in the 1-hexylbenzene molecule at B3LYP/6-311++G(3d,3p).

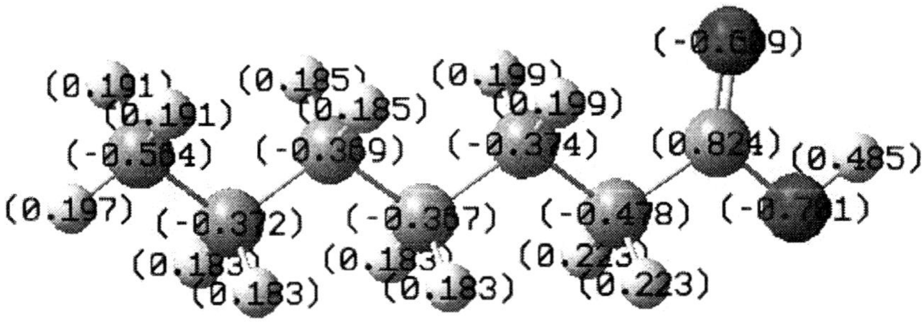

Figure 16. Natural atomic charges in the molecule of 1-heptanoic acid at B3LYP/6-311++G(3d,3p).

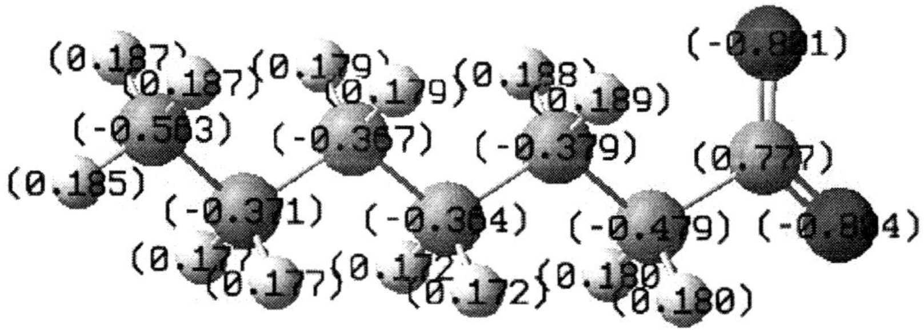

Figure 17. Natural atomic charges in the 1-heptanoate anion at B3LYP/6-311++G(3d,3p).

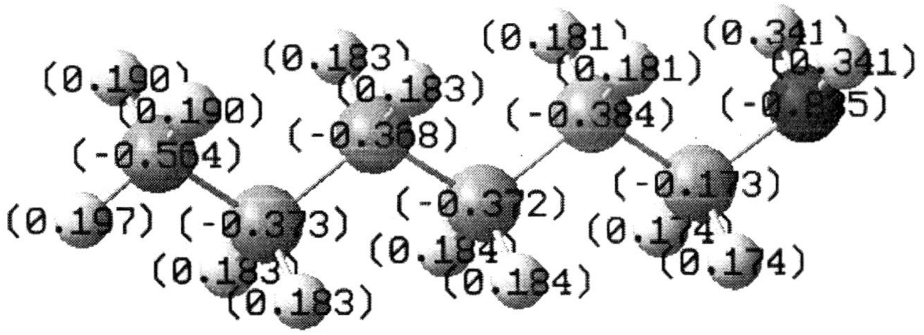

Figure 18. Natural atomic charges in the molecule of 1-aminohexane at B3LYP/6-311++G(3d,3p).

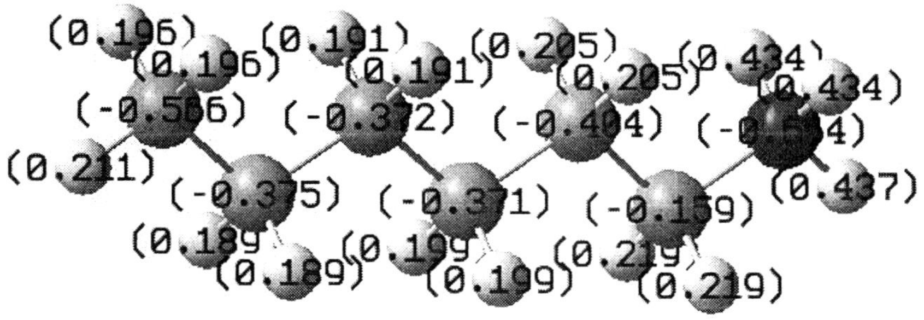

Figure 19. Natural atomic charges in the 1-hexylammonium cation at B3LYP/6-311++G(3d,3p).

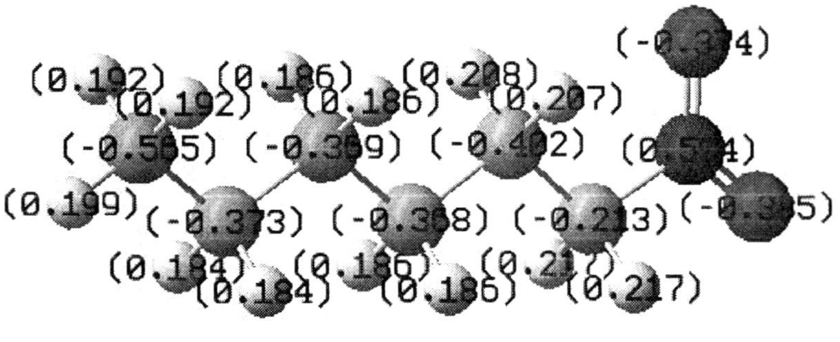

Figure 20. Natural atomic charges in the molecule of 1-nitrohexane at B3LYP/6-311++G(3d,3p).

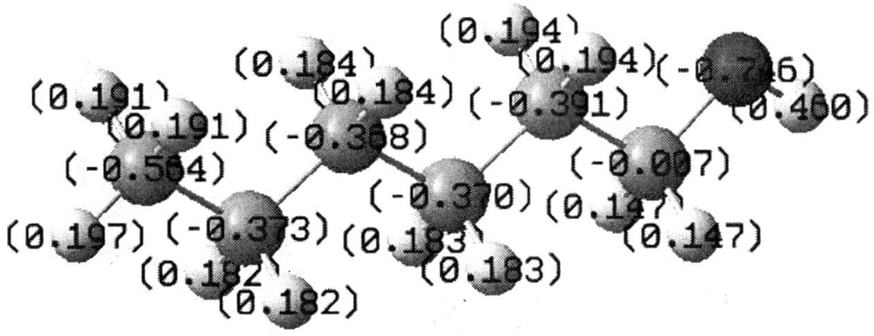

Figure 21. Natural atomic charges in the molecule of 1-hexanol at B3LYP/6-311++G(3d,3p).

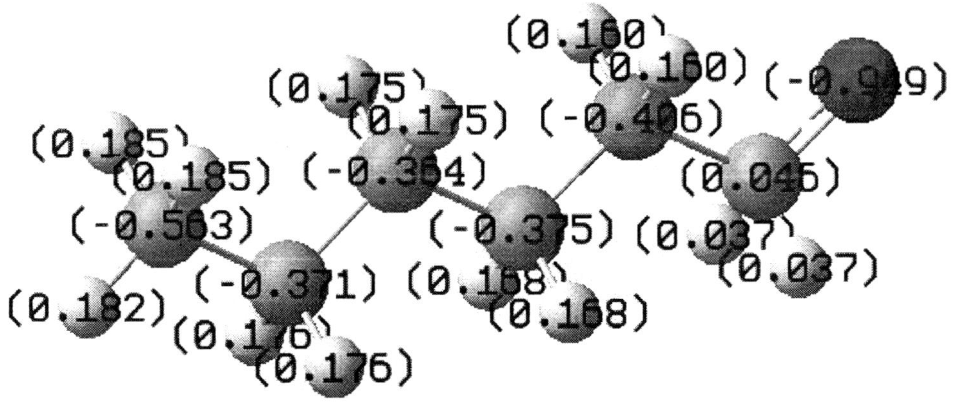

Figure 22. Natural atomic charges in the 1-hexanolate anion at B3LYP/6-311++G(3d,3p).

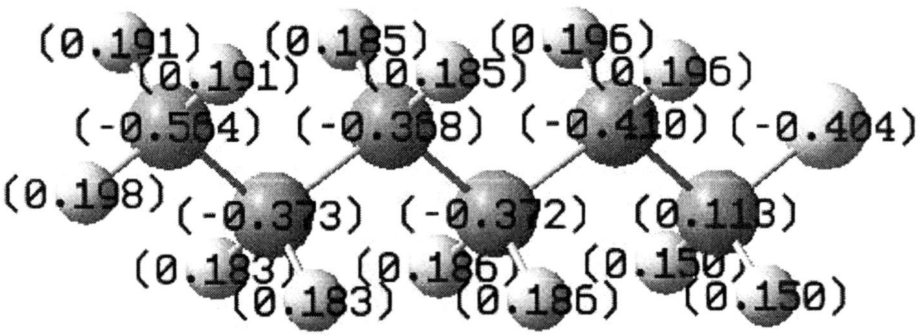

Figure 23. Natural atomic charges in the molecule of 1-fluorohexane at B3LYP/6-311++G(3d,3p).

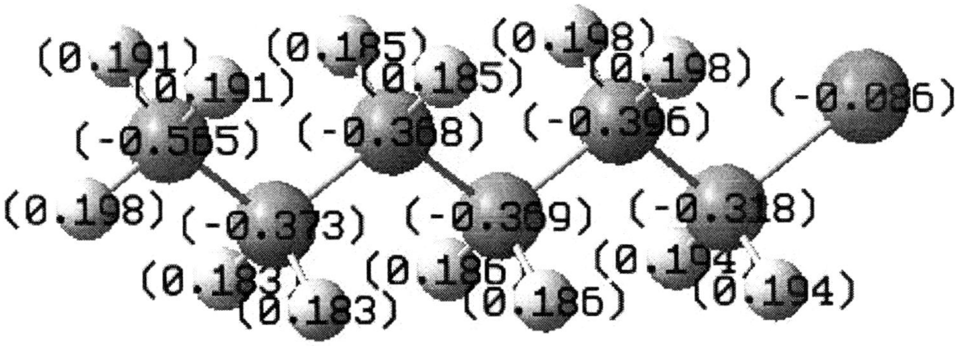

Figure 24. Natural atomic charges in the molecule of 1-chlorohexane at B3LYP/6-311++G(3d,3p).

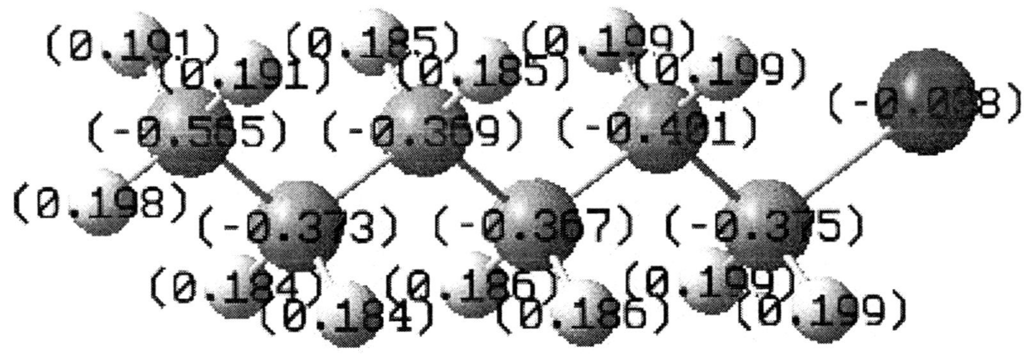

Figure 25. Natural atomic charges in the molecule of 1-bromohexane at B3LYP/6-311++G(3d,3p).

The NBO charge distribution in the molecules of *n*-hexane, *n*-heptane and *n*-octane is in agreement with the ^{13}C NMR data [72-74]. For 1-aminohexane, the ^{13}C chemical shifts [74] correlate linearly with the natural charges on atoms. Thus, the correctness of our evaluations of the electron distribution in molecules is confirmed.

Following the order of changes in the values of charge on the α-C atom, the atomic groups are arranged in the series just below (from left toward right, the charge shifts into negative region):

$$\begin{gathered}
\mathrm{F} \gg \mathrm{O^-} > \mathrm{O{-}C{\equiv}N} \gtrapprox \mathrm{OCH_3} \gtrapprox \mathrm{OH} > \mathrm{OPO_3H_2} > \mathrm{OCOCH_3} \gtrapprox \mathrm{OSO_3H} > \mathrm{ONO_2} > \\
> \mathrm{ONO} \gg \mathrm{{}^+N(CH_3)_3} > \mathrm{N(CH_3)_2} \gtrapprox \mathrm{{}^+NH_3} \gtrapprox \mathrm{NHCH_3} \gtrapprox \mathrm{N{=}C{=}O} \gtrapprox \mathrm{NH_2} \gtrapprox \mathrm{N{=}C{=}S} \gtrapprox \\
\gtrapprox \mathrm{N_3} \gtrapprox \mathrm{NHCOCH_3} > \mathrm{{}^+N{\equiv}C^-} > \mathrm{{}^+N{\equiv}N} \gtrapprox \mathrm{NO_2} > \mathrm{NO} \gg \mathrm{Cl} > \mathrm{C(CH_3)_3} \gtrapprox \mathrm{CH(CH_3)_2} \gtrapprox \\
\gtrapprox \mathrm{CH_2CH_3} \gtrapprox \mathrm{CH_3} \gtrapprox \mathrm{Br} > \mathrm{C_6H_5} > \mathrm{CH{=}CH_2} > \mathrm{CCl_3} > \mathrm{CBr_3} > \mathrm{CF_3} \gtrapprox \mathrm{S{-}C{\equiv}N} > \\
> \mathrm{CONH_2} \gtrapprox \mathrm{COCH_3} \gtrapprox \mathrm{SH} \approx \mathrm{COOCH_3} > \mathrm{CHO} \approx \mathrm{COOH} \gtrapprox \mathrm{COO^-} \gtrapprox \mathrm{CN} \gtrapprox \\
\gtrapprox \mathrm{{}^+S(CH_3)_2} \gtrapprox \mathrm{COCl} \gtrapprox \mathrm{COBr} \gtrapprox \mathrm{SCH_3} \gtrapprox \mathrm{S^-} \gtrapprox \mathrm{COF} > \mathrm{C{\equiv}CH} > \mathrm{SOCH_3} > \mathrm{H} > \\
> \mathrm{SO_2CH_3} \approx \mathrm{SO_3H} > \mathrm{SO_3^-} > \mathrm{PH_2} > \mathrm{PHCH_3} > \mathrm{{}^+PH_3} \gtrapprox \mathrm{P(CH_3)_2} > \mathrm{{}^+P(CH_3)_3} \gg \\
\gg \mathrm{PO_3H_2} \gtrapprox \mathrm{BH_2} \gtrapprox \mathrm{B(CH_3)_2} \gg \mathrm{SiH_3} > \mathrm{Si(CH_3)_3}.
\end{gathered}$$

The substituents positioned on the left from H in the above series, result in the shift of natural charge on α-C atom toward positive side as compared with the *n*-hexane molecule. Therewith only the substituents F, O^- and O-C≡N change a sign of charge on α-C atom into a positive one. Contrary, the groups positioned on the right from hydrogen enhance a negative charge on α-carbon atom as compared with the n-C_6H_{14} molecule.

The range of charge values on α-C atom falls from 0.1126 (X = F) to –0.6975 (X = $^+P(CH_3)_3$), –0.7558 (X = PO_3H_2), –0.7623 (X = BH_2), –0.7687 (X = $B(CH_3)_2$) and even –0.8236 (X = SiH_3) and –0.8397 (X = $Si(CH_3)_3$). At X = H the said quantity consists –0.5642.

The charge on carbon atom in β-position, when varying the substituent X, changes within a fairly narrow range: from –0.4103 for X = F to –0.3527 on X = C≡CH, therewith in the case of X = H a charge is equal to –0.3725. The values of natural charge on β-C atom follow the series below (a negative charge being diminished from left toward right):

$$\mathrm{F} \gtrapprox \mathrm{NO} \gtrapprox \mathrm{O^-} \gtrapprox \mathrm{{}^+N(CH_3)_3} \gtrapprox \mathrm{{}^+NH_3} \gtrapprox \mathrm{NO_2} \gtrapprox \mathrm{SOCH_3} \approx \mathrm{Br} \gtrapprox \mathrm{O\text{-}C{\equiv}N} \gtrapprox \mathrm{SO_3^-} \approx$$
$$\approx \mathrm{{}^+S(CH_3)_2} \gtrapprox \mathrm{SO_3H} \approx \mathrm{SO_2CH_3} \approx \mathrm{Cl} \gtrapprox \mathrm{N(CH_3)_2} \gtrapprox \mathrm{S\text{-}C{\equiv}N} \approx \mathrm{OSO_3H} \gtrapprox \mathrm{ONO_2} \gtrapprox$$
$$\gtrapprox \mathrm{ONO} \approx \mathrm{OPO_3H_2} \gtrapprox \mathrm{OH} \gtrapprox \mathrm{OCOCH_3} \gtrapprox \mathrm{SH} \approx \mathrm{S^-} \approx \mathrm{OCH_3} \approx \mathrm{NHCH_3} \gtrapprox \mathrm{SCH_3} \approx$$
$$\approx \mathrm{{}^+PH_3} \approx \mathrm{NH_2} \approx \mathrm{CBr_3} \gtrapprox \mathrm{CCl_3} \gtrapprox \mathrm{{}^+N{\equiv}N} \gtrapprox \mathrm{PO_3H_2} \approx \mathrm{P(CH_3)_2} \gtrapprox \mathrm{N_3} \approx \mathrm{CF_3} \approx \mathrm{COO^-} \approx$$
$$\approx \mathrm{CHO} \approx \mathrm{{}^+P(CH_3)_3} \approx \mathrm{PHCH_3} \gtrapprox \mathrm{NHCOCH_3} \approx \mathrm{PH_2} \gtrapprox \mathrm{CH(CH_3)_2} \gtrapprox \mathrm{COCH_3} \approx$$
$$\approx \mathrm{CH{=}CH_2} \gtrapprox \mathrm{N{=}C{=}O} \gtrapprox \mathrm{COCl} \approx \mathrm{COF} \approx \mathrm{COBr} \approx \mathrm{COOH} \approx \mathrm{SiH_3} \approx \mathrm{CONH_2} \approx$$
$$\approx \mathrm{C(CH_3)_3} \approx \mathrm{COOCH_3} \gtrapprox \mathrm{C_6H_5} \approx \mathrm{H} \approx \mathrm{N{=}C{=}S} \gtrapprox \mathrm{{}^+N{\equiv}C^-} \gtrapprox \mathrm{Si(CH_3)_3} \gtrapprox \mathrm{CH_3} \gtrapprox$$
$$\gtrapprox \mathrm{CH_2CH_3} > \mathrm{BH_2} \gtrapprox \mathrm{CN} \gtrapprox \mathrm{B(CH_3)_2} \approx \mathrm{C{\equiv}CH}.$$

As for the charge on γ-carbon atom, its values fall into even narrower interval compared with β-C: from –0.3754 (X = O^-) to –0.3624 (X = BH_2), including the value of –0.3692 on X = H. The sequence of shifts in natural charges on γ-C atom (from left to right) toward less negative region appears to be as follows:

$$\mathrm{O^-} \gtrapprox \mathrm{{}^+N{\equiv}N} \gtrapprox \mathrm{N{=}C{=}S} \gtrapprox \mathrm{N{=}C{=}O} \approx \mathrm{F} \approx \mathrm{{}^+N{\equiv}C^-} \gtrapprox \mathrm{NH_2} \approx \mathrm{NHCOCH_3} \gtrapprox \mathrm{CN} \approx$$
$$\approx \mathrm{{}^+NH_3} \approx \mathrm{OPO_3H_2} \gtrapprox \mathrm{OH} \approx \mathrm{OSO_3H} \approx \mathrm{O\text{-}C{\equiv}N} \approx \mathrm{N_3} \approx \mathrm{OCH_3} \approx \mathrm{NHCH_3} \approx \mathrm{OCOCH_3} \approx$$
$$\approx \mathrm{C{\equiv}CH} \gtrapprox \mathrm{ONO} \approx \mathrm{Cl} \approx \mathrm{CF_3} \approx \mathrm{ONO_2} \approx \mathrm{H} \gtrapprox \mathrm{N(CH_3)_2} \approx \mathrm{PH_2} \approx \mathrm{{}^+PH_3} \approx \mathrm{COF} \approx$$
$$\approx \mathrm{{}^+P(CH_3)_3} \approx \mathrm{NO_2} \approx \mathrm{Br} \approx \mathrm{COBr} \approx \mathrm{COCl} \approx \mathrm{{}^+S(CH_3)_2} \approx \mathrm{S^-} \approx \mathrm{PHCH_3} \approx \mathrm{COOH} \approx \mathrm{CH_3} \gtrapprox$$
$$\gtrapprox \mathrm{COOCH_3} \approx \mathrm{PO_3H_2} \approx \mathrm{CH_2CH_3} \approx \mathrm{CONH_2} \approx \mathrm{CHO} \approx \mathrm{{}^+N(CH_3)_3} \approx \mathrm{SO_3H} \approx \mathrm{NO} \approx \mathrm{SH} \approx$$
$$\approx \mathrm{SCH_3} \approx \mathrm{S\text{-}C{\equiv}N} \approx \mathrm{COCH_3} \approx \mathrm{P(CH_3)_2} \approx \mathrm{SiH_3} \approx \mathrm{Si(CH_3)_3} \approx \mathrm{CCl_3} \approx \mathrm{CBr_3} \approx$$
$$\approx \mathrm{CH(CH_3)_2} \approx \mathrm{COO^-} \approx \mathrm{CH{=}CH_2} \approx \mathrm{SO_2CH_3} \gtrapprox \mathrm{C(CH_3)_3} \approx \mathrm{C_6H_5} \approx \mathrm{SOCH_3} \gtrapprox$$
$$\gtrapprox \mathrm{B(CH_3)_2} \approx \mathrm{SO_3^-} \gtrapprox \mathrm{BH_2}.$$

A character of influence of atomic groups upon the electron density distribution on carbon atoms in α- and β-positions of aliphatic chain is so that the electronegativity effect leads in the cases of X = $CH(CH_3)_2$, $C(CH_3)_3$, CF_3, CCl_3, CBr_3, CH=CH_2, C_6H_5, CHO, $COCH_3$, COOH, $CONH_2$, $COOCH_3$, COO^-, COF, COCl, COBr, NH_2, $NHCH_3$, $N(CH_3)_2$, $^+N(CH_3)_3$, $NHCOCH_3$, $^+N{\equiv}N$, N=C=O, N_3, NO, NO_2, OH, O^-, OCH_3, O-C≡N, $OCOCH_3$, ONO, ONO_2, OPO_3H_2, OSO_3H, SH, S^-, SCH_3, $^+S(CH_3)_2$, S-C≡N, $SOCH_3$, F, Cl, Br to alternating the signs of charge changes (but usually not charges themselves; solely in going from X = H to X = F, O^-, O-C≡N a charge on α-C atom becomes positive). The above fact is in agreement with a knowledge on α- and β-effects in ^1H NMR spectroscopy [75]. It is known [75] that for the fragment CH-CHX, the increase in electronegativity of the group X leads to the displacement of signal of the proton at α carbon atom toward weak field, and of the proton at β carbon atom toward strong field.

As seen, α- and β-effects are widespread, but not universal, since the substituent introduction not always leads to the decrease in electron density on α-C atom and to its increase on β-C atom.

It is important to emphasize that under the effect of substituent, the most essential (in the first significant digit) changes occur with the charges on α-C atom. The charges on the second and third carbon atoms vary within the second or third significant digit. Thus, the electronegativity effect in the aliphatic chain readily attenuates.

The electronegativity effect distance over aliphatic chain usually does not exceed three atoms.

Using sum of the charges of fragment C_6H_{13} as a criterion, for the above series of the n-$C_6H_{13}X$ molecules, we obtain the following sequence of weakening (from left to right) electron-acceptor ability of substituents X:

$^{+}N{\equiv}N \gg F \gtrapprox O\text{-}C{\equiv}N > OSO_3H \gtrapprox ONO_2 > OPO_3H_2 \gtrapprox {}^{+}NH_3 \gtrapprox {}^{+}N(CH_3)_3 \gtrapprox OCOCH_3 >$
$> ONO > OCH_3 > OH > {}^{+}N{\equiv}C^{-} > NO_2 \gtrapprox N{=}C{=}S > N{=}C{=}O \gtrapprox NHCOCH_3 > N_3 >$
$> N(CH_3)_2 > NO > NHCH_3 > NH_2 > Cl \gtrapprox {}^{+}S(CH_3)_2 > CCl_3 > CBr_3 > CN \gtrapprox Br > COBr >$
$> COCl \gtrapprox C_6H_5 > CF_3 \gtrapprox C(CH_3)_3 > COF \gtrapprox COOH \gtrapprox CH{=}CH_2 \gtrapprox COOCH_3 \gtrapprox S\text{-}C{\equiv}N \gtrapprox$
$\gtrapprox CH(CH_3)_2 \gtrapprox CH_2CH_3 \gtrapprox CH_3 \gtrapprox CONH_2 \gtrapprox CHO \gtrapprox COCH_3 > C{\equiv}CH > O^{-} \gtrapprox {}^{+}PH_3 >$
$> SH > SO_3H \gtrapprox SCH_3 > SO_2CH_3 \gtrapprox SOCH_3 > {}^{+}P(CH_3)_3 > COO^{-} > H > PH_2 > SO_3^{-} \gtrapprox$
$\gtrapprox PHCH_3 \gtrapprox PO_3H_2 > P(CH_3)_2 > S^{-} > BH_2 > B(CH_3)_2 \gg SiH_3 > Si(CH_3)_3.$

For X = $^{+}N{\equiv}N$ we have $\Sigma = 0.5083$; for X = F, $\Sigma = 0.4027$; for X = H, $\Sigma = -0.1963$; for X = S^{-}, $\Sigma = -0.2873$, for X = BH_2, $\Sigma = -0.3192$, for X = $B(CH_3)_2$, $\Sigma = -0.3448$, for X = SiH_3, $\Sigma = -0.3910$, for X = $Si(CH_3)_3$, $\Sigma = -0.4218$.

As obvious from the discussion presented above, the distance and specificity of electronegativity effect do not depend directly upon the overall electron-acceptor ability of atomic groups.

A position of SO_2CH_3 substituent in the last series seems at first glance to be contradictory to such well-known facts as the electron-acceptor character of SO_2 fragment, that assists the appearance of superacidic properties of substances. Obviously, the electron-acceptor properties of SO_2CH_3 group are displayed by the stabilization of anions formed as a result of acidic dissociation of the corresponding molecules.

The Σ values do not correlate with the constants σ^*, σ_I and other inductive constants following from the linear free energy relationships principle. Since the series we constructed display the effect of substituents on the electronic structure of molecules over a chain of bonds, but not through space, thus the absence of such-kind correlations testifies indirectly to the fact that the inductive σ-constants involve the information on field (*via* space) interaction.

At the same time, with the Σ values for atomic groups X = H, CH_3, CH_2CH_3, $CH(CH_3)_2$, $C(CH_3)_3$, $CH{=}CH_2$, C_6H_5, CN, CHO, $COCH_3$, COOH, $CONH_2$, $COOCH_3$, COCl, $Si(CH_3)_3$, NH_2, $NHCH_3$, $N(CH_3)_2$, $NHCOCH_3$, N=C=O, NO_2, OH, OCH_3, $OCOCH_3$, SH, SCH_3, F, Cl, the inductive parameter *iota* (ι) does correlate linearly [76]:

$$\iota = (z^* + 1)n^*.$$

This characteristic reflects (by means of two parameters: z^*, the effective charge of valence shell, and n^*, the effective principal quantum number) the effect of substituents over the bonds chain: $\iota = f + g\Sigma$, $f = 2.328 \pm 0.048$, $g = 1.657 \pm 0.258$, correlation coefficient r = 0.9330, adequacy dispersion $S_0^2 = 0.01302$.

Besides, the index Σ is related linearly to the atomic groups electronegativity values obtained by various experimental and theoretical methods and expressed on different scales, with the correlation coefficients from 0.95 to 1.

Therefore, the Σ value can serve as a measure of electronegativity of the substituent X in molecule. Earlier, W.F. Reynolds, S. Marriott, R.W. Taft and R.D. Topsom [77, 78] assumed a criterion of electronegativity to be a value $1 - Q_H$, where Q_H is a Mulliken charge on hydrogen atom in the HX molecule, computed at the HF/6-31G(d) level of theory. The advantage of the criterion Σ compared to the analogous index $1 - Q_H$ consists in the adequate allowance, by the Σ value, for the possibility of electron density dissimilation within the limits of sufficiently extended σ-core fragment C_6H_{13} (core σ-relaxation).

In the course of quantum chemical studies, including those of reaction mechanisms, hydrogen bonding interactions and electronic absorption spectra, the aqueous and non-aqueous media effect on the reactivity has been considered, doing so by different methods. The following approaches are used therewith.

1. *The explicit allowance for hydration in the supermolecule approximation.*

The corresponding clusters contain more than 100 water molecules. Therewith a density of water molecules distribution approaches to this value for liquid state [79]. Thus, actual aqueous solutions are simulated (*e.g.*, Figure 26).

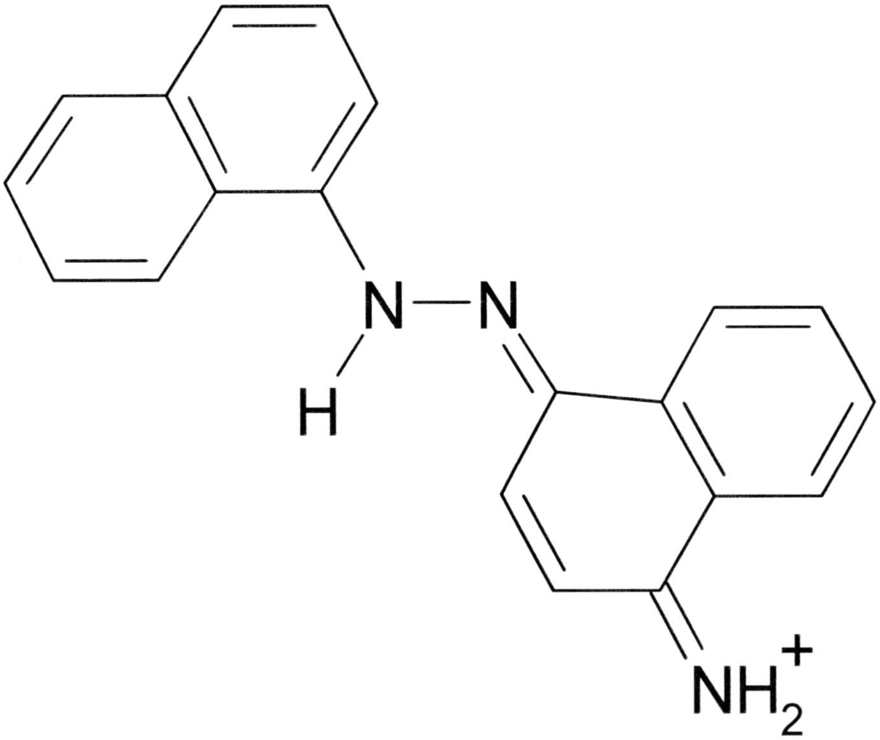

Figure 26. Product (analytical form) of the azo coupling reaction of naphthalene-1-diazonium cation with naphthalene-1-amine in the environment of water molecules (continued on next page).

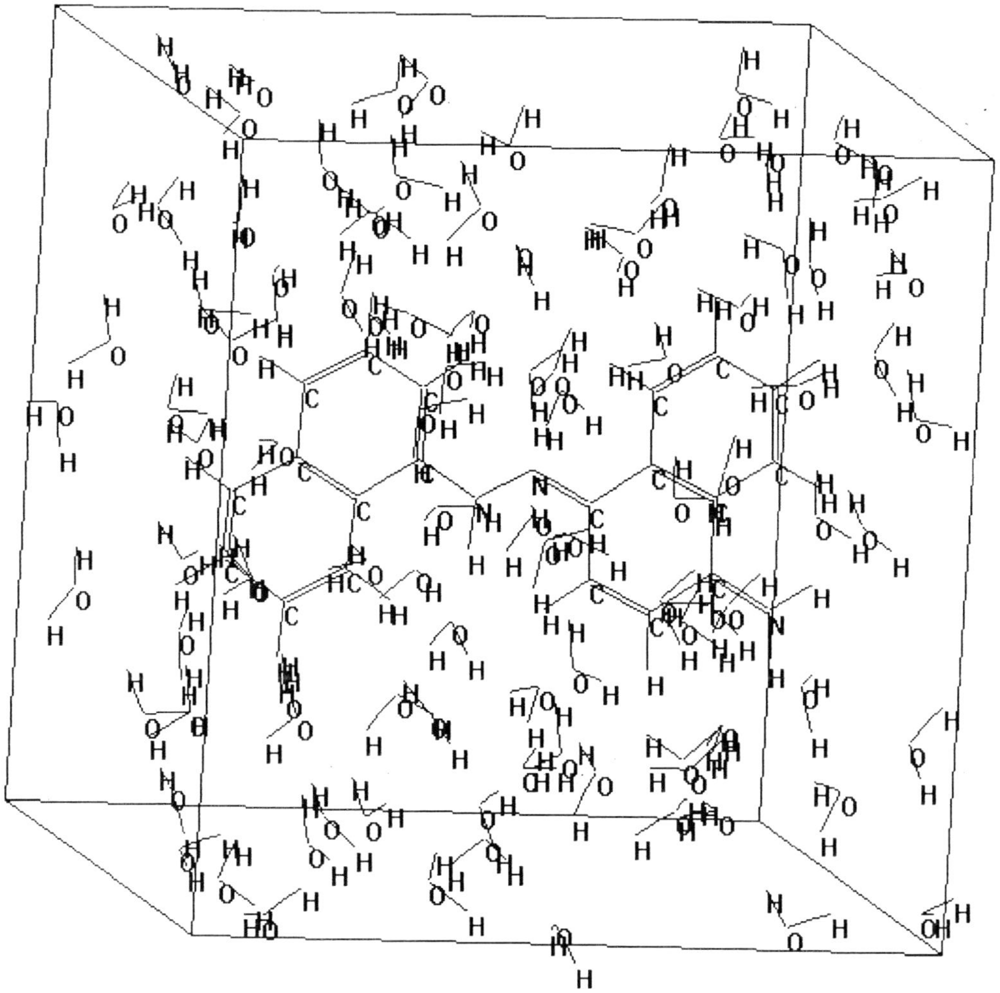

Figure 26. Product (analytical form) of the azo coupling reaction of naphthalene-1-diazonium cation with naphthalene-1-amine in the environment of water molecules (completed).

Scientific bases of the analytical characteristics prediction for azo coupling reactions have been created [80, 81]. The reactions have been found to be frontier-controlled, and their regioselectivity is unequivocally determined by thermodynamics of the intermediates - σ-complexes, as well as by the effect of aqueous medium. The latter defines high positional selectivity, providing the decision between two possible routes of reaction predicted on the basis of quantum chemical computations for separate molecular systems (*e.g.*, on Figure 27, the coupling proceeds *via* position 4 of the naphthalene-1-amine molecule). Furthermore, aqueous medium makes the selection of analytical forms - protonated azo coupling products with quinonehydrazonic structure (as for the example given on Figure 27, within the set of possible protonation centres - nitrogen atom (*p*) of primary amino group, α and β azo nitrogens - namely β nitrogen atom undergoes protonation):

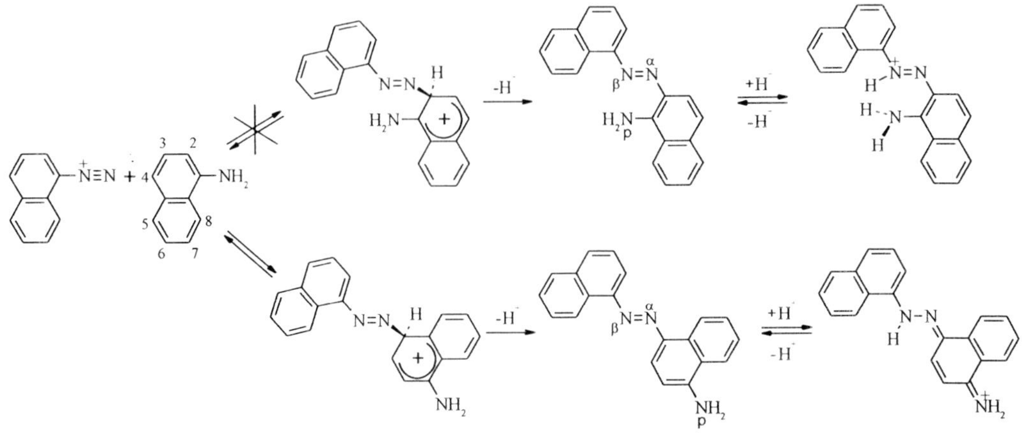

Figure 27. Azo coupling: possible directions of the reaction between naphthalene-1-amine and naphthalene-1-diazonium cation; regioselectivity of aminoazonaphthalenes protonation.

Aqueous medium has been established to exert a considerable influence upon the reaction rate for 4-carboxybenzenediazonium, 4-nitrobenzenediazonium and 4-sulphobenzenediazonium cations with diphenylamine. Furthermore, this medium stabilizes predominantly the σ-complex occurring on electrophilic attack by 4-sulphobenzenediazonium of *para* position in the diphenylamine molecule, and in doing so it chooses between two alternative routes of azo coupling reaction, which are predicted by the quantum chemical computations of isolated σ-adducts. Thus, the reaction product has the following structure:

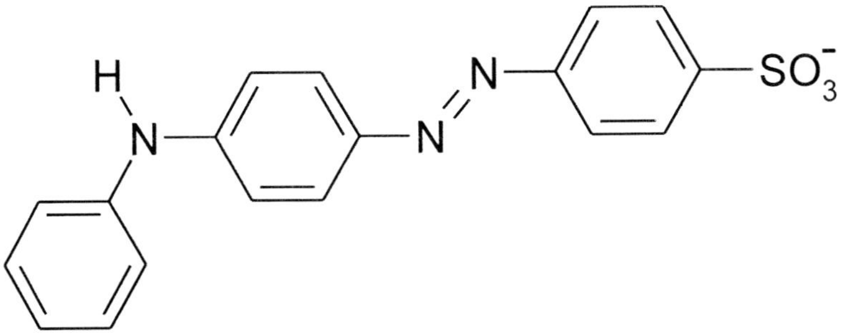

Hydration enthalpies of phenothiazine, N-acetylphenothiazine and their conjugate acids have been evaluated. The difference between the amino compounds basicity in gaseous phase and in aqueous media is connected with differentiating hydrating ability of water in relation to amines and ammonium cations.

For aromatic amines, carbonyl and azo compounds, 2-pyrrolidone, acridone, phenothiazines, it has been established that the protonation promotes hydration. It is natural, since the hydrophilic hydration is contributed mainly by electrostatic interaction which increases w-hen molecular system gains charge.

It has been cleared up that the protonation of anilines proceeds *via* the amine nitrogen atom, even if other possible centres of electrophilic attack do exist in the molecules [82].

On the basis of quantum chemical study it has been shown that for isolated molecules of sulphophthalexons (Figure 28), the heterocyclic form (H) is more thermodynamically stable than the open zwitterionic one (Z). The latter, on the contrary, is prevalent in aqueous medium [83].

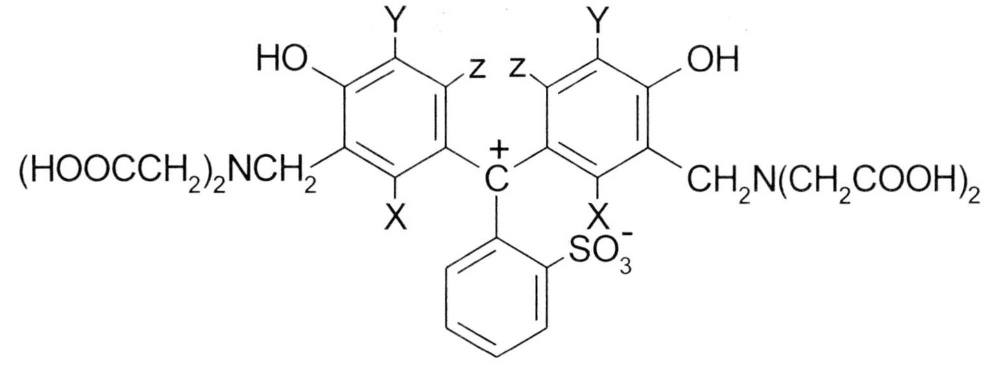

Z

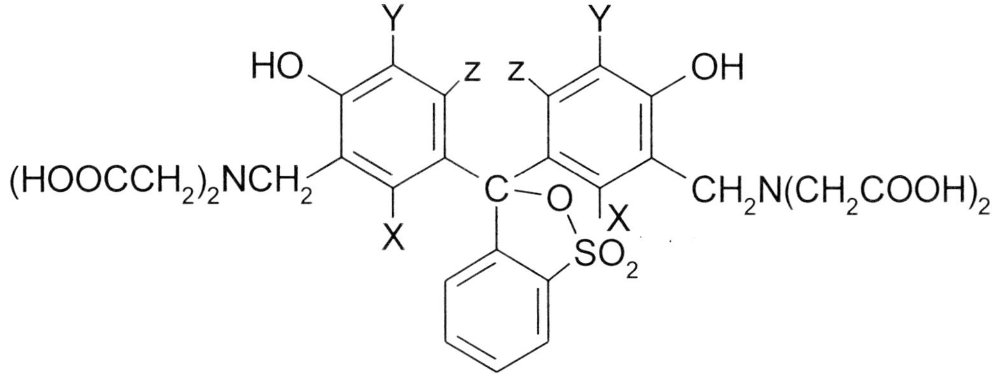

H

a) X =Y = Z = H; b) X = Z = H, Y = Cl; c) X = Z = H, Y = Br; d) X = Z = H, Y = CH_3;
e) X = Y = CH_3, Z = H; f) X = CH_3, Y = $CH(CH_3)_2$, Z = H; g) X = Y = H, Z = CH_3

Figure 28. Zwitterionic (Z) and heterocyclic (H) forms of sulphophthalexons.

2. *Study of the environment influence by the self-consistent reaction field (SCRF) method.*

IHB interactions in 8-hydroxyquinoline in relation to its zwitterionic tautomer and the rotamer without IHB have been studied by the B3LYP and MPW1K density functionals. The rotation of O-H bond and intramolecular proton transfer reactions were studied theoretically. The following theory levels have been applied: B3LYP/6-31G(d,p), B3LYP/6-311++G(d,p), MPW1K/6-311++G(d,p), MPW1K/6-311++G(2d,3p)//MPW1K/6-311++G(d,p). The effect of environment (benzene, chloroform, tetrahydrofuran, 1,2-dichloroethane, acetone, water) was simulated by the self-consistent reaction field (SCRF) method within the polarizable continuum model (PCM) at the MPW1K/6-311++G(d,p) theory level. The evolution of geometry, relative energies, rotation (around the O-H bond) and tautomerization barriers heights, IHB energies and ΔG(solv) have been systematically investigated. The results obtained have shown the failure to neglect some changes of the aforesaid characteristics in polar media with respect to the gaseous phase. The IHB energy diminishes with the solvent polarity growth. The NBO analysis at the B3LYP/6-311++G(d,p) theory level has shown that the O-H···N interaction satisfies the indicative criteria for hydrogen bonding. The charge transfer occurs from the lone pairs of nitrogen and oxygen to the aromatic system. Electron density diminishes on the OH hydrogen and increases on the N atom, *i.e.* is redistributed between the atoms - direct participants of hydrogen bonding. Our calculations have demonstrated also that the structure and properties of 8-hydroxyquinoline can be adequately discussed within the framework of hybridization. The zwitterions formation process has a barrier of ~20 kcal/mol (MPW1K/6-311++G(d,p)) both in the gaseous phase and in the presence of solvent. The reverse proton transfer reaction has no barrier almost at all in the gaseous phase and a considerable barrier (~5-10 kcal/mol) in polar solvents. The results obtained indicate that the zwitterization reaction is difficult to be realized experimentally in any medium [84].

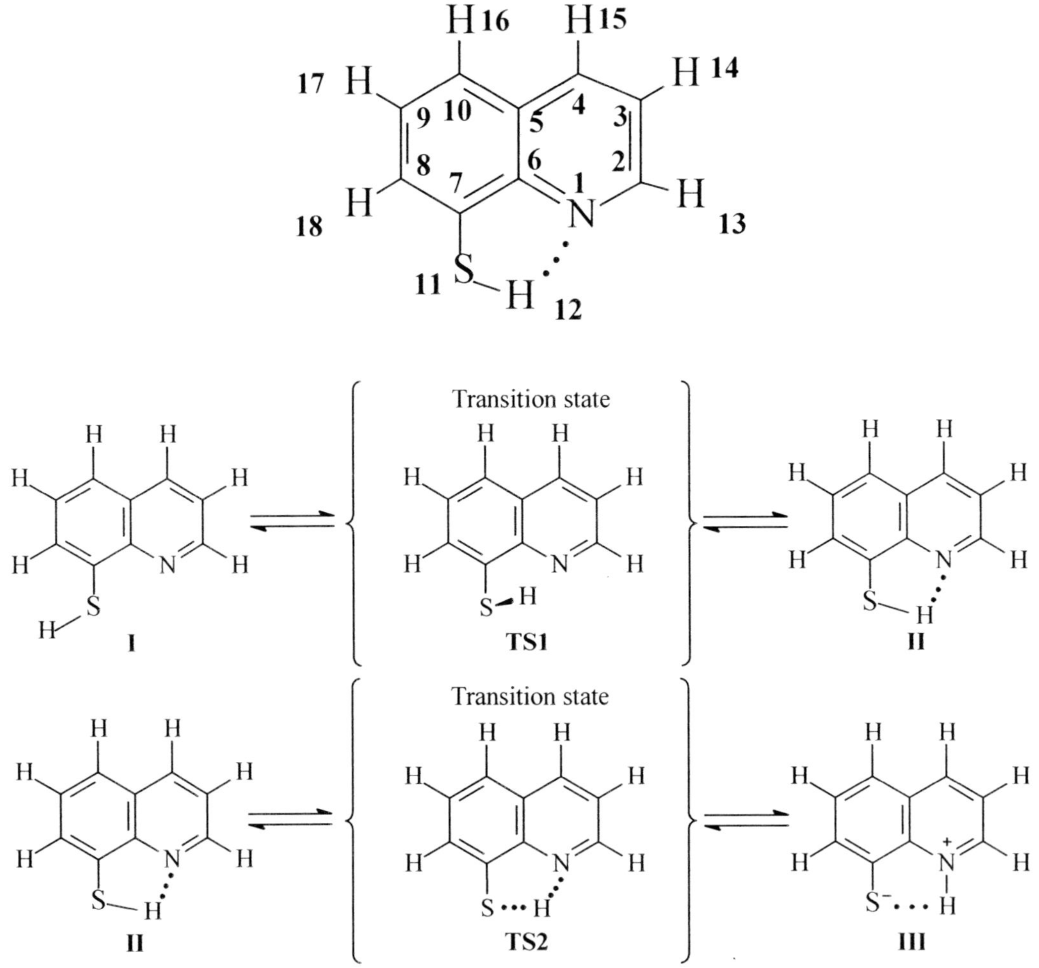

Figure 29. Rotamers and tautomers of 8-mercaptoquinoline.

A theoretical quantum chemical study of the IHB interactions in 8-mercaptoquinoline has been carried out. Special attention has been paid to the rotation (I $\rightleftarrows$ II) of S-H bond and intramolecular proton transfer (II ⇄ III) reactions (Figure 29). Therefore B3LYP/6-311++G(d,p), B3LYP/6-31+G(2d,2p), MPW1K/6-311++G(d,p), MPW1K/6-31+G(2d,2p), BH&HLYP/6-311++G(d,p), G96LYP/6-311++G(d,p) methods have been used. The most successful result of the IHB energy about 2.3 kcal/mol (difference of sums of electronic and thermal enthalpies) was computed using MPW1K/6-311++G(d,p) level. The Onsager and PCM reaction field methods at the MPW1K/6-311++G(d,p) и B3LYP/6-311++G(d,p) theory levels have been applied to study solvent (benzene, 1,2-dichloroethane, ethanol, nitromethane, water) effects on hydrogen bond energies, conformational equilibria, rotational barriers and tautomerism. Natural bond orbital (NBO) analysis has been performed to study the IHB in the gaseous phase and in aqueous medium.

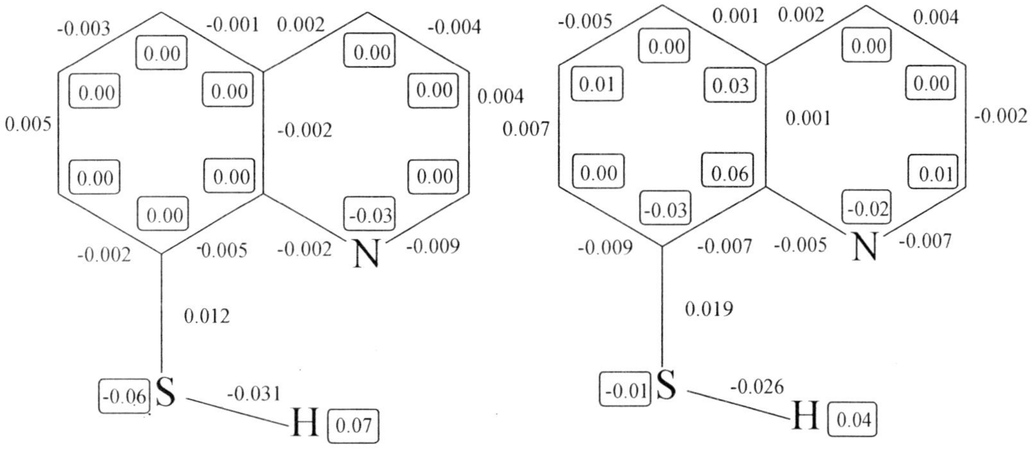

Figure 30. Changes in natural charges (circled, in atomic units) and Wiberg bond orders in II with respect to I from NBO analysis at the B3LYP/6-311++G(d,p) level in the gaseous phase (A) and in aqueous solution (B) using the PCM method.

The effect of the IHB has been studied by comparing two conformers I and II. We assume that the differences in the charges obtained by the NBO analysis, as well as in Wiberg bond orders, reflect the effect of IHB on the electron density distribution in the molecule. The changes in the natural charges and in the bond orders of II with respect to I from B3LYP/6-311++G(d,p) calculations in the gaseous phase and in aqueous solution are depicted in Figure 30. The charge distribution of II shows that the weak (as compared to I) character change of the bonds involved in the hydrogen-bonded five-membered quasicycle takes place. The electron density on sulphur and nitrogen atoms increases, on labile hydrogen - diminishes, the smaller change being in aqueous solution. In the latter, the greatest increase in the electron density is observed on C7, and some diminution on C5 and C6. The involvement of C5, C6 and C7 atoms in the electron density changes in solution, in contrast to the gaseous phase, is in agreement with significant BD*(C5-C6) → BD*(C7-C8) donor-acceptor interaction occurring only in the aqueous medium. The latter interaction may be a result of a strong augmentation of the attractive electrostatic terms and solvent polarization. The charges on the other ring carbons does not alter under the above intramolecular interaction. Among the parameters of II, the S11-H12 bond order shows a significant decrease, and C7-S11 bond order increases appreciably as a result of the IHB formation. These tendencies occur in the both media.

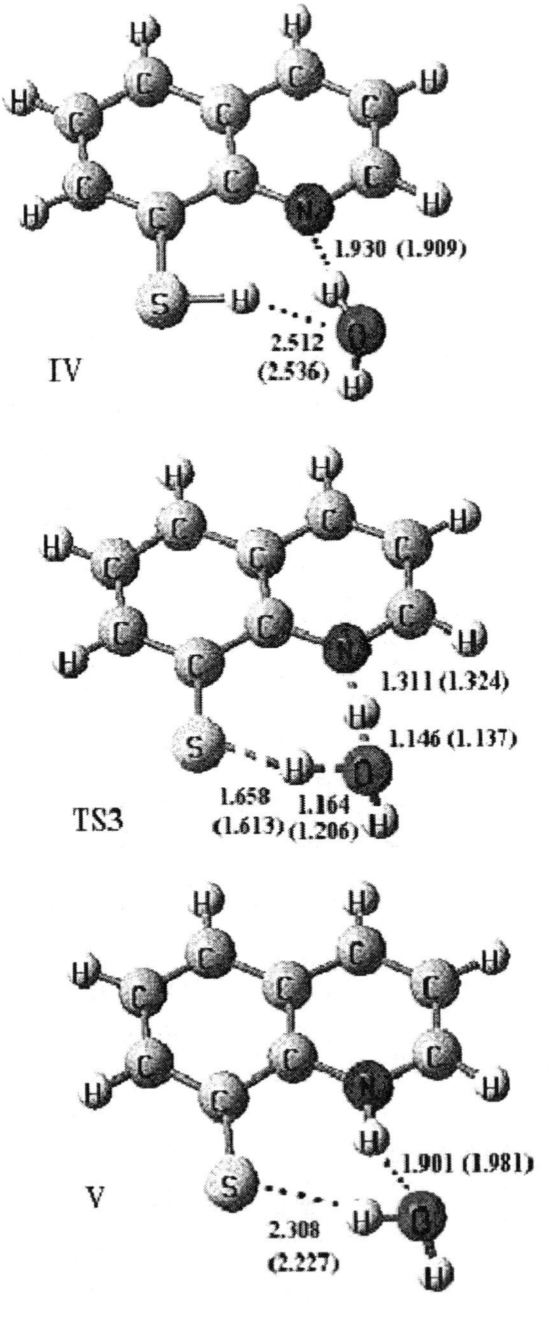

Figure 31. Some interatomic distances (Å) of tautomers IV, V and transition state TS3 at the MPW1K/6-311++G(d,p) level for the gaseous phase and aqueous solution (in parentheses).

The proton transfer on zwitterization is featured by a gaseous-phase barrier about 10.5 kcal/mol, and the reverse proton transfer nearly has no barrier [85].

Double proton transfer in the prototropic tautomerization of 8-mercaptoquinoline complex with one water molecule (IV $\rightleftarrows$ V) (Figure 31) in the gaseous phase and in solution has been studied.

To clear up a question whether any high-energy intermediate exists along the reaction path, we have computed the energetic profile (intrinsic reaction coordinate) for double proton transfer at the B3LYP/6-311++G(d,p) level starting from the transition state TS3. The results are shown in Figure 32. There is no intermediate along the intrinsic reaction coordinate, and the reaction proceeds smoothly from reactant to product. It is safe to say in this connection that two protons in the gaseous phase transfer concertedly and synchronously. In the transition state TS3 (Figure 31), both protons of the water molecule are disposed closely to the oxygen atom.

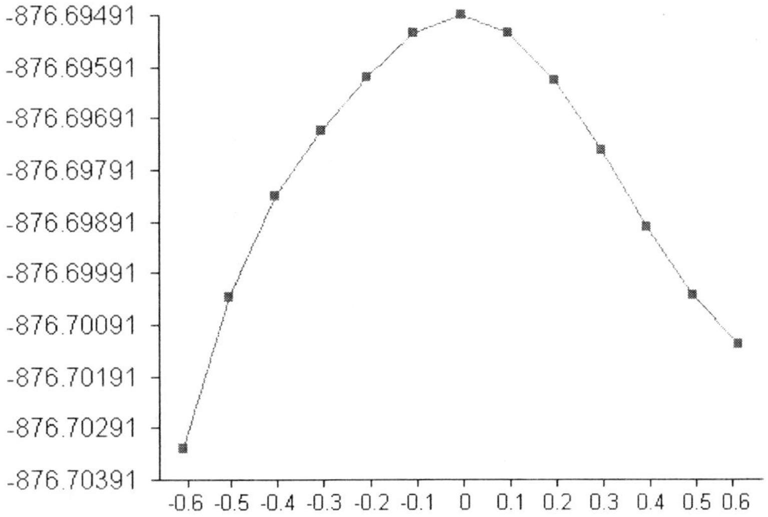

Figure 32. Energetic profile (intrinsic reaction coordinate) for 8-mercaptoquinoline water-assisted tautomerization IV $\rightarrow$ V calculated for the gaseous phase at the B3LYP/6-311++G(d,p) level. The vertical axis is for the total energies in Hartree/particle and the horizontal axis is for reaction coordinate in the units of Bohr.

The water-assisted tautomerization is kinetically less and thermodynamically more favourable than the single proton transfer. The tautomerization energy and barrier height are decreased as the dielectric constant is increased both for single proton transfer and water-assisted reaction, which imply that the tautomerization of 8-mercaptoquinoline occurs faster and more completely in polar solvent. The stability of studied forms in solution, on the contrary to that for 8-hydroxyquinoline, does not coincide with one in the gaseous phase, implying that the electrostatic influence of solvent is very important [85].

There are the manifest principal differences in the extent of influence of the electrostatic solvent effects on the relative stability of various forms of 8-hydroxyquinoline and 8-mercaptoquinoline. For the former of the above reagents, the contribution to stability, which is intrinsic for the conformers and tautomers as themselves, occurs to be determining.

Contrary, in the case of 8-mercaptoquinoline, the polar medium plays a decisive role for governing the relative stability of different forms.

It is known [86] that the principal consequences of the specific interaction with solvent could be taken into account provided that the complex of the studied molecule with one solvent molecule is explicitly considered. When regarding the water-assisted proton transfer and placing the 8-mercaptoquinoline complex with water molecule in the dielectric continuum of the solvent, we simultaneously take into account the effect of specific and universal solvation on the reactivity within the discrete-continuum approach. On studying the water-assisted reaction we have shown a possibility of a by-step proton transfer as that occurs in real water with its complex structure, the profound distant-order elements in a liquid state. The regularities elucidated therewith in respect to the effect of complexation with water molecule on the mechanism, kinetics and thermodynamics of tautomerization could compose the basis for the directive regulation of the reactions with the aim of obtaining the compounds with desirable properties.

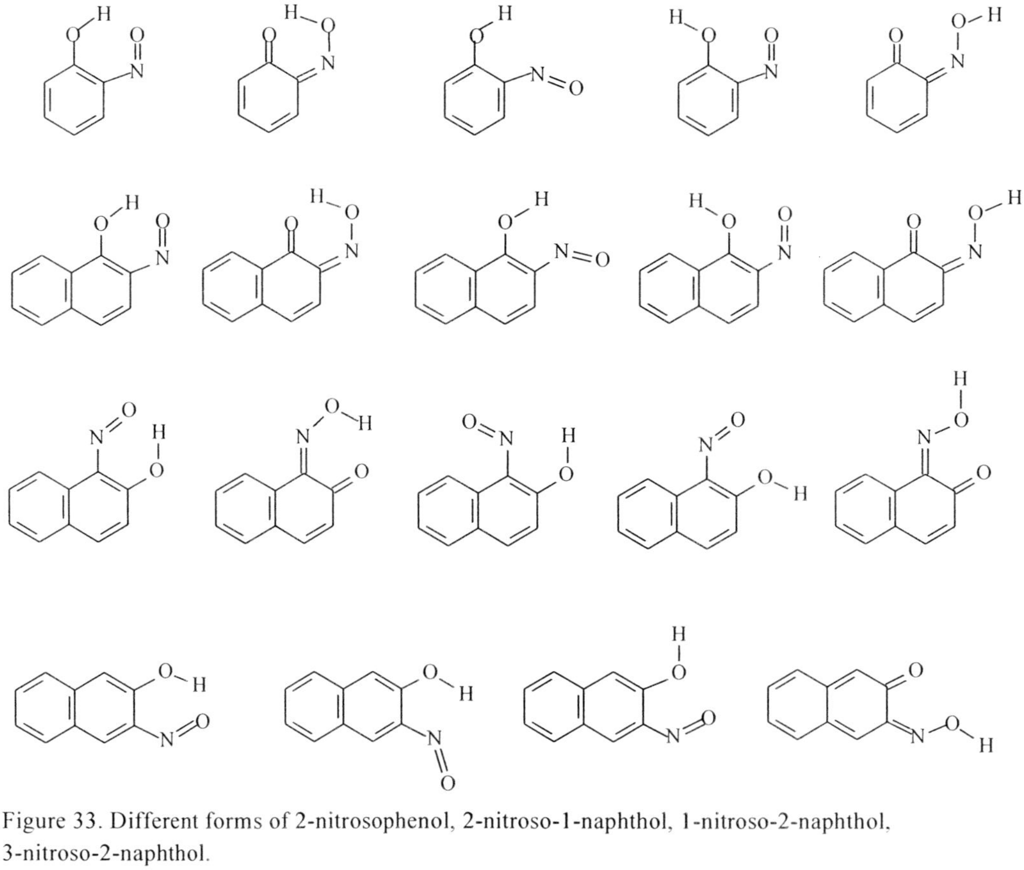

Figure 33. Different forms of 2-nitrosophenol, 2-nitroso-1-naphthol, 1-nitroso-2-naphthol, 3-nitroso-2-naphthol.

Intramolecular hydrogen bonding (IHB) interactions and molecular structures of 2-nitrosophenol, nitrosonaphthols and their quinone-monooxime tautomers (Figure 33) were investigated at *ab initio* (HF/6-31G(d,p), MP4/6-311G(d)//MP2/6-31G(d,p)) и DFT (B3LYP/6-311++G(d,p), MPW1K/6-311++G(d,p)) and density functional theory (B3LYP/6-311++G(d,p), MPW1K/6-311++G(d,p)) levels. Possible conformations with different types of the IHB of the tautomers were considered to understand the nature of the hydrogen bonding among these conformers. The effect of solvent on hydrogen bond energies, conformational equilibria, and tautomerism in aqueous solution were studied by means of SCRF within the PCM model. Natural bond orbital analysis was performed to study the IHB in the gaseous phase and in aqueous medium. The NMR ^{1}H, ^{13}C, ^{15}N, ^{17}O chemical shifts in the gaseous phase and in solution for the studied compounds were calculated using GIAO approach. The optimized geometrical parameters and ^{1}H NMR chemical shifts are in good agreement with previous experimental data [87].

It has been elucidated that the hydrogen bond energy formation for nitroso tautomeric forms is a larger than for oxime ones. The analysis of charge density has shown that all the systems studied here satisfy the indicative criteria of hydrogen bonding interactions.

Careful conformational analyses of given molecules show that both that the size and charge differences between oxygen atoms in these compounds are very important factors not only for comparing their conformational behaviour but also for understanding the nature of hydrogen bonding.

Besides, good correlations between O···H distance and the difference of charge density ($Q_H - Q_A$, where A is the atom - acceptor in the H···A hydrogen bond), electron density, Laplacian density at bond critical point, and ellipticity has been found.

The results of NBO analysis indicate that interactions in nitroso forms are stronger than those in oxime ones (except for 2-nitrosophenol). The hydrogen bond is stronger in the *syn*-oxime structure of 2-nitrosophenol than that of the others, especially in aqueous medium.

The results of this study also show that the topological parameters may be applied to estimate the H-bond strength. The NMR ^{1}H chemical shifts may be useful descriptors for the strength of intramolecular H-bonds.

By means of the B3LYP, MPW1K and B3PW91 methods with the 6-311++G(d,p) and AUG-cc-pVDZ basis sets, the equilibrium geometries and vibrational frequencies of stable conformers of the feasible monomers and dimers of 2-pyrrolidone have been studied [88].

The problem of 2-pyrrolidone hydrogen-bonded dimers formation (Figure 34) is of great interest, since all hypothalamus hormones are low-molecular peptides, many of them carry a fragment of pyroglutamic acid [89], *i.e.*, in essence, the fragment of 2-pyrrolidone. Such hormones are thyroliberin, gonadoliberin, melanostatin, *etc.* The sites of synthesis of hypothalamus hormones are nerve endings - synaptosomes of hypothalamus. The large amounts of hormone associates have been discovered just here. It cannot be excluded that the aforementioned associate species are stabilized by hydrogen bonds of the pyroglutamic acid fragment.

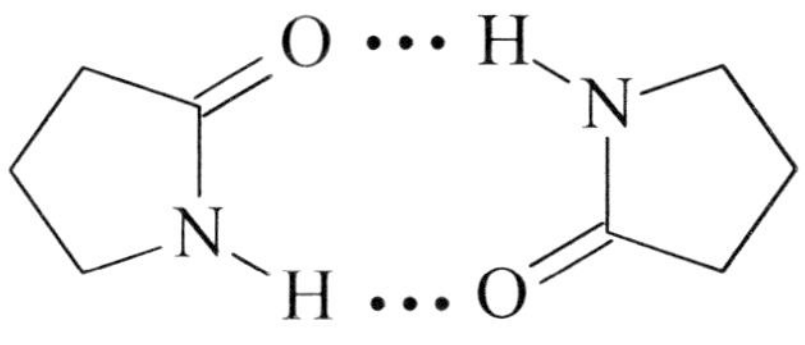

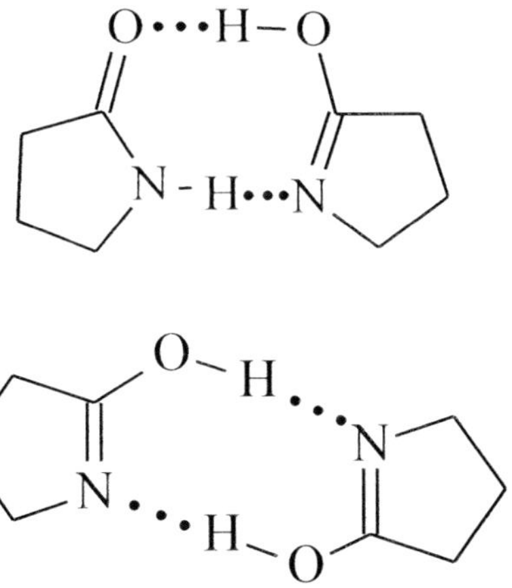

Figure 34. Theoretically possible dimers of 2-pyrrolidone based on its different tautomers.

As we have shown [88], the intramolecular proton transfer in the monomer, 2-pyrrolidone, has a barrier of ~50 kcal/mol and is essentially independent on the solvent. Double proton transfer in the hydrogen-bonded dimers proceeds concertedly and synchronously with a low barrier. Therefore, the prototropic tautomerization of 2-pyrrolidone:

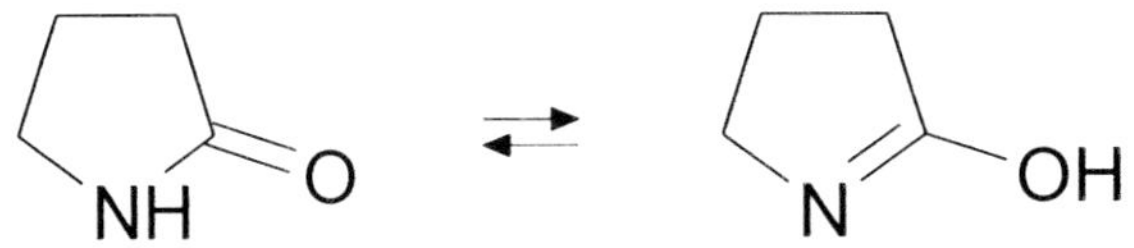

occurs *via* the intermolecular mechanism.

The dimers have been investigated by performing the topological analysis of electron density at bond critical points within the framework of the Bader's atoms in molecules (AIM) quantum theory, the analyses of NBO, orbital interactions, electrostatic potential. As a whole, the undertaken studies have shown that all the systems under consideration can meet the criteria for hydrogen bonding. The dependences between the N-H (O-H) bond's stretching vibration frequency and this bond's length; between the intermolecular hydrogen bond length in dimer, the dimerization energy and the electron density at bond critical point have been found. The NBO analysis has led to a conclusion that all the complexes incorporate the strong intermolecular interactions of the type of charge transfer from the lone electron pairs of heteroatoms to the σ^*(N-H) and σ^*(O-H) antibonds. As a consequence, the occupancy of these antibonding orbitals is fairly high.

3. *Indirect elucidation of the medium contribution by observing deviations from linear dependences between the experimental values of molecular properties in solutions (e.g., pK_a values) and gaseous-phase characteristics obtained by the quantum chemical methods (in particular, proton affinities).*

Revealing the character (specific or conditionally universal solvation) and value of medium effect from equations linking some properties of substances in solution and in the gaseous phase.

Predictive linear interrelations linking the benzenethiols (thiophenols) pK_a values with the quantum chemically evaluated enthalpies and free energies of the benzenethiolate anions proton affinity (*PA*) have been established. Differential influence of organo-aqueous media on the benzenethiols acidity has been estimated.

For the anions (A^-) formed on 27 benzenethiols (AH) dissociation, we have computed the proton affinity (Table 7) represented by enthalpy [90-93] and free energy of reactions of proton detachment in gaseous phase $AH \rightarrow A^- + H^+$:

$$\Delta\Delta H_f = \Delta H_f(A^-) + \Delta H_f(H^+) - \Delta H_f(AH),$$
$$\Delta\Delta G_f = \Delta G_f(A^-) + \Delta G_f(H^+) - \Delta G_f(AH).$$

The objects under study were the XC_6H_4SH compounds, where X = H (VI), 2-CH_3 (VII), 3-CH_3 (VIII), 4-CH_3 (IX), 3-$COCH_3$ (X), 4-$COCH_3$ (XI), 2-COOH (XII), 4-COOH (XIII), 3-$COOCH_3$ (XIV), 4-$COOCH_3$ (XV), 4-NH_2 (XVI), 4-CH_3CONH (XVII), 3-NO_2 (XVIII), 4-NO_2 (XIX), 4-OH (XX), 3-OCH_3 (XXI), 4-OCH_3 (XXII), 3-CH_3SO_2 (XXIII), 4-CH_3SO_2 (XXIV), 4-F (XXV), 3-Cl (XXVI), 4-Cl (XXVII), 3-Br (XXVIII), 4-Br (XXIX), 3-I (XXX), 4-I (XXXI), as well as 2,4-dichlorobenzenethiol (XXXII).

Starting from the modified equation of isobar of chemical reaction, $\Delta G^0 = -RT\ln K_p$ (ΔG^0 is the change in standard free energy in the course of reaction, K_p is the equilibrium constant), we have the right to predict the linear interrelation between pK_a and free energy of proton affinity: $pK_a = h + i\cdot\Delta\Delta G_f$. Therewith for perfect system (gaseous phase), the intercept (*h*) and the slope (*i*) have the following values: $h = 0$, $i = (\ln 10\cdot RT)^{-1} = 0.7330$ mol/kcal. The change in entropy remains near-constant in the similar-type reactions with the participation of relative compounds of close molecular geometry. Therefore, the linear dependence between pK_a and enthalpy of proton affinity, $pK_a = h + i\cdot\Delta\Delta H_f$, could also exist.

We have found that the literature pK_a values of benzenethiols in the aqueous-acetone, aqueous-ethanol media and in methanol do correlate linearly with the gaseous-phase proton affinity $PA = \Delta\Delta H_f$, $\Delta\Delta G_f$ (Table 7; see the example in Figure 35; *m* is number of points in correlation equation, *r* is correlation coefficient).

Table 7. Correlations of the benzenethiols pK_a values with the benzenethiolate anions proton affinities (*PA*) (continued on next page)

Conditions of pK_a determination*	*PA*	Set of compounds	*m*	*h*	*i*, mol/kcal	*r*
1	$\Delta\Delta H_f$	VI-IX, XVII, XXV, XXVII, XXX	8	-71.84 ± 33.50	0.2395 ± 0.0998	0.9234
1	$\Delta\Delta G_f$	VI-IX, XVII, XXV, XXVII, XXX	8	-93.70 ± 21.74	0.3134 ± 0.0667	0.9787

Table 7. Correlations of the benzenethiols pK_a values with the benzenethiolate anions proton affinities (PA) (completed)

Conditions of pK_a determination*	PA	Set of compounds	m	h	i, mol/kcal	r
2	$\Delta\Delta H_f$	VI, VIII-XV, XIX-XXII, XXVI, XXVII, XXIX-XXXII	19	–49.22 ± 9.89	0.1690 ± 0.0298	0.9453
2	$\Delta\Delta G_f$	VI, VIII-XV, XIX-XXII, XXVI, XXVII, XXIX-XXXII	19	–49.45 ± 10.73	0.1743 ± 0.0332	0.9368
3	$\Delta\Delta H_f$	VI, VIII-XV, XIX-XXII, XXVI, XXVII, XXIX-XXXII	19	–53.05 ± 11.14	0.1849 ± 0.0335	0.9423
3	$\Delta\Delta G_f$	VI, VIII-XV, XIX-XXII, XXVI, XXVII, XXIX-XXXII	19	–53.28 ± 12.08	0.1907 ± 0.0374	0.9335
4	$\Delta\Delta H_f$	VI, VIII, IX, XVIII-XXIV, XXVI, XXVII, XXIX-XXXII	16	–47.10 ± 8.81	0.1627 ± 0.0265	0.9619
4	$\Delta\Delta G_f$	VI, VIII, IX, XVIII-XXIV, XXVI, XXVII, XXIX-XXXII	16	–46.87 ± 9.48	0.1665 ± 0.0293	0.9558
5**	$\Delta\Delta H_f$	VI, IX, XVI, XIX, XXII, XXVII, XXVIII	7	–49.06 ± 30.23	0.1709 ± 0.0907	0.9080
5**	$\Delta\Delta G_f$	VI, IX, XVI, XIX, XXII, XXVII, XXVIII	7	–50.30 ± 28.91	0.1797 ± 0.0892	0.9180
5***	$\Delta\Delta H_f$	VI, IX, XIX, XXII, XXVII, XXVIII	6	–42.03 ± 21.90	0.1493 ± 0.0659	0.9531
5***	$\Delta\Delta G_f$	VI, IX, XIX, XXII, XXVII, XXVIII	6	–43.03 ± 23.81	0.1567 ± 0.0737	0.9472

*1 - Determined by potentiometric titration in water - acetone mixture (1 : 3 by volume) at 27 °C [94]
2 - Determined by potentiometric titration in 48.9 vol. % aqueous ethanol at 20-22 °C [95]
3 - Determined by potentiometric titration in 95 vol. % aqueous ethanol at 20-22 °C [95]
4 - Determined by spectrophotometric technique in 48 vol. % aqueous ethanol at 25 °C [96]
5 - Determined by potentiometric titration in methanol [97]

**Correlation series involves the PA and pK_a values for all the compounds, for which the pK_a values in methanol have been measured

***Data for the species XVI (X = NH_2) are excluded

The interrelations found are indicative of the regularity - the approximately constant contribution of medium to the protolytic properties of benzenethiols.

A measure of the influence of constant component of medium contribution on the benzenethiols acidity can be: 1) the deviation of i value in Table 7 from the theoretical slope of dependence pK_a *vs* $\Delta\Delta G_f$, the latter being equal to $(\ln 10 \cdot RT)^{-1}$ = 0.7330 mol/kcal and assigned to gaseous phase; 2) the absolute h value equal to zero in the case of gaseous phase.

We have estimated differential effect of solvent for aqueous-ethanol media with varied ethanol content. There is the equation $pK = j + k/\varepsilon$ valid for the constants of various equilibria [98]. That is why the universal (electrostatic) or conditionally universal (corresponding to the conservation of energy of specific interactions) solvation, that depends upon the dielectric constant ε, should contribute at least to the coefficient i in the equalities $pK_a = h + i \cdot PA$, therewith $i \sim \varepsilon^{-1}$. The higher the alcohol content in aqueous-ethanol medium, the greater is the pK_a value, *i.e.* the weaker are the acidic properties of benzenethiols. The relationship of pK_a with ε has an antibate character. At the same time, the h value in going from 48.9 vol. % ethnol to 95 vol. % ethanol decreases slightly. Therefore, the conditionally universal solvation, that depends on ε, does not determined by the h value. For the explicit proton-donors ArSH, the obvious is the occurrence of specific interactions with the aqueous-ethanol media, that diminish pK_a compared to the gaseous phase. Provided that the above interactions would be displayed by the coefficient h, thus the increase in h value with the increase in the ethanol content in the aqueous-ethanol mixture should be assumed, since water assists the proton ejection from the molecule of benzenethiol ArSH to a somewhat greater extent (the values of pK_b for water and ethanol are 15.77 and 16, respectively [99]) and stabilizes more effectively the anion ArS^- formed as a result of ArSH dissociation (the acceptor numbers of water and ethanol are 54.8 and 37.3, respectively [98]) as compared to ethanol. The actual coefficient h is somewhat more negative in the case of medium enriched with ethanol.

The increase in the i value, as we observed in going from 48.9 vol. % ethanol to 95 vol. % ethanol, agrees with the corresponding effect on pK_a exerted by both the polarity (the ε value for water and ethanol are 78.39 and 24.55, respectively [98]) and the specific interactions. Consequently, the influence of both conditionally universal and specific solvation is comprised into the coefficient i. Differential effect of medium manifests itself in the component $i \cdot PA$ of the equality $pK_a = h + i \cdot PA$. The coefficient h likely has no physical meaning, but presents a scaling factor [37].

It would be reasonable to stress that the coefficient i values for the solutions of benzenethiols in 48.9 vol. % and 95 vol. % aqueous ethanol differ only slightly, that is natural for two media close in composition. However, it is difference in the coefficient i, being even small, that provides nevertheless the predomination of $i \cdot PA$ component over the h value in correlation equations. As a consequence, the experiments show the increase in pK_a in going from 48.9 vol. % ethanol to 95 vol. % ethanol.

With a series of solvents with different dielectric constants at hand, the conditionally universal and specific contributions of medium to the slope of the proton-affinity dependences of pK_a could be separated, and the system's "solute - solvent" shift to imperfection estimated ("gaseous phase", or an isolated molecule of solute) separately due to each type of solvation.

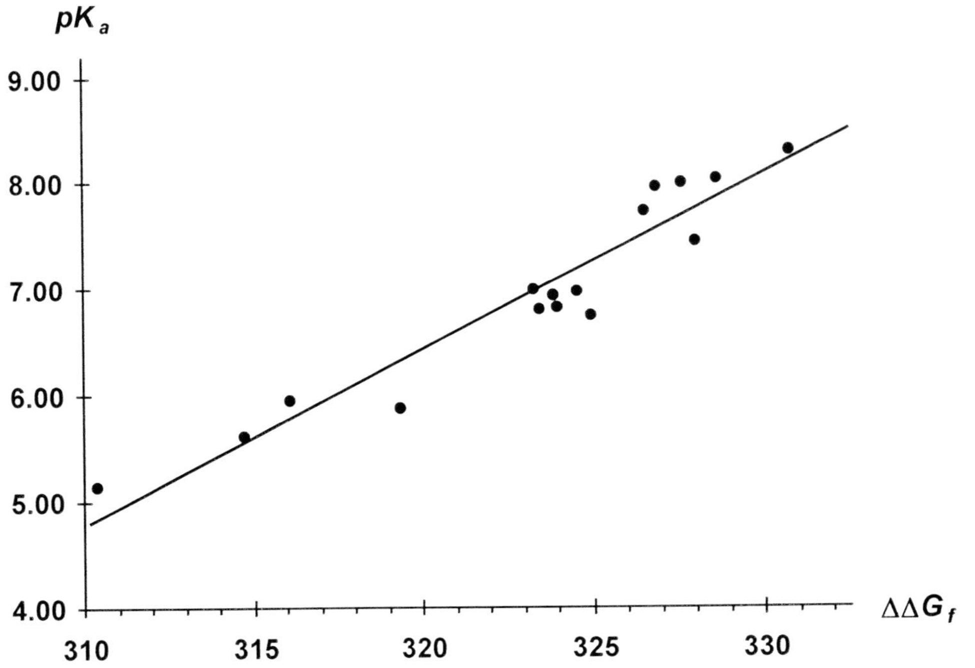

Figure 35. Dependence of benzenethiols XC_6H_4SH pK_a values determined by spectrophotometric technique in 48 vol. % aqueous ethanol at 25 °C [95] *vs* $\Delta\Delta G_f$ (kcal/mol). X = H, 3-CH_3, 4-CH_3, 3-NO_2, 4-NO_2, 4-OH, 3-OCH_3, 4-OCH_3, 3-CH_3SO_2, 4-CH_3SO_2, 3-Cl, 4-Cl, 3-Br, 4-Br, 3-I, 4-I. Correlation coefficient $r = 0.9558$.

Let us also offer another example. We have computed enthalpies and free energies of gaseous-phase proton affinity (*PA*) for aniline and its 62 derivatives with different kinds of electron-donor and electron-acceptor substitution in aromatic ring and at nitrogen atom. Linear correlations of the type pK_a *vs PA* have been found. Deviations of the data for some *ortho* substituted anilines from the above relationships have been discussed in view of possible hydrophobic hydration [100] of the molecular fragments adjacent to the protonation centre.

4. *Indirect knowledge on a character of hydrate shell developed from the information on charge localization or delocalization in molecular system.*

The effect of electrostatic, hydrophobic factors, as well as, using (1), (2) and (4) approaches, of hydration on reactivity, physicochemical and analytical properties is studied. In particular, in terms of the analysis of the mentioned effects, the regularities of phenothiazine fluorescence in micelles of cationic, anionic, non-ionic surfactants, the formation and extraction of ionic associates of 2,6-diphenyl-4-(4-dimethylamino)styrylpyrilium (XXXIII) and -thiopyrilium (XXXIV) cations with anionic surfactants [101] have been substantiated.

In Table 8, presented are new (obtained at a higher level of theory) data, as compared to those discussed in [101], on the electronic structure of the XXXIII, XXXIV systems, along with there selenium analog XXXV.

In the cation XXXIV, the overall charge of hetero ring is somewhat shifted to the positive side as compared to XXXIII, and positive charge is localized in significant extent on sulphur atom, whereas in the system XXXIII the charge of hetero ring is delocalized, and oxygen atom possesses even negative charge. Therefore the cation XXXIV characterized by higher charge density is evidently surrounded by more closer hydrate shell in aqueous solutions compared to the molecular system XXXIII. As a consequence, the cation XXXIII with anions of surfactants and inorganic ions in aqueous media forms the ionic associates, which are more stable and capable of being easier extracted by toluene as compared to the cation XXXIV. At the same time, in toluene extract, ionic associates and their components are not hydrated, and in accordance with the aforesaid peculiarities of cycles' charge distribution, ionic associates of the cation XXXIV with counterions are more stable than ionic pairs containing the cation XXXIII. The value and localization extent of charge in the cations of organic reagents XXXIII and XXXIV influence the magnitudes of stability constants and associates' distribution constants, as well as the reagents selectivity with respect to inorganic ions.

Table 8. Natural charges on atoms in chalcogenopyrilium cations at B3LYP/6-311G(d)

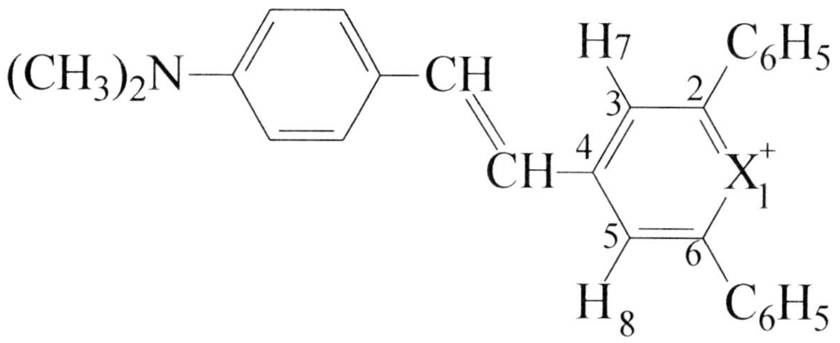

XXXIII X = O; XXXIV X = S; XXXV X = Se

Cation	X	Natural charge on the atom								
		1	2	3	4	5	6	7	8	Σ*
XXXIII	O	–0.463	0.426	–0.260	0.052	–0.251	0.427	0.225	0.228	–0.069 (0.384)
XXXIV	S	0.513	–0.088	–0.214	0.040	–0.207	–0.089	0.221	0.225	–0.045 (0.401)
XXXV	Se	0.638	–0.131	–0.225	0.042	–0.218	–0.132	0.219	0.222	–0.026 (0.415)

*Sum of charges on atoms of the heterocycle: the first number - without H7 and H8, in parentheses - including H7 and H8

From the viewpoint of charge distribution in the heterocycle, and due to slightly more positive total charge of the ring, for the selenium-containing cation XXXV one can expect more pronounced incapability of forming close ionic pairs in aqueous solutions, more difficult extraction by inert solvents, but, at the same time, occurrence of more stable ionic associates in the extracts as compared to the sulphur analog XXXIV.

5. *A specific kind of indirect medium consideration within the problem of quantum chemical calculation of electronic absorption spectra.*

The possibility of quantitative predictions of absorption maxima in electronic spectra of unsaturated organic compounds by means of the Pariser - Parr - Pople (PPP) method with adaptation of parameters has been examined. In this viewpoint, a schematic classification of quantum chemical methods has been proposed which is based on the idea of extrapolating: 1) from elementary particles (physical constants) to molecule; 2) from atoms to molecule; 3) from fragments to molecule [102, 103].

To reproduce quantitatively the wavelenghths of long-wave absorption bands, a possible way out is decreasing the "height" of the leap involved in the quantum chemical method which now may be designated as "fragments - molecule" (the case 3 among mentioned ones). Its advantages consist in the following:

[i] Fragments, being separate molecules, may be placed into the same medium as the initial molecule. It is of importance, since an explicit consideration of the solvation effects in electronic spectroscopy represents a complicated problem (*e.g.* [104] and Ref. therein), it is rarely possible to attain a quantitative agreement of calculated transition energies with experimental data.

[ii] The choice of the parameters is made using electronic molecular spectroscopy which is more accessible than atomic.

[iii] Fragments, in contrast to atoms and elementary particles, allow for a gradual complexation which provides for possible correction of errors at each step; the total error is the error of the last step, thus appearing minimal.

For realization of the above-described idea it is advisable that at a minimal complexation of the molecule only one unknown parameter would appear. In the case of the PPP method, the parameter to be corrected is two-electron one-centre Coulomb integral (its evaluation is the main object of criticism in the whole method [105]). The correction implies that deviations exceeding their conventional values by more than 30 % are highly undesirable.

Practical use of the method consists in the following. A series of compounds is built, in the beginning of which there is an appropriate conjugated hydrocarbon (it corresponds to the first parameter in the scheme to be corrected, *i.e.* the Coulomb integral of the carbon atom), and the terminal is the molecule under study. In going from one compound in the series to the neighbouring one, one new type of π-centre corresponding to the latter must appear (note that it may be several identical π-centres). The spectra of the compounds in this series should, as far as possible, be recorded under similar conditions. After that the parameters of each π-centre, by one at each step, are selected with fixation of all the parameters determined at preceding steps.

Thus the calibrated parameter "remembers" all the factors accounting for deviations of the experimental spectrum from the theoretical one and reproduces them at each following step.

For correct reproduction of spectral wavelengths, the construction of complete series of compounds, beginning with a hydrocarbon, with the appearance of a new type of π-centre at each step is not always necessary; spectra may be calculated according to a "cut-down" scheme using one or more model systems for calibration of the parameters.

The PPP method with adaptation of parameters has been used to calculate the electronic spectra of some unsaturated, aromatic and heterocyclic compounds, such as: N,N'-diphenyl-*p*-diphenoquinonediimine dication (oxidation product of diphenylamine in acidic media) [66] and derivatives of ethylene, benzene, naphthalene, quinone, pyridine (substituted and condensed), pyrimidine, some other azaheterocycles, thiopyranone and furan [102, 103].

The third group of methods (extrapolating from fragments to molecule) can hardly be expected to expand on account of all-valence methods, since the latter do not provide for the appearance of the only one parameter to be calibrated, and not several ones simultaneously, after a minimal complexation of the molecule structure. In going from one group to the neighbouring one with a higher number, the amount of computational work exhibits a clear tendency towards its decrease, while that of experimental (spectroscopic) work tends to grow.

Let us discuss the relation between the methods of the first and the second groups from one hand, and the third group, from the other hand. There is an important advantage that the *ab initio* and DFT approaches, as well as the all-valence semiempirical models as CNDO, INDO, PNDO, NDDO *et al.* have over the PPP, the latter even with adaptation of parameters. Through automatic core (sigma) relaxation they need not distinguish between different kinds of the same atom, *e.g.* aniline (pyrrole) and pyridine type nitrogen, phenol (furan) and carbonyl type oxygen, *etc.* The above-mentioned all-valence semiempirical methods, unlike the PPP method with adaptation of parameters, use unique sets of parameters for all molecular systems. Therefore, they are, in general, less accurate in prediction of electronic transition energies, especially for large molecules. The fact that the PPP approach used deals with different sets of parameters for different kinds of molecules reduces the predictive power of the method. However, it gives an opportunity to reproduce in many cases real absorption wavelengths of rather complicated organic compounds within minimal CI both without use of any combination of single and double configurations selected on the basis of some threshold criteria with the possibility of making additional corrections due to excluded configurations by means of the perturbation theory [106], and without explicit consideration of medium effects.

It is necessary to note that on calibration, the parameters become somewhat (in the cases of benzoic acid, aniline, phenol, furan - by 3-5 eV) greater compared to the original ones. That should lead to the undesirable enhancement of energies of the excited states disposed higher than the first one. Being within the framework of the formulated problems, the aforesaid seems appropriate.

The methods of the third group seem to have some advantages at present, in the cases when the quantitative reproducing of any characteristics (in particular, a wavelength in electronic spectrum, which is necessary, for example, at chemical substances identification) is aimed. For indirect confirmation of structure of reaction products, especially for those which can not be isolated and characterized by means of IR, NMR, mass-spectrometric, X-ray and other investigations, computations of electronic spectra and comparison with experiments are expedient. Detailed analysis of energetics and nature of molecular systems stationary states, the establishing of the corresponding regularities in the series of compounds and solvents, require the methods belonging to the first or the second group.

Further, we mention some other interpretational tasks of chemistry which we have dealt with.

Basing on the analysis of comparative thermodynamic stability of various configurations, the consideration of experimental and computed dipole moments, the allowance for polarity of medium, it has been concluded that the molecules $(C_6H_5)_3AsCl_2$, $(C_6H_5)_3SbCl_2$, $(C_6H_5)_3Sb(OH)Cl$, $(C_6H_5)_3BiCl_2$, $(C_6H_5)_3Bi(ONO_2)_2$ exist as solutes in the form of trigonal bipyramid with two axial inorganic ligands [25, 33, 35]. An example has been displayed on Figure 36.

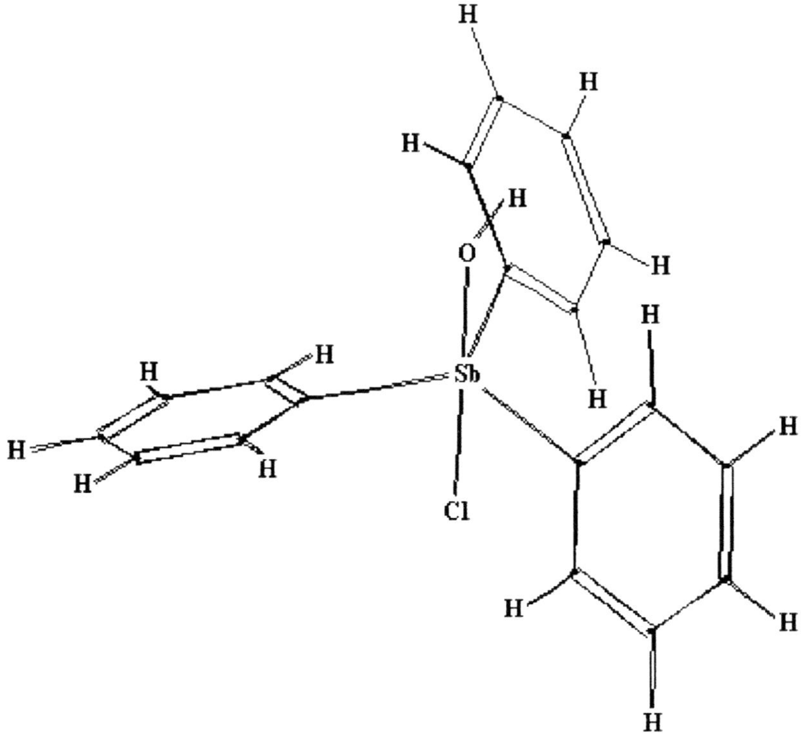

Figure 36. Optimal configuration and conformation of triphenylstibinehydroxychloride $(C_6H_5)_3Sb(OH)Cl$.

It has been established that in gaseous phase and aqueous solution, the most thermodynamically stable tautomer of 2-pyrrolidone is lactam. According to quantum chemical evaluatons with an explicit account for aqueous medium, the state of tautomeric equilibrium serves as a prerequisite to participation of 2-pyrrolidone's lactim tautomer (pyrroline-2-ol) in complexation with palladium (II) in aqueous solution. 2-Pyrrolidone protonation in gaseous phase and aqueous medium has been shown to proceed *via* oxygen atom, that corresponds to knowledge on mesomeric displacement of electron density in amide fragment. Aqueous medium stabilizes 2-pyrrolidone's lactim tautomer to the greater extent compared to lactam, and O-protonated cyclic amide compared to N-protonated one. Stereodirective character of palladium (II) complexation with chloride ion and pyrroline-2-ol has been explained. The initially formed tetragonal-pyramidal adduct with axial organic ligand then regroups into precursor of *cis* product - intermediate with extracoordinated axial chlorine atom. Less thermodynamically stable *cis* isomer of the complex $[PdCl_2(Prol)_2]$ (Prol = pyrroline-2-ol) appears due to the fact that its structural precursor is the most preferable, in respect to energy, intermediate of associative nucleophilic substitution. At a supermolecular level, *cis* product is capable of being stabilized by means of intermolecular dipole-dipole association in a crystal [107]. The said reaction stereoselectivity treatment is of importance, since namely *cis* complexes of palladium (and also platimum) show pronounced antitumour activity.

Spatial and electronic structure of the molecules of aluminon, chromoxane violet B, chromoxane purely-blue B, eryochrome cyanine R, sulphochrome, chromazurol S

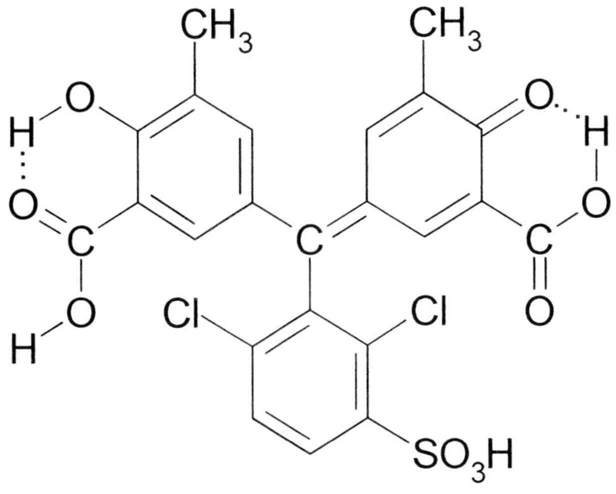

Chromazurol S

was considered. In the context of the analysis of electrostatic potential (ESP) of the reagents molecular systems, primary dissociation of carboxylic group in *ortho* carboxyphenolic ring was substantiated. The ESP character allowed us to present in more detail the steps of reagents complexing with metals cations at different pH levels. Favourable topology of nucleophilic interaction of non-ionic surfactants' hydroxyethyl fragments with protons of carboxylic and hydroxy groups of phenolcarboxylic acids of triphenylmethane series, serving as a prerequisite to the reagents solubilization in non-ionic surfactants micelles, was shown. Much the same contribution of medium to acid-base properties of phenolcarboxylic acids of triphenylmethane series, that conceives the similar characteristics of hydrophilic and hydrophobic hydration of the molecular systems of reagents and their conjugate bases, was established. The substrate specificity of the reagents chromoxane purely-blue B and chromazurol S to non-ionic surfactants is evidently caused by local properties of chloro substituents and atomic basins in the vicinity of the latter (volume and hydrophobicity of chlorine atom combined with a comparatively high polarity of the C-Cl bonds) [108].

Ab initio, electronic structure of the molecules HOCN, HSCN, HNCO, HNCS and the anions $[OCN]^-$, $[SCN]^-$ has been investigated. The higher thermodynamic stability of the HNCO and HNCS molecules as compared to HOCN and HSCN, respectively, has been shown. The series of the protolytes strength alteration: HSCN > HOCN, HNCS > HNCO, HOCN > HNCO, HSCN > HNCS have been substantiated. The computations taking into account the electronic correlation reflect the general sequence of change in proton-donor properties: HSCN > HOCN > HNCS > HNCO, coinciding with the raw of descending hydrophobicity of the compounds. A higher degree of the stabilization-by-hydration for the oxygen-containing anion compared to the sulphur-containing analog has been confirmed, however, that does not govern the sequence of changes in acidic properties of the different forms of hydrogen oxonitridocarbonate (IV) and hydrogen sulphidonitridocarbonate (IV). Comparative proton-donor ability of the above acids in aqueous solutions is determined basically by the electronic structure and size of their molecules and anions $[OCN]^-$, $[SCN]^-$, but not by the medium effects [109].

Of great inportance in chemistry are reactions involving a step that may be conceived formally as hydride ion transfer [110]. Transformations of 1,5-dicarbonyl compounds and chalcogenopyrans into the mixtures of chalcogenopyrilium salts and chalcogenacyclohexanes or chalcogenopyrilium salts and other reduced products serve as examples [111].

For a wide range of chemical compounds, the following mechanisms of "hydride" transfer are discussed [110].

1. Direct splitting-out of a hydride ion ($-$ H^-).
2. Consequent withdrawing of an electron, a proton and yet another electron ($-$ e^- $-$ H^+ $-$ e^-).
3. Extraction of an electron and a hydrogen atom ($-$ ē $-$ $H^\bullet$).
4. Transfer of a hydrogen atom and an electron ($-$ $H^\bullet$ $-$ ē).
5. Deprotonation followed by a transfer of two electrons ($-$ H^+ $-$ ē $-$ e^-).

The mechanism of "minus electron minus electron minus proton" ($-$ ē $-$ ē $-$ H^+) type is not realizable, since there is little likelihood of a dication radical formation.

Outside the domain of this discussion, one more mechanism probably exists which involves primary protonation of heteroring and final step of H_2 [112] or another electroneutral molecule [113] liberation. Feasibility of heterocycle protonation at a chalcogen atom has been unambiguously proved in the works by J.B. Lambert and S.I. Featherman [114].

Table 9. Correspondence between steps of "hydride" transfer and reactivity indices describing the above steps

Mechanism	Steps of transfer	To be analyzed	
		Index	For molecular system
1	$BH \rightarrow B^+ + H^-$	*HIA*	B^+
2	$BH \rightarrow BH^{\bullet+} + e^-$	*IP*	BH
	$BH^{\bullet+} \rightarrow B^\bullet + H^+$	*PA*	$B^\bullet$
	$B^\bullet \rightarrow B^+ + e^-$	*IP*	$B^\bullet$
3	$BH \rightarrow BH^{\bullet+} + e^-$	*IP*	BH
	$BH^{\bullet+} \rightarrow B^+ + H^\bullet$	*HAA*	B^+
4	$BH \rightarrow B^\bullet + H^\bullet$	*HAA*	$B^\bullet$
	$B^\bullet \rightarrow B^+ + e^-$	*IP*	$B^\bullet$
5	$BH \rightarrow B^- + H^+$	*PA*	B^-
	$B^- \rightarrow B^\bullet + e^-$	*IP*	B^-
	$B^\bullet \rightarrow B^+ + e^-$	*IP*	$B^\bullet$

We have proposed [115] a strategy of the quantum chemical study of "hydride ion" transfer based on the analysis of correspondence of the sequences of changes in ionization potentials (*IP*), enthalpies and free energies of the affinities to hydride ion (*HIA*), to hydrogen atom (*HAA*) and to proton (*PA*) of substrates molecules, as well as their derivatives as cations, cation radicals, radicals, anions, to the experimentally substantiated series of "hydride" mobility (Table 9).

We have shown that for six chalcogenopyrans based on "semicyclic" 1,5-diketones:

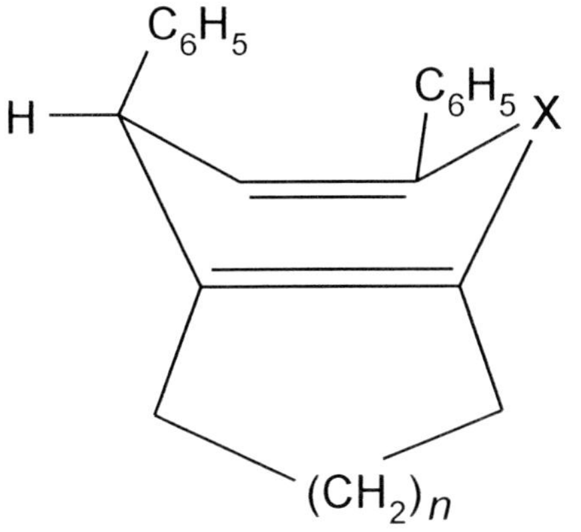

X = O, S, Se
n = 1, 2

the experimental series of "hydride" mobility S > O ≈ Se [116-118] is not inconsistent with: i) ionization potentials (*IP*) of the molecules studied; ii) hydrogen atom affinities (*HAA*) of radicals formed by abstraction of hydrogen atom in heterocycle position 4 from the initial molecular systems.

Let us compare that to the data from the Table 9. The *IP* index describes an electron extraction from the substrate's molecule BH - the first step of the mechanisms 2 and 3. The ejection of a hydrogen atom from the initial molecule BH (the *HAA* value corresponds just to this process) presents the first step of the mechanism 4. Provided that the first step is rate-determining one, the experimental results' correlation with the ionization potentials implies the mechanisms 2 or 3 realization, and the experiments' agreement with the radicals $B^{\bullet}$ affinity to hydrogen atom means the mechanism 4.

Therefore the mechanisms 1 and 5 are to be rejected. The mechanisms 2 to 4 are feasible provided that the first step is limiting one in every case.

Thus, the direct splitting-out of hydride ion and primary deprotonation of substrates followed by two electrons withdrawing have been elucidated to be unlikely. Feasible are the mechanisms of "hydride" mobility, the first step of which consists in electron or hydrogen atom transfer from chalcogenopyrans molecules [115].

The results obtained allow the experimenting researchers to focus their attention on the problem of substantiating the actually possible mechanisms of "hydride" transfer in the chemistry of chalcogenopyrans.

Molecular spatial parameters and polarity of acrylic acid proposed as an additive to methylsilicone OV-1 appear to work well in fitting the corresponding characteristics of stationary phase (SP), that has been proved by quantum chemical computations. The modified SP formed demonstrates a complementary gas chromatographic behaviour in the cases of test hydroxylic compounds used. Using packed columns with OV-1 as a SP certainly modified with appropriate amounts of acrylic acid, a simplified gas chromatographic technique for the rapid determination of benzoic and heptanoic acids directly in aqueous solutions has been elaborated [119].

On the basis of quantum chemical modeling, the interpretation of silica sorption properties has been presented: the most active ion-exchange centres are silanolic groups, but not a mineral structure defect, related to the isomorphous substitution of silicon by aluminum; the closer the defect to the sorbent surface, the more acidic are protons of silanolic groups [120].

A considerable amount of investigations in the area of physical chemistry of morphogenic proteins of higher fungi has been carried out in collaboration with Colleagues from Laboratory of Microbiology and Mycology, Institute of Biochemistry and Physiology of Plants and Microorganisms, Russian Academy of Sciences, Saratov, Russia (above all, Senior Researcher Olga M. Tsivileva and the Head of the said Laboratory Professor Valentina E. Nikitina).

The dependence of the extracellular lectin activity of *Lentinus edodes* (Berk.) Sing [*Lentinula edodes* (Berk.) Pegler] and the formation of a pigmented mycelial film in the submerged culture upon the presence of essential amino acids in a synthetic medium has been studied. The pronounced effect of asparagine in the presence of Ca^{2+} or Mn^{2+} cations on the formation of the above vegetative structure has been found. Quantum chemical *ab initio* calculations of the electronic structure of zwitterions of asparagine and its nearest structural analog have been performed. The van der Waals surface and volume of molecular systems, as well as refraction, polarisability and lg$P$ (P is a distribution coefficient in the system of 1-octanol - water) have been computed. On the basis of the research carried out, the supposition was made that the different character of the interaction of amino acids with the aforementioned ions is due to differences in the hydrophobicity (lipophilicity) of the amino acids rather than to differences in the electronic structure of the amino acids zwitterions [121, 122].

The activity of extracellular lectins of the basidiomycete *Lentinus edodes* (Berk.) Sing [*Lentinula edodes* (Berk.) Pegler] has been shown to be related to the presence of definite double-charged metal cations in the synthetic liquid medium of cultivation. The effect of Fe^{2+}, Co^{2+}, Ni^{2+} on the lectin activity has been demonstrated to change with a symbate character in respect to the energies of chelating reactions of metals hexaaqua complexes with the model ligand ethylene glycol, computed *ab initio*. At the initial steps of the lectins biosynthesis, the above metals cations exert a much greater influence upon the lectin activity of *Lentinus edodes*, as compared to the results of subsequent interactions between the cations and proteins under study in the culture medium [123].

An invited review [124] in the area of mycology, including our scientific results on quantum chemical interpretations in mycology, has been published.

Within the area of our attention are the classification aspects and the analysis of Web resources, too.

Classing generalizations on reactions in homogeneous and heterogeneous systems (theories of acids and bases, intramolecular factors effecting protolytes strength, types of complex compounds, kinds of co-precipitation, precipitate aging mechanisms) have been drawn [63, 125, 126].

The alternative classification of the methods of quantum chemistry has been proposed [102, 103].

The data on information resources on natural sciences, on ecology, have been systematized and generalized. Information on a large amount of Publishers of scientific literature (publishing houses, scientific institutions and organizations, educational schools, societies, associations, federations, unions and companies) contributing greatly to publishing printed and electronic production, as well as of electronic libraries, search systems, databases and catalogs of natural-scientific and ecological literature, referred to Web-addresses, has been collected. Valuability of the work carried out is high the more so as the sites of a number of publishing houses, search systems, databases, catalogs offer the information search by authors, paper title, name of journal, of book, by research subject, science area, key words, *etc.* Many electronic libraries provide an access to the Publishers Web sites, that facilitates substantially a process of obtaining the detailed information on the corresponding journals (addresses, Editorial Board, aims and scope, audience, subscription information, guide for authors, Copyright information, Tables of contents, abstracts, full text papers, news, *etc.*) on Web sites. The information on status changes for many scientific journals is presented. The creation of journals (newly or on the basis of the old ones), discontinuation of issues, journals incorporation in other ones, combined and divided journals are mentioned [127-130].

In general, the investigations carried out are concerned with a voluminous scope of scientific interests, among which are physical organic, inorganic, organoelement and coordination chemistry, structure and reactivity, quantitative structure - property relationships, reaction mechanisms, medium role in chemical processes, applied quantum chemistry and molecular dynamics, theory of analytical reagents action, hydrogen bond, physical chemistry of morphogenic proteins of higher fungi, Internet in natural sciences and in ecology.

A considerable component of the studies conducted is presented by the methods of quantum chemistry.

The quantum chemical consideration of the chemical problems is not solely the application of powerful up-today research methods. That is also new ideology of chemistry science, its modern language, the achievement of qualitatively distinct, true scientific, *i.e.* predictive, level.

The use of quantum chemical methods combined with the physical, physicochemical, chemical research methods is key for gaining the principally important information on the reactivity, chemical reaction mechanisms, analytical properties of reagents, biological activity of substances, for establishing the prognostic quantitative structure - property relationships.

Theoretical, including quantum chemical, modeling of chemical compounds with virtual testing of their various properties, the study of substances' reactivity, reaction mechanisms in relation to the electronic structure of molecules, render chemistry a predictive science, *i.e.* permit its contemporary level and the trends of development.

Therewith the general (with numerous fine details at closer consideration) methodology of investigations regardless of the essence of subjects (inorganic, organic, organoelement, coordination, high-molecular compounds, including biopolymers, as well as molecular complexes, exciplexes, excimers, overexcited Rydberg molecules of up to 1000 Å sizes, ionic and molecular crystals, clusters (supermolecules, nanostructures), permolecular, supramolecular, surface structures, adsorption, catalytic systems, by-electrode layers, *etc.*), of the structure peculiarities (charge and spin multiplicity of the molecular system, ground and excited electronic, vibrational, rotational and other levels), of the types of interaction in molecules or between them (electrostatic, van der Waals, "guest - host", hydrophobic, chemical), of the character of chemical bond (ionic, covalent, donor-acceptor, hydrogen, dative, agostic, metallic; banana, multicentral, interstitional; two- and three-dimensional electronic delocalization [131]; aromaticity and related phenomena, including antiaromaticity, quasiaromaticity, or non-classic aromaticity, heteroaromaticity, homoaromaticity, Y-aromaticity, spiroaromaticity, bicycloaromaticity and bicycloantiaromaticity, intra-plane, or radial, aromaticity, cross-aromaticity, metalloaromaticity, three-dimensional, in plane, spherical aromaticity, σ-aromaticity and σ-antiaromaticity [132], superaromaticity, induced aromaticity [133]), of aggregation state, of a type of solvation (universal, specific), demonstrates the universality of chemistry (and quantum chemistry) as a science.

Quantum chemistry is as unbounded and inexhaustible as the whole chemistry, equally as the surrounding reality, just reflected in chemistry science. It is not accident that the specialists have troubles understanding their colleagues working in different branches of quantum chemistry and its applications. So, the researcher solving a wide set of interpretation tasks with the use of quantum chemical methods, does not rank below the scientist who use a combination of different experimental methods, in multilateralism and variety of aspects of his scientific activity.

Nevertheless, along with the inevitable differentiation and divergence rising from the variability of subjects under investigation and of special approaches to their description, high integration potential occurs in quantum chemistry alike in the whole chemistry, the basis for which being underlined by the unique chemical form of the matter movement and, as a consequence, the unified general methodology of the matter knowledge.

The information potentialities of quantum chemical methods (even taking into account the restrictions, which are well-known and allow evaluations of the computations accuracy) are essentially greater, than of the experimental ones [134].

Quantum chemical computations enable one to determine simultaneously the assemblage of principal characteristics of molecules and substances, which can not be extracted from any experimental method [134]. Among the above properties are geometry, symmetry point group, electron density distribution (including charges on atoms, bond orders) (for the systems with open shells - also spin density distribution), topology of electron density (bond path, location of bond critical point, electron density in the latter, Laplacian of the above value, ellipticity, other parameters of AIM-analysis), dipole, quadrupole, octapole, hexadecapole moments, polarisability, hyperpolarisabilities, composition, hybridization, population, symmetry and energies of molecular orbitals, characteristics of natural bond orbitals (NBO), superdelocalisabilities, electrostatic potential of molecule, total energy, thermodynamic functions and their components, heat of formation, entropy, free energy of formation, proton affinity, ionization energy (potential), electron affinity, rotational constants, frequencies and intensities of bands in vibrational spectra, wavelengths and oscillator strengths of bands in electronic spectra, chemical shifts and spin-spin coupling constants in NMR spectra of various nuclei, parameters of NQR and Mössbauer spectra, hyperfine constants in EPR spectra, *et al.* By means of quantum chemical methods, the data of X-ray and ultraviolet photoelectron spectroscopy, Auger spectroscopy, mass spectrometry, gas electron diffraction, *etc.*, are interpreted.

On the basis of quantum chemical considerations, one could judge on the reactivity of different positions of molecules in the charge-controlled and frontier-controlled processes, when the reaction rate is determined by the charges on atoms and the electron density of frontier orbitals, respectively. There is a possibility to reveal the character of chemical bonds, different intra- and intermolecular interactions, charge transfer, to calculate the thermodynamic parameters of reactions, to compute the transition states, to evaluate the reaction barriers, to study the topology of potential energy surface and to construct the reaction paths. Along with the aforesaid, the effect of medium on all the calculated characteristics becomes possible to be taken into account.

The information provided by the quantum chemical methods is substantially multidimensional. In this sense, among the experimental methods, approaches and techniques, those are only partially similar to the Hyphenated Techniques created on the basis of methods of separation and diagnostics.

Sometimes [134] the quantum chemical methods are referred to as being comparable with NMR in their information-supply ability. As Professor Boris A. Ershov (Faculty of Chemistry, St. Petersburg State University) tells, "NMR could make everything".

Like any science and methodology, quantum chemistry encounters problems in connection with both its own internal development and application of the corresponding methods.

Nevertheless, in the same sense as for NMR, the quantum chemistry should be referred to as capable of "making everything" even more reasonably at present. There are no doubts that quantum chemistry will answer any challenge and will overcome possible difficulties at the steps of further development.

Quantum chemistry determines, to a considerable extent, a progress of the whole chemistry as a science, which moves in the course of its development toward the sophistication of both the research methods (physical, mathematical ones) and the subjects mentioned above, the contemporary knowledge of the chemistry of the latter being impossible without the quantum chemical treatment.

For realizing the quantum chemical methods, any unambiguous and unclear-in-sense constants are not required usually. The methods of quantum chemistry ([2, 86, 105, 134-144] *et al.*) either do not use the postulates outside the scope of fundamental points, principles and mathematical tools of quantum mechanics, theoretically substantiated approximations of quantum chemistry (adiabatic, single-electron, MO LCAO), procedures for taking into account the relativistic, non-adiabatic effects and electron correlation, strict rules of composing the basis sets (*ab initio*), or exploit the simplified non-empirical schemes involving the effective core potential (ECP), or pseudopotential (ECP method), or use the reasonable approach of zero differential overlap, consider in the explicit form the valence electrons only, and introduce the parameters, which already prove their efficiency for predicting different properties (semiempirical ones). Besides, the parameters could be fitted for solving the given problems (*e.g.*, [102, 103]). The promising package of essentially hybride methods, that combines the strict non-empirical computations with the semiempirical account for the electron correlation energy, is based on the Density Functional Theory (DFT) [138]. For computing the large systems, including high-molecular compounds, biopolymers, the approaches QM/MM and ONIOM have been developed [139]. The principally novel general avenue of inquiry in computer-assisted quantum chemistry, compared to the methods MO and VB, free of requiring computational mega resources, is presented by the method of multiparticle wavefunctions using the hypothesis of equipotential surfaces prohibition (Professor Eugeny A. Smolenskii and Professor Nickolai D. Chuvylkin, N. D. Zelinsky Institute of Organic Chemistry, Russian Academy of Sciences, Moscow) [137, 140-144]. The aforesaid does not exhaust the variety of methods and approaches of the current quantum chemistry.

The experience shows that commonly the properly selected semiempirical methods, provided that each of those corresponds to a sufficiently adequate model of definite-kind-molecules and describes a part of molecular properties, are not worse, in respect to efficiency, than one complicated method. That obeys the law of technical systems development: the efficiency of instrument increases with the increase in degree of its narrow specialization.

Up to the present, for solving the numerous interpretational problems of chemistry, the semiempirical quantum chemical methods hold true an importance.

Their use with respect to "classical" systems may be tolerated provided that those are capable of explaining and predicting the most important features of structure and reactivity of large molecules, macromolecules, periodic structures, crystal lattices, aggregates of molecules, clusters, therewith capable also of reproducing the physicochemical properties of molecular systems (microspecies) and substances (assemblages of microspecies) from the test-kit. At least for the preliminary estimations followed by the considerations at a higher level of theory, their application is encouraged.

The semiempirical methods would apparently find their further application for modeling the specific and non-specific solvent effects, as well as for computation of electronic spectra when it is necessary to involve thousands or even millions configurations to CI.

In order to justify the economical reasoning and expended time required for the establishment of predictive quantitative structure - property relationships aimed at the molecular design of chemical compounds, such the relationships should be obtained sufficiently simply and fast, and in this respect semiempirics also finds its "niche".

At the same time, for studying the electronic structure of molecular systems in fine detail, the electronic effects in molecules, the nature of chemical bonds, dispersion and other weak interactions, adsorption, catalysis, for analyzing the potential energy surfaces of the majority of reaction types, the *ab initio* and DFT methods are required. Moreover, these methods allow the highly-precise simultaneous computations for the whole complex of properties to be done. As applied to "non-classical" systems (with non-standard stereochemistry, hypercoordination, *etc.*), the use of sophisticated approaches for solving any problems necessitating the quantum chemical consideration, is alternativeless.

As for the term "non-classical systems", *e.g.*, for the organic compounds, non-classical structures are defined according to the presence of non-tetrahedral tetracoordinated or/and hypercoordinated carbon atoms in their molecules or ions [145, 146].

The future of applied quantum chemistry may undeniably lie with the rigorous, sophisticated methods. A considerable progress in solving the various problems of chemistry, biology and interdisciplinary areas at high level of theory (*ab initio*, DFT) will be favoured by the further development of computing devices, including the improvement of workstations-supercomputers, fast-operating clusters, which realize the distributed parallel computations, as well as by a probable appearance, within near-future, of the quantum computers with an about terabyte memory and a rate of data transfer at a gigabyte level.

It should be remembered, however, that the appearance, development and fruitfulness of the semiempirical methods of quantum chemistry follow inevitably from the fundamental principle of complementarity (Niels H.D. Bohr), in consequence of which the clarity of knowledge and description of the molecular objects properties is accomplished by means of model insight and imagines [147].

Ab initio computational methods are not in fact deduced completely "from the first principles", but are also essentially semiempirical, the empirics being transferred to the level of basis choice [147]. The empirics is intrinsic, even to a greater extent, in hybrid DFT methods, that are not at a disadvantage in relation to *ab initio* and often superior in their potentialities.

No the most perfect experimental and theoretical methods allow one to overcome the accuracy limits of determination of molecules' parameters: 10^{-2} Å for the geometric parameters, 2-3 degrees for valence angles, $(2\text{-}3)\cdot10^{4}$ cm^{-2} for force constants [147]. An experience of *ab initio* computations application in molecular spectroscopy has shown that the resulting values of parameters could be used only as the initial approximations to be refined by means of the inverse tasks [147].

As Professor Lev A. Gribov (V. I. Vernadsky Institute of Geochemistry and Analytical Chemistry, Russian Academy of Sciences, Moscow) [148] notes, semiempirical and *ab initio* approaches and methods are not mutually exclusive, but complement each other, and the most rational avenue of the further development of quantum chemistry lies in searching a "golden middle".

In our mind, the latter would be achieved on solving the interpretational chemical problems (including those described in the present chapter) by means of the quantum chemical methods provided that: i) the methods of different hierarchy and theory level are adequate to the problems to be solved; ii) the hybrid methods of Density Functional Theory (DFT) are used.

As for solving the given scientific problems requiring the quantum chemical methods application, so the most intriguing and promising results should be expected within the interdisciplinary areas of science, when the work is conducted in cooperation, by theoretically- and experimentally-oriented researchers (spectroscopists, electrochemists, photochemists, radiochemists, plasmachemists, cryochemists, syntheticians, analyticians, material scientists, astronomists, astrophysicists and cosmochemists, geochemists and geologists, lithologists, hydrologists, oceanologists, limnologists, agrochemists, soil scientists, biochemists, enzymologists, immunochemists, biophysicists, biologists, microbiologists, biotechnologists, pharmaceuticians, pharmacologists, toxicologists, medicians, ecologists, specialists in stereochemistry, structural chemistry, thermodynamics, phase equilibria and transitions, kinetics, chemistry of combustion, shock compression, detonation and explosion, chemistry of high energies and extremal attacks (reactions in plasma, in the course of photolysis and radiolysis, under the action of shock waves, of high temperatures and pressures, of electric and magnetic fields, in electric discharge, under the electro thermal conditions, under the effect of high-frequency electromagnetic vibrations, ultrasound, electrohydraulic hammer, *etc.*), in catalysis, solution chemistry, physics and chemistry of disperse systems, condensed states, in surface phenomena, topochemistry, physics of semiconductors, dielectrics, segnetoelectrics, piezoelectrics, ferromagnetics, auxetics, supercritical and non-Newton fluids, in the problems of superfluidity and fermionic condensates, Bose - Einstein condensates, supersolids, high-temperature superconductivity, in magnetochemistry, dielectrometrics, optics, laser chemistry, electrooptics, acoustooptics, electronics, spintronics, optoelectronics, photonics, vacuum engineering and instrumentation, photographic chemistry, chemistry of crystals and crystallography, mineralogy, physics and chemistry of surface, films and coatings, adhesion, corrosion and protection against it, mechanochemistry and rheology, tribochemistry and tribology, sonochemistry, electrohydraulic effect, radiation and nuclear chemistry, biophysical chemistry, molecular biology, bionics, genomics, transcriptomics, proteomics, metabolomics, in environmental science, in extraction, flotation, adsorption and sorption at all, ion exchange, chromatography, diffusion, electromigration, membrane processes, chemical engineering, metallurgy, powder technique, welding, nanochemistry and nanotechnology, chemical and electrochemical energetics, including chemical, biochemical current sources and energetic substances, diffraction and ionization research methods in chemistry, *etc.*).

Therewith the peculiarities of experimental methods, the data of which are subject to treat by means of quantum chemistry, should be taken into consideration.

Every physical method has its own characteristic time.

Besides, the methods of direct determination of the molecules geometry (X-ray structure analysis, neutron diffraction, gas electron diffraction, microwave spectroscopy) yield, strictly speaking, values of spatial parameters different-in-sense (distinct are r_e-, r_0-, r_z-, r_s-, r_g- and another types of structures) [149].

How should the data from different physical methods be adjusted to a unified basis, and what could the values computed by the quantum chemical methods be referred to?

As seen from the list of scientific fields presented above, the quantum chemistry is applicable for interpreting the results of experimental studies and for prognosis at both microscopic and macroscopic levels.

One could separate the following approaches to the quantum chemical description of processes and phenomena capable of being registered at a macroscopic level [2, 86, 105, 135, 136].

1) Heats [136] and free energies of formation, heat capacities, entropies are inherent to the assemblages of species. However, the semiempirical methods of quantum chemistry estimate these quantities with a rather well accuracy, considering the isolated molecules [136]. That is related to the compensation for the affecting leap "molecule - assemblage of species" by means of the appropriate parameterization. For instance, the formula for heat of formation is contributed from the enthalpies of atoms formation - the values gained from experiments [150]. The relevant expressions have been also chosen for computing the potentials of core-core repulsion [150].
2) For improving the adequacy of models underlying the basis of quantum chemical computations, the latter are performed not for the separate molecules, but for voluminous clusters (supermolecules, nanostructures), to simulate the real or near-real state of substance.
3) The alternative to the second approach is indirect account for surrounding within the framework of continuum models.
4) There are discrete-continuum approaches involving the essential features of the points 2 and 3.

Presented above are the examples of all the four mentioned approaches realization.

We believe that the development in the century XXI lies ahead primarily in the fields among which are computer-assisted chemistry (quantum chemistry, molecular dynamics, molecular design), the study of mechanisms of chemical and electrochemical reactions in inorganic and organic chemistry, chemical material science, ecological chemistry [151]. Our assessments coincide sometimes with the commonly accepted ones. Thus, in questionnaire [151], the first ten of scientists-chemists of the century XX included John A. Pople, who was soon awarded (along with Walter Kohn) with Nobel Prize-1998 in chemistry [152] for a fundamental contribution to the development of quantum chemical methods. The Nobel Prize-1999 in chemistry was gained by Ahmed Zewail for his fundamental works in the field of femtochemistry, that open up possibilities for studying the course of reactions in real-time mode, for obtaining the direct information on the transition states, for providing the selectivity of yields of the desirable reaction products [153].

The further development of the scientific avenue discussed in this chapter is planned as being related to the studies of fine peculiarities of the molecules electronic structure, the reactivity in chemistry and biology, the mechanisms of action of analytical reagents, the molecular drug design. Therewith in accordance with the whole-world tendency, the role of the *ab initio* and DFT quantum chemical methods increases, the same is true for molecular dynamics, QSAR (QSPR) modeling, molecular docking, *de novo* design.

Acknowledgement

The author would like to thank Professor Frank Columbus (Nova Science Publishers, Inc., President), Dr. Nadya Columbus (Nova Science Publishers, Vice President) for kind invitation to contribute to this book, Dr. Donna Dennis (Nova Science Publishers, Inc., Editorial Production Manager) and other Colleagues for careful proceeding with the manuscript, Dr. Olga M. Tsivileva (Laboratory of Microbiology and Mycology, Institute of Biochemistry and Physiology of Plants and Microorganisms, Russian Academy of Sciences, Saratov, Russia) and Igor A. Kargin, MS (Department of Chemistry, N. G. Chernyshevskii Saratov State University, Saratov, Russia) for technical assistance.

References

[1] Chemical Encyclopaedia (in Russian); Knunyants, I.L.; Editor-in-Chief; Zefirov, N.S.; Vice Editor-in-Chief; Kulov, N.N.; Vice Editor-in-Chief; Scientific Publishing House "Big Russian Encyclopaedia" (in Russian: "Bol'shaya Rossiiskaya Entsiklopediya"): Moscow, vol. 3, 1992, 639 pp.

[2] Minkin, V.I.; Simkin, B.Ya.; Minyaev, R.M. Quantum Chemistry of Organic Compounds. Reaction Mechanisms (in Russian); Khimiya: Moscow, 1986, 248 pp.

[3] Tobe, M.L. Inorganic Reaction Mechanisms; Thomas Nelson and Sons Ltd.: London, 1972. Russian Edition: Mechanisms of Inorganic Reactions. Popular Essay of Contemporary Views on Reaction Kinetics of Inorganic Complex Compounds (in Russian); Mir: Moscow, 1975, 277 pp.

[4] Jackson, R.A. Mechanism: An Introduction to the Study of Organic Reactions; Clarendon Press: Oxford, 1972. Russian Edition: Introduction to the Mechanism Study of Organic Reactions (in Russian); Khimiya: Moscow, 1978, 192 pp.
[5] Volovik, S.V.; Staninets, V.I.; Zefirov, N.S. *Doklady Akad. Nauk SSSR* 1992, 324, 354-358.
[6] Volovik, S.V.; Staninets, V.I.; Zefirov, N.S. *Doklady Akad. Nauk* 1993, 330, 321-323.
[7] Urusov, V.S. *Zhurn. Strukt. Khimii* 1994, 35, No. 1, 111-127.
[8] Rozen, A.M.; Krupnov, B.V. *Uspekhi Khimii* 1996, 65, 1052-1079.
[9] Cherkasov, A.R.; Galkin, V.I.; Zueva, E.M.; Cherkasov, R.A. *Uspekhi Khimii* 1998, 67, 423-441.
[10] Zueva, E.M.; Galkin, V.I.; Cherkasov, A.R.; Cherkasov, R.A. *Rossiisk. Khim. Zhurn.* (*Zhurn. Rossiisk. Khim. Obshchestva im. D.I. Mendeleeva*) 1999, 43, No. 1, 39-48.
[11] *Poisk* 2003, August 29, No. 35 (745), 5-6.
[12] *Poisk* 2002, April 19, No. 16 (674), 8-10.
[13] Buchachenko, A.L. *Uspekhi Khimii* 1999, 68, 99-118.
[14] Alfimov, M.V.; Novikov, V.D. *Poisk* 2000, May 26, No. 21 (575), 5-6.
[15] 1000 Most Cited Chemists 1981-1997. Web: www.ill.fr/dif/citations/chemists.html.
[16] Zolotov, Yu.A. *Zhurn. Anal. Khimii* 2003, 58, 1013-1014.
[17] Berezkin, V.G.; Sidorenko, N.A.; Arkhipov, D.B. *Zhurn. Anal. Khimii* 2007, 62, 100-110.
[18] Pankratov, A.N. *J. Mol. Struct. Theochem* 1998, 453, 7-15.
[19] Pankratov, A.N.; Shchavlev, A.E. *Monatsh. Chem.* 1998, 129, 1007-1017.
[20] Pankratov, A.N. *Afinidad* 1999, 56, 257-262.
[21] Ovchinsky, V.A.; Pankratov, A.N.; Shchavlev, A.E.; Uchaeva, I.M.; Kulapina E.G. In Physicochemical Methods in Coordination and Analytical Chemistry: Proceedings of Volga Region Confer. in Honour of 80th Anniversary of A.A. Popel Birthday. Kazan, Oct.26, 1999; Ulakhovich, N.A.; Zakharov, A.V.; Ed.; Kazan State University: Kazan, 1999, pp. 126-129.
[22] Pankratov, A.N.; Shchavlev, A.E. *Zhurn. Strukt. Khimii* 1999, 40, 1059-1066.
[23] Pankratov, A.N.; Shchavlev, A.E. *Canad. J. Chem.* 1999, 77, 2053-2058.
[24] Pankratov, A.N.; Tsivileva, O.M.; Nikitina, V.E. *J. Biochem. and Mol. Biol.* 2000, 33, 37-42.
[25] Pankratov, A.N.; Uchaeva, I.M. *J. Mol. Struct. Theochem* 2000, 498, 247-254.
[26] Pankratov, A.N. *J. Mol. Struct. Theochem* 2000, 507, 239-244.
[27] Pankratov, A.N. *Zhurn. Strukt. Khimii* 2000, 41, 696-700.
[28] Pankratov, A.N.; Shchavlev, A.E. *Zhurn. Anal. Khimii* 2001, 56, 143-150.
[29] Pankratov, A.N. *Zhurn. Neorgan. Khimii* 2001, 46, 791-794.
[30] Pankratov, A.N. *Zhurn. Strukt. Khimii* 2001, 42, 822-826.
[31] Pankratov, A.N. *Heteroatom Chemistry* 2002, 13, 229-241.
[32] Pankratov, A.N. *J. Serb. Chem. Soc.* 2002, 67, 339-346.
[33] Pankratov, A.N.; Uchaeva, I.M. *Zhurn. Neorgan. Khimii* 2003, 48, 91-96. *Eng. Transl.: Russian J. Inorg. Chem.* 2003, 48, 83-88.
[34] Pankratov, A.N. *Khimiya Prirod. Soedin.* 2003, 457-464.
[35] Pankratov, A.N.; Uchaeva, I.M. *Zhurn. Neorgan. Khimii* 2004, 49, 1520-1525. *Eng. Transl.: Russian J. Inorg. Chem.* 2004, 49, 1403-1408.
[36] Pankratov, A.N. *Afinidad* 2004, 61, 256-260.

[37] Pankratov, A.N.; Shalabay, A.V. *Zhurn. Strukt. Khimii* 2004, 45, 800-806.
[38] Pankratov, A.N. *Zhurn. Anal. Khimii* 2005, 60, 149-156. *Eng. Transl.: J. Anal. Chem.* 2005, 60, 130-136.
[39] Pankratov, A.N. *Khimiya Geterotsikl. Soedin.* 2005, 391-402.
[40] Pankratov, A.N.; Borodulin, V.B.; Chaplygina, O.A. *Koord. Khimiya* 2005, 31, 696-702. *Eng. Transl.: Russian J. Coord. Chem.* 2005, 31, 660-666.
[41] Pankratov, A.N. *Afinidad* 2006, 63, 404-410.
[42] Duchowicz, P.R.; Castro, E.A.; Fernández, F.M.; Pankratov, A.N. *J. Argentine Chem. Soc.* 2006, 94, No. 4/6, 31-45.
[43] Pankratov, A.N. *Zhurn. Strukt. Khimii* 2007, 48, 478-484.
[44] Stull, D.R.; Westrum, E.F.; Sinke, G.C. The Chemical Thermodynamics of Organic Compounds; John Wiley & Sons, Inc.: New York, London, Sydney, Toronto, 1969. Russian Edition: Mir: Moscow, 1971, 807 pp.; Suppl.
[45] MacDonald, A.L.; Trotter, J. *J. Chem. Soc. Perkin. Trans.* Part 2 1973, 476-480.
[46] van Bolhuis, F.; Kiers, C.Th. *Acta Crystallogr. B* 1978, 34, 1015-1016.
[47] Shchukarev, S.A.; Lobaneva, O.A.; Ivanova, M.A.; Kononova, M.A. *Vestnik Leningrad. Univ. Ser. Fiz. i Khim.*, 1961, No. 10, issue 2, 152-155.
[48] Biryukov, A.A.; Shlenskaya, V.I. *Zhurn. Neorgan. Khimii* 1964, 9, 813-816.
[49] Burger, K.; Dyrssen D. *Acta Chem. Scand.* 1963, 17, 1489-1501.
[50] Burger, K. *Acta Chim. Acad. Scient. Hung.* 1964, 40, Sectio 1, 261-273.
[51] Burger, K. *Magyar Kém. Folyóirat* 1964, 70, No. 4, 179-184.
[52] Nikonorov, P.G.; Kozhina, L.F.; Mushtakova, S.P. In New Achievements in Organic Chemistry: Collection of Scientific Works (in Russian); Saratov University Press: Saratov, 1997, pp. 75-76.
[53] Dneprovskii, A.S.; Temnikova, T.I. Theoretical Bases of Organic Chemistry. Structure and Reaction Mechanisms of Organic Compounds (in Russian); Khimiya. Leningrad Branch: Leningrad, 1991, 560 pp.
[54] Leatham, G.F.; Stahmann, M.A. *J. Gen. Microbiol.* 1981, 125, 147-157.
[55] Pimentel, G.C.; McClellan, A.L. The Hydrogen Bond; Pauling, L.; Ed.; W.H. Freeman: San Francisco and London, 1960, 475 pp.
[56] Scheiner, S. Hydrogen Bonding: A Theoretical Perspective (Topics in Physical Chemistry); Oxford University Press: Oxford, 1997, 394 pp.
[57] Golubev, N.G.; Denisov, G.S.; Shraiber, V.M. In Hydrogen Bond; Sokolov, N.D.; Responsible Ed.; Nauka: Moscow, 1981, pp. 212-254.
[58] El-Nahas, A.; Hirao, K. *J. Mol. Struct. Theochem* 1999, 459, 229-237.
[59] Pankratov, A.N.; Shalabay, A.V. *J. Serb. Chem. Soc.* 2007, 72, 151-157.
[60] Pankratov, A.N.; Shalabay, A.V. *J. Serb. Chem. Soc.* 2007, 72, 265-272.
[61] Pankratov, A.N.; Shalabay, A.V. *Zhurn. Strukt. Khimii* 2007, 48, 472-477.
[62] Shalabay, A.V.; Pankratov, A.N.; Nechaev, V.V.; Berezin, K.V. In Molecular Modeling in Chemistry, Biology and Medicine: II Russian School-Confer. (in Russian). Saratov, Russia, 13-16 October, 2004; Saratov University Press: Saratov, 2004, pp. 44-45.
[63] Pankratov, A.N. Acids and Bases in Chemistry (in Russian); Saratov University Press: Saratov, 2006, 196 pp.
[64] Pankratov, A.N.; Stepanov, A.N. *Croatica Chem. Acta* 1997, 70, 585-598.
[65] Pankratov, A.N.; Bilenko, O.A.; Mushtakova, S.P. *Afinidad* 2000, 57, 201-208.

[66] Pankratov, A.N. *Zhurn. Anal. Khimii* 2001, 56, 161-163.
[67] Pankratov, A.N. *J. Mol. Struct. Theochem* 1994, 315, 179-186.
[68] Pankratov, A.N.; Shchavlev, A.E.; Osokin, D.Ya.; Bozhenko, K.V. In New Achievements in Organic Chemistry: Collection of Scientific Works (in Russian); Saratov University Press: Saratov, 1997, pp. 92-93.
[69] Pankratov, A.N. *Indian J. Chem.* 1995, 34B, 689-694.
[70] Vereshchagin, A.N. Inductive Effect (in Russian); B.A. Arbuzov; Responsible Ed.; Nauka: Moscow, 1987, 326 pp.
[71] Minkin, V.I. *Pure and Applied Chemistry* 1999, 71, 1919-1981.
[72] Grant, D.M.; Paul, E.G. *J. Amer. Chem. Soc.* 1964, 86, 2984-2990.
[73] Lindeman, L.P.; Adams, J.Q. *Anal. Chem.* 1971, 43, 1245-1252.
[74] Levy, G.C.; Nelson, G.L. Carbon-13 Nuclear Magnetic Resonance for Organic Chemists; Wiley-Interscience: New York, 1972. Russian Edition: Guide for Nuclear Magnetic Resonance of Carbon-13 for Organic Chemists (in Russian); Mir: Moscow, 1975, 295 pp.
[75] Sergeev, N.M. NMR Spectroscopy (for Organic Chemists) (in Russian); Moscow University Press: Moscow, 1981, 279 pp.
[76] Inamoto, N.; Masuda, Sh. *Tetrahedron Letters* 1977, 3287-3290.
[77] Reynolds, W.F.; Taft, R.W.; Marriott, S.; Topsom, R.D. *Tetrahedron Letters* 1982, 23, 1055-1058.
[78] Marriott, S.; Reynolds, W.F.; Taft, R.W.; Topsom, R.D. *J. Org. Chem.* 1984, 49, 959-965.
[79] Naberukhin, Yu.I.; Luchnikov, V.A.; Malenkov, G.G.; Zheligovskaya, E.A. *Zhurn. Strukt. Khimii* 1997, 38, 713-722.
[80] Pankratov, A.N. *Helvetica Chim. Acta* 2004, 87, 1561-1573.
[81] Pankratov, A.N. *Zhurn. Anal. Khimii* 2005, 60, 1036-1046. *Eng. Transl.: J. Anal. Chem.* 2005, 60, 920-929.
[82] Pankratov, A.N.; Uchaeva, I.M. *J. Serb. Chem. Soc.* 2002, 67, 111-113.
[83] Pankratov, A.N.; Mustafin, A.I. *Zhurn. Anal. Khimii* 2005, 60, 455-457. *Eng. Transl.: J. Anal. Chem.* 2005, 60, 397-399.
[84] Shchavlev, A.E.; Pankratov, A.N.; Shalabay, A.V. *Int. J. Quantum Chem.* 2006, 106, 876-886.
[85] Shchavlev, A.E.; Pankratov, A.N.; Shalabay, A.V. *J. Phys. Chem. A* 2005, 109, 4137-4148.
[86] Minkin, V.I.; Simkin, B.Ya., Minyaev, R.M. Theory of Molecules Structure (in Russian); Phoenix: Rostov-on-Don, 1997, 560 pp.
[87] Shchavlev, A.E.; Pankratov, A.N.; Enchev, V. *J. Phys. Chem. A* 2007, 111, 7112-7123.
[88] Shchavlev, A.E.; Pankratov, A.N.; Borodulin, V.B.; Chaplygina, O.A. *J. Phys. Chem. A* 2005, 109, 10982-10996.
[89] Berezov, T.T.; Korovkin, B.F. In Biological Chemistry, Chapter 8: Textbook (in Russian); Medicine (in Russian: Meditsina): Moscow, 1998, pp. 248-297.
[90] Kabachnik, M.I. *Uspekhi Khimii* 1979, 48, 1523-1547.
[91] De Kock, R.; Jasperse, C. *Inorg. Chem.* 1983, 22, 3839-3843.
[92] Olivella, S.; Urpi, F.; Vilarrasa, J. *J. Comput. Chem.* 1984, 5, 230-236.
[93] Ozment, J.L.; Schmiedekamp, A.M. *Int. J. Quantum Chem.* 1992, 43, 783-800.
[94] Dmuchovsky, B.; Zienty, F.B.; Vredenburgh, W.A. *J. Org. Chem.* 1966, 31, 865-869.

[95] Schwarzenbach, G.; Rudin, E. *Helvetica Chim. Acta* 1939, 22, 360-376.
[96] Bordwell, F.G.; Andersen, H.M. *J. Amer. Chem. Soc.* 1953, 75, 6019-6022.
[97] David, J.G.; Hallam, H.E. *Trans. Faraday Soc.* 1964, 60, 2013-2016.
[98] Fialkov, Yu.Ya. Solvent As a Facility for Managing the Chemical Process (in Russian); Khimiya. Leningrad Branch: Leningrad, 1990, 240 pp.
[99] Blumenthal, G.; Engels, S.; Fitz, I.; Haberditzl, W.; Heckner, K.-H.; Henrion, G.; Landsberg, R.; Schmidt, W.; Scholz, G.; Starke, P.; Wilke, I.; Wilke, K.-Th. Anorganikum. Lehr- und Praktikumsbuch der anorganischen Chemie mit einer Einführung in die physikalische Chemie; Kolditz, L.; Ed.; VEB Deutscher Verlag der Wissenschaften: Berlin, 1984, vol. 1. Russian Edition: Anorganicum. Structure of Matter. Chemical Kinetics. Thermodynamics. Electrochemistry. Inorganic Chemistry: Compounds and Reactions (in Russian); Mir: Moscow, 1984, vol. 1, 672 pp.
[100] Samoilov, O.Ya. *Zhurn. Fiz. Khimii* 1978, 52, 1857-1862.
[101] Pankratov, A.N.; Yastrebova, N.I. In X All-Russia Workshop on Quantum Chemistry, Coordinating Workshop on Quantum Chemistry: Abstr. (in Russian). Kazan, Russia, September 16-21, 1991; Institute of Org. and Phys. Chemistry, Kazan Scientific Centre, Academy of Sciences of the USSR, Kazan Physical-Technical Institute, Kazan Scientific Centre, Academy of Sciences of the USSR, Kazan State University: Kazan, 1991, p. 258.
[102] Pankratov, A.N.; Shmakov, S.L.; Shchavlev, A.E. *Croatica Chem. Acta* 1995, 68, 325-334.
[103] Pankratov, A.N.; Shmakov, S.L.; Shchavlev, A.E. *Zhurn. Strukt. Khimii* 1996, 37, 1016-1022.
[104] Karelson, M.M.; Zerner, M.C. *J. Phys. Chem.* 1992, 96, 6949-6957.
[105] Semiempirical Methods of Electronic Structure Calculation; Segal, G.A.; Ed.; Plenum Press: New York, London, 1977, Part A: Techniques; Part B: Applications. Russian Edition: Mir: Moscow, 1980, vol.1, 328 pp.; vol.2, 373 pp.
[106] Del Re, G. *Int. J. Quantum Chem.* 1990, 37, 497-508.
[107] Pankratov, A.N.; Borodulin, V.B.; Chaplygina, O.A. *Koord. Khimiya* 2005, 31, 523-529. *Eng. Transl.*: *Russian J. Coord. Chem.* 2005, 31, 494-500.
[108] Pankratov, A.N. In Organic Reagents in Organized Media: Collected Scientific Articles of Higher Educational Institutions (in Russian); Scientific Book (in Russian: Nauchnaya Kniga): Saratov, 2003, issue 7, pp. 188-193.
[109] Pankratov, A.N.; Khmelev, S.S. *Zhurn. Strukt. Khimii* 2005, 46, 421-426. *Eng. Transl.*: *J. Struct. Chem.* 2005, 46, 404-408.
[110] Okhlobystin, O.Yu. Electron Transfer in Organic Reactions (in Russian); Zhdanov, Yu.A.; Responsible Ed.; Rostov University Press: Rostov-on-Don, 1974, 120 pp.
[111] Kharchenko, V.G.; Chalaya, S.N. Thiopyrans, Thiopyrilium Salts and Related Compounds (in Russian); Saratov University Press: Saratov, 1987, 160 pp.
[112] Detty, M.R. *Organometallics* 1988, 7, 1122-1126.
[113] Blinokhvatov, A.F.; Berberova, N.T.; Archegova, A.S.; Klimov, E.S.; Shpakov, A.V.; Okhlobystin, O.Yu. *Khimiya Geterotsikl. Soedin.* 1991, 900-904.
[114] Lambert, J.B.; Featherman, S.I. *Chemical Reviews* 1975, 75, 611-626.
[115] Pankratov, A.N.; Drevko, B.I. J. *Serb. Chem. Soc.* 2004, 69, 431-439.
[116] Degani, I.; Fochi, R.; Vincenzi, C. *Boll. scient. Fac. chim. industr. Bologna*, 1965, 23, 21-30 (1965).

[117] Degani, I.; Fochi, R.; Spunta, G. *Boll. scient. Fac. chim. industr. Bologna*, 1965, 23, 243-244 (1965).

[118] Berberova, N.T.; Blinokhvatov, A.F.; Archegova, A.S.; Klimov, E.S.; Shpakov, A.V.; Okhlobystin, O.Yu. *Khimiya Geterotsikl. Soedin.* 1991, 47-50.

[119] Pankratov, A.N.; Tsivileva, O.M. *Rev. Roumaine Chim.* 2003, 48, 491-498.

[120] Nikiforov, I.A.; Shchavlev, A.E.; Pankratov, A.N.; Sevost'yanov, V.P. In All-Russia Confer. of Young Scientists "Contemporary Problems of Theoretical and Experimental Chemistry": Abstr. (in Russian). Saratov, Russia, June 25-26, 1997; Saratov University Press: Saratov, 1997, pp. 44-45.

[121] Tsivileva, O.M.; Pankratov, A.N.; Nikitina, V.E.; Garibova, L.V. *Mikrobiologiya* 2004, 73, 486-490. *Eng. Transl.*: *Microbiology* 2004, 73, 410-413.

[122] Tsivileva, O.M.; Pankratov, A.N.; Nikitina, V.E.; Garibova, L.V. *Food Technol. and Biotechnol.* 2005, 43, 227-234.

[123] Tsivileva, O.M.; Pankratov, A.N.; Loshchinina, E.A.; Nikitina, V.E. *Vestnik Moskovsk. Universiteta. Ser. 2: Khimiya* 2006, 47, No. 2, 91-96.

[124] Nikitina, V.E.; Tsivileva, O.M.; Pankratov, A.N.; Bychkov, N.A. *Food Technol. and Biotechnol.* 2007, 45, 230-237.

[125] Pankratov, A.N.; Shchavlev, A.E.; Uchaeva, I.M.; Sedanov, M.Yu. *Zhurn. Anal. Khimii* 1995, 50, 399-400.

[126] Pankratov, A.N. *Zhurn. Anal. Khimii* 1997, 52, 889-893.

[127] Pankratov, A.N. *Online Information Review* 2005, 29, No. 2, 168–192.

[128] Pankratov, A.N.; Tsivileva, O.M.; Nikitina, V.E. *Online Information Review* 2006, 30, 43-52.

[129] Pankratov, A.N. *Zhurn. Anal. Khimii* 2006, 61, 566-580. *Eng. Transl.*: *J. Anal. Chem.* 2006, 61, 520-534.

[130] Information Resources on Natural Sciences and Ecology. Web: chem.saratov.ru/pankratov; www.chem.saratov.ru/pankratov; sgu.ru/faculties/chemical/pankratov; www.sgu.ru/faculties/chemical/pankratov.

[131] Gambaryan, N.P.; Stankevich, I.V. *Uspekhi Khimii* 1989, 58, 1945-1970.

[132] Gorelik, M.V.; Efros, L.S. Bases of Chemistry and Technology of Aromatic Compounds (in Russian); Khimiya: Moscow, 1992, 640 pp.

[133] Minyaev, R.M.; Minkin, V.I.; Starikov, A.G.; Gribanova, T.N. *Izv. Akad. Nauk. Ser. Khim.* 2001, 2219-2228.

[134] Clark, T. A Handbook of Computational Chemistry. A Practical Guide to Chemical Structure and Energy; A Wiley Interscience Publication, John Wiley and Sons: New York, Chichester, Brisbane, Toronto, Singapore, 1985. Russian Edition: Computational Chemistry (in Russian), Mir: Moscow, 1990, 383 pp.

[135] Foresman, J.B.; Frisch, A.E. Exploring Chemistry with Electronic Structure Methods; Gaussian: Pittsburgh, 1996, 302 pp.

[136] Stepanov, N.F. Quantum Mechanics and Quantum Chemistry (in Russian); Mir: Moscow, 2001, 520 pp.

[137] Chuvylkin, N.D. *Zhurn. Vsesoyuz. Khim. Obshchestva im. D.I. Mendeleeva* 1990, 35, 660-669.

[138] Kohn, W. *Uspekhi Fiz. Nauk* 2002, 172, 336-348.

[139] Dapprich, S.; Komaromi, I.; Byun, K.S.; Morokuma, K.; Frisch, M.J. *J. Mol. Struct. Theochem* 1999, 461-462, 1-21.

[140] Smolenskii, E.A. *Doklady Akad. Nauk SSSR* 1982, 266, 160-164.
[141] Smolenskii, E.A. *Zhurn. Fiz. Khimii* 1988, 62, 3273-3278.
[142] Smolenskii, E.A. *Izv. Akad. Nauk. Ser. Khim.* 1995, 1666-1673.
[143] Chuvylkin, N.D.; Smolenskii, E.A.; Zefirov, N.S. *Izv. Akad. Nauk. Ser. Khim.* 2005, 1118-1131.
[144] Chuvylkin, N.D.; Smolenskii, E.A.; Chernykh, A.V. *Khim. Fizika* 2006, 25, 55-61.
[145] Minkin, V.I.; Minyaev, R.M. Non-Classical Structures of Organic Compounds (in Russian); Kletskii, M.E.; Responsible Ed.; Rostov University Press: Rostov-on-Don, 1985, 168 pp.
[146] Minkin, V.I.; Minyaev, R.M.; Hoffmann, R. *Uspekhi Khimii* 2002, 71, 989-1014.
[147] Gribov, L.A. *Zhurn. Strukt. Khimii* 1994, 35, No. 4, 122-134.
[148] Gribov, L.A. *Zhurn. Fiz. Khimii* 2005, 79, 688-692.
[149] Vilkov, L.V.; Mastryukov, V.S.; Sadova, N.I. Determination of Geometrical Structure of Isolated Molecules (in Russian); Khimiya. Leningrad Branch: Leningrad, 1978, 224 pp.
[150] Voityuk, A.A. *Zhurn. Strukt. Khimii* 1988, 29, No. 1, 138-162.
[151] Pankratov, A.N. Chemistry (Weekly Application to the Newspaper "First of September") (in Russian); 1997, June, No. 24, p. 13.
[152] Verkhovskii, L.I.; Chuvylkin, N.D. *Khimiya i Zhizn - XXI Vek* 1999, No. 1, 12-13.
[153] Verkhovskii, L.I. *Khimiya i Zhizn - XXI Vek* 2000, No. 1, 14-15.

In: Quantum Chemistry Research Trends
Editor: Mikas P. Kaisas, pp. 127-145
ISBN 978-1-60021-620-6

Chapter 3

A NOVEL GRID BASED APPROACH TO THE ELECTRONIC STRUCTURE PROBLEM: INTERPOLANTS AND DERIVATIVES

***James S. M. Anderson, Juan I. Rodriguez, David C. Thompson*[*] and Paul W. Ayers*[†]**
Department of Chemistry, McMaster University, Hamilton, ON L8S 4M1, Canada

Abstract

The electronic structure problem is introduced and briefly reviewed within the context of density-functional theory. The use of numerical grid based schemes, circumventing the use of a basis set, will be discussed. We will briefly introduce our numerical grid methodology for use in electronic structure theory. This method, which is based upon a combination of the Smolyak cubature construction and a novel nonlinear transformation, has been shown to be efficacious when applied to cubature of physically relevent functions. In this paper we extend these results to interpolation and differentiation; these being the other fundamental operations essential for a numerical solution of the electronic structure problem. We shall investigate the performance of our method as a function of spatial dimension, number of grid points and, across a wide range of relevant functions.

1. Introduction

The fundamental problem of molecular electronic structure determination, or the solution of the underlying equations governing the behaviour of electrons, traditionally involves computing accurate approximations to the ground state electronic wavefuction $\Psi(\mathbf{x}_1, \ldots, \mathbf{x}_N)$. Here, $\mathbf{x}_i$ represents the spin-resolved spatial coordinate, $\mathbf{x}_i = \mathbf{r}_i s_i$. Once determined, all observable quantities are accessible from the wave function via direct integration:

[*]Present address: Department of Structural Biology and Computational Chemistry, Chemical & Screening Sciences, Wyeth Research, 200 Cambridge Park Drive, Cambridge, MA 02140
[†]E-mail address: ayers@mcmaster.ca

$$Q[\Psi] \equiv \sum_{s_i} \int \int \cdots \int \left[\Psi^*(\mathbf{x}_1, \ldots, \mathbf{x}_N) \hat{Q} \Psi(\mathbf{x}_1, \ldots, \mathbf{x}_N) \right] d\mathbf{r}_1 \ldots d\mathbf{r}_N, \tag{1}$$

with the quantum-classical correspondance principle providing a link between the property Q and its Hermitian operator. The Hamiltonian operator is of great importance, allowing us to determine the energy spectrum of the system. For a system of N electrons and M nuclei, in atomic units, and within the Born-Oppenheimer approximation, this operator is written as:

$$\hat{H} \equiv \sum_{i=1}^{N} \left(-\frac{1}{2}\nabla_i^2 + \sum_{\alpha=1}^{M} -\frac{Z_\alpha}{|\mathbf{r}_i - \mathbf{R}_\alpha|} + \sum_{i<j} \frac{1}{|\mathbf{r}_i - \mathbf{r}_j|} \right). \tag{2}$$

One can then obtain the total energy of the system through

$$\begin{aligned} W[\Psi] &\equiv E_{nucl} + E_{elec}[\Psi] \\ &= \sum_{\alpha=1}^{M-1} \sum_{\beta=\alpha+1}^{M} \frac{Z_\alpha Z_\beta}{|\mathbf{R}_\alpha - \mathbf{R}_\beta|} + \langle \Psi | \hat{H} | \Psi \rangle. \end{aligned} \tag{3}$$

To determine Ψ a variational construct is employed: we know that no 'trial' wave function, $\tilde{\Psi}$, has a lower energy than the exact ground state, Ψ_{gs}:

$$\langle \Psi_{gs} | \hat{H} | \Psi_{gs} \rangle \leq \langle \tilde{\Psi} | \hat{H} | \tilde{\Psi} \rangle. \tag{4}$$

Thus, minimization of $E_{elec}[\Psi]$, with respect to Ψ, such that it is constrained to be normalized and anti-symmetric with respect to exchange of spatial and spin coordinates, yields the exact ground state energy.

Unfortunately, application of the Hamiltonian operator to the electronic wave function results in a second-order partial differential equation in 3 N real valued coordinates and N dichotomic spin coordinates. The ground state is thus inaccessible except for small or model systems [1, 2, 3, 4, 5, 6]. Moreover, the scaling of the computational cost is exponential with respect to N. Given this limitation, most chemically relevant processes of biological activity are outside the scope of this traditional wave function based methodology. Clearly the root of this problem lies in the high dimensionality of the wave function. Should a molecular system, and its properties, be accessible through a function (or functional) with fewer coordinates, then some of these problems encountered when determining the electronic structure of large systems could be circumvented. It is in this light that we now turn to density-functional theory (DFT), wherein the key descriptor of the system is the ground state electronic density.

1.1. Density-Functional Theory

In the encompassing theorem of Hohenberg and Kohn it is demonstrated that, like the wave function, the ground state's electron density determines all of the properties of an electronic system in its ground state [7]. This remarkable result is achieved as follows: we recall that the number of electrons is determined directly from the electron density:

$$N[\rho] \equiv \int \rho(\mathbf{r}) d\mathbf{r}. \tag{5}$$

Next, it is demonstrated that the external potential, $v(\mathbf{r})$, is determined directly from $\rho(\mathbf{r})$. With N and $v(\mathbf{r})$ one can determine the electronic Hamiltonian and solve Schrödinger's equation for the wave function; all observable properties are then easily obtained through Eqn. 1. Within density-functional theory the independent variable has changed from $\Psi(\mathbf{x}_1, \ldots, \mathbf{x}_N)$ to $\rho(\mathbf{r})$. We have reduced the space of the solution to three variables (from $3N$ spatial and N dichotomic in the wave function formulation). However, the following problem remains: the Hohenberg-Kohn theorem is one of existence and it gives no clue as to how to determine the unique ground state electronic density. Indeed, the key result of the so-called 'second' theorem of Hohenberg and Kohn, and the foundation of all practical approaches, is suitably vague:

$$E_{gs} = E_v[\rho_{gs}] = \underbrace{\min}_{\text{all } N-\text{electron}\rho(\mathbf{r})} E_v[\rho]; \tag{6}$$

the problem of dealing with the $3N$ spatial and N dichotomic variables is transferred into the lack of knowledge of the energy functional. The core problem thus becomes finding suitable approximations to the Hohenberg-Kohn functional, $F[\rho]$:

$$F[\rho] = E[\rho] - \int \rho(\mathbf{r}) v[\rho; \mathbf{r}] d\mathbf{r}. \tag{7}$$

A practical solution was presented by Kohn and Sham in Ref. [8]. One constructs the electron density from one-particle orbitals:

$$\rho(\mathbf{r}) = \sum_{i=1}^{N} |\phi_i(\mathbf{r})|^2. \tag{8}$$

Applying the variational principle, Eqn. 6, to the resulting Euler-Lagrange equations yields:

$$\left(-\frac{1}{2}\nabla^2 + v(\mathbf{r}) + \int \frac{\rho(\mathbf{r}')}{|\mathbf{r} - \mathbf{r}'|} d\mathbf{r}' + \mu_{xc}(\rho) \right) \phi_i(\mathbf{r}) = \epsilon_i \phi_i(\mathbf{r}). \tag{9}$$

These are the so-called Kohn-Sham equations. Here, μ_{xc} is the exchange-correlation potential, the functional derivative of the exchange-correlation energy functional with respect to the density: $\mu_{xc} = \frac{\delta E_{xc}[\rho]}{\delta \rho}$. The total energy is then:

$$E^{KS} = \sum_i \epsilon_i - \frac{1}{2} \int \frac{\rho(\mathbf{r})\rho(\mathbf{r}')}{|\mathbf{r} - \mathbf{r}'|} d\mathbf{r} d\mathbf{r}' - \int \mu_{xc}(\mathbf{r})\rho(\mathbf{r}) d\mathbf{r} + E_{xc}[\rho]. \tag{10}$$

The Kohn-Sham equations defined through Eqn. 9, are to be solved self-consistently. The difficulty in determining the exact ground state density from this formulation has been recast into the unknown form of the *exact* exchange-correlation energy functional E_{xc}. Progress has been made, and the functional 'zoo' containing functionals of varying sophistication, continues to grow [9, 10, 11, 12, 13]. The Kohn-Sham formalism has been successfully applied to a wide variety of problems in solid state physics, chemistry, biochemistry, and within the field of drug discovery [14, 15, 16, 17, 18, 19, 20]. A fuller description of DFT, its theoretical development, and its usage within quantum chemistry can be found in Ref. [21].

A naive implementation of the Kohn-Sham method would result in an $\mathcal{O}(N^3)$ scaling tool, and in such cases DFT may only be successfully applied to systems of around a few hundred atoms; clearly this limits its applicability as a method for use in medicinal chemistry, or biophysics, where the number of particles to treat can be orders of magnitude greater. A successful treatment of these systems requires methods which scale linearly in particle size (and with early onset and a small prefactor). To date, research in this area has focussed on avoiding explicitly constructing the Kohn-Sham orbitals, ϕ_i, and directly constructing the density instead [22, 23, 24, 25, 26, 15]. Our own research is related to the field of linear scaling DFT calculations and this forms the content of the rest of this chapter. In the following section we discuss the plane wave and basis set approaches for solving the Kohn-Sham equations of DFT; these are two commonly used methods for representing orbitals within the Kohn-Sham formalism. Basic background on each method will be presented and we shall focus on the relative strengths and weaknesses of each approach whilst not discussing the merits/demerits of any specific implementation. In section 3 we introduce real space methods, specifically adaptive coordinate approaches. In section 4 we present an overview of our own method for incorporating coordinate adaptability into a real space formalism and then in section 5 we show how this efficiently adapted real space grid can be used to perform all of the fundamental numerical operations neccessary for performing an electronic structure calculation; we concentrate our discussion on the operations interpolation and differentiation. In section 6 we present results and discussion and in the final section our concluding remarks.

2. Plane Wave and Basis Set Methods

In the plane wave approach orbitals are expanded in a non-local plane wave basis. Core states are modeled using pseudopotentials and allow for a reasonable number of plane waves to represent most of the chemical elements found in a materials simulation [27]. The use of Fast Fourier transform (FFT) techniques when dealing with the orbitals and electrostatic potential is efficacious, and there is no dependence of the basis on atom positions. Also, subsequent algorithmic improvements have led to a reduction in the number of steps required to attain self consistency [28, 29]. Rigorous control of the numerical convergence is attained through a few parameters, principally the wave vector of the highest frequency Fourier mode, k. However, there are a number of drawbacks to the approach. Even with advances in pseudopotential methods, strong variations in the potential occur within the core region; these are difficult to resolve with a plane wave representation. Moreover, such a representation would extend over the whole domain, into areas where such resolution would most certainly not be neccessary. This is particularly troublesome for localized, highly inhomogeneous systems like clusters, molecules, or surfaces. Here a huge amount of effort is expended in modeling the vacuum accurately. The correct description of charged systems is also somewhat involved, as a uniform neutralizing background needs to be correctly added in order to correctly compute total energies. Finally, without a localized orbital representation, the orthogonalization step is seen to scale as N^3.

In contrast to the use of non-local plane wave methods, one can use a localized basis set representation, typically with either Slater or Gaussian basis functions [30]. Here molecular orbitals are constructed from a linear combination of atomic orbitals (the LCAO approximation) and < 30 Gaussians can provide an accurate representation of a typical first row chemical element. The use of the localized basis is somewhat more chemically intuitive than the plane wave approach which heralds from condensed matter physics. Both choice of localized function have their strengths and weaknesses [31]: Slater functions have an improved form both close to, and far from, the nuclei; they are also more sharply peaked than Gaussian functions. The use of Slater orbitals avoids the cumbersome use of contractions of Gaussian functions. Unfortunately, most integrals involving Slater functions have to be done numerically, while most integrals involving a Gaussian basis are analytically tractable. Gaussian functions also have the attractive property that they can be easily differentiated any number of times, whilst Slater functions have the complication that this repeated differentiation leads to the introduction of factors of x/r. Irrespective of which function is used, algorithmic advances have led to significant advances in the acceleration of the convergence behaviour of basis set self-consistent methods [32, 33, 34]. However, for basis set methods in general, care must be taken to account for basis set superposition error (BSSE). This arises through the overlap of non-orthogonal atom centered functions of composite systems. Finally, we note that scaling in basis set methods can vary but, with recent developments, linear scaling for large systems is accessible.

There also exist hybrid methods, combining the strengths of both the basis set, and plane wave approaches. In the Gaussian/Plane wave method (GPW) the wave function is described using an atom centered Gaussian basis, but an auxillary plane wave basis is used to describe the density [35, 36]. This description of the density allows for the use of FFT methods to solve the Poisson equation and to obtain the Hartree energy in a linear scaling fashion. Whilst the GPW method requires a pseudopotential representation of the core region, the susequent Gaussian and augmented plane wave (GAPW) method does allow for all electron calculations [37].

3. Real Space Methods

There are a number of key advantages to a real space representation of an atomic or molecular problem, as opposed to either the plane wave, or basis set representations, described in the preceding section. Firstly, the potential term is diagonal in coordinate space, whilst the kinetic term, the Laplacian, is nearly local. The near-locality is ideal for linear scaling algorithms and facillitates parallelization. Within the real space formalism it is trivial to deal with both finite or charged systems. In strict contrast to the basis set, or plane wave methods, the real space approach produces structured highly banded matrices which are efficiently solved using multiscale approaches [38, 39]. For an excellent introduction to real space methods and electronic structure calculations we refer the interested reader to Ref. [40].

Real space methods are perhaps closest in structure to the plane wave approach: both are fully 'numerical' with a few parameters controlling convergence (the grid spacing, h, and the wave vector, k, described above). Indeed, the problem of resolution within plane wave methods (the difficulty of representing inhomogenous systems, or the vacuum) is also found within the real space method and is referred to as *adaptability* of the grid. For a highly efficient description of an inhomogenous system it is beneficial to allow the resolution of the grid to match the requirements of the physical system. This is achieved by an increase in resolution in the grid near the nuclei (moving the points in the grid to where the density, or some other such representative function, is large). Within the literature there are two strategies which have been used thus far: that of 'local refinement' or that of 'grid curving'. The local refinement approach has been performed within the context of wavelets [41, 42], finite elements [43, 44] and multigrid methods [45, 46]. As our method is conceptually related to the grid curving approach and we shall discuss this in somewhat more detail.

In the ACRES method of Modine *et al.* [47] standard Cartesian coordinates $\mathbf{r}$ are related to curvilinear coordinates through a coordinate transformation $\mathbf{r}(\xi)$. The change of coordinates maps the regular grid in curvilinear coordinates to an adaptive mesh in Cartesian coordinates with a finer resolution where it is needed. This transformation of coordinates is different from the more 'classical' coordinate transformations (eg. log. spherical) in a number of key ways. Firstly, the grid is adapted to an arbitrary arrangement of atoms through a linear superposition of the grid displacements of each atom. Secondly, the transformation is smooth and continuous everywhere. The method implemented in the ACRES program was extended into the HARES program although the form of the grid curving remains the same [48]. We note that this form of grid curving was initially implemented by Gygi [49] using a plane wave basis and within this context has been extended by a number of groups [50, 51, 52, 53]. Gygi and Gallo have also recently studied real space approaches using curvilinear coordinates and pseudopotentials [54].

4. Grid Curving and the Smolyak Construct

We shall now briefly summarize our recent research on grid curving methods in electronic structure theory. We shall follow the formalism of Ref. [55] and introduce our methodology in terms of multi-dimensional numerical integration, or cubature.

In one dimension, Gaussian quadrature is the *best possible integration method* for integrals of the form:

$$\int f(x)w(x)dx, \tag{11}$$

where $f(x)$ has any of the following properties: it is analytic; smooth (derivatives of all orders exist); is a polynomial (or is well approximated by polynomials) or has derivatives up to some order, n, with $n \geq 0$. For the first three cases, or when $n > 0$ *no other method is comparable*. There are numerous closed form Gaussian quadrature formulae in one-dimension, however, whilst special cases are known of formulae in many dimensions, in general the problem is regarded as one of immeasurable complexity [56]. Notable exceptions to this rule are rules for integration on the unit sphere and unit hypercube, with the Smolyak construction being an example of the latter; the recent work of Rees and Hall represents another special case with an 'atom-like' spherical grid [57]

Before discussing the Smolyak construction in detail, let us recap. The problem of numerical integration is such that we seek good approximations to the functional Q_d where

$$Q_d(f) = \int_{[0,1]^d} f(x)dx. \tag{12}$$

A near optimal choice was suggested by Smolyak [58] and we quote without proof the following formulae, henceforth referred to as the Smolyak construction:

$$A(q,d) = \sum_{q-d+1 \le |\mathbf{i}| \le q} (-1)^{q-|\mathbf{i}|} \binom{d-1}{q-|\mathbf{i}|} (U^{i_1} \otimes \cdots \otimes U^{i_d}). \tag{13}$$

In the above expression d is the dimension of the problem and q, the effort, can be thought of as a parameter akin to the grid spacing. Here U^i refers to the one dimensional quadrature formulae:

$$U^i(f) = \sum_{j=1}^{m_i} f(x_j^i).a_j^i \tag{14}$$

such that $m_i \in \mathbb{N}$ is known. The a_j^i and x_j^i are the *weights* and *nodes* of our integration scheme. Thus, the Smolyak construction allows us to define an efficacious grid for Gaussian cubature on the unit hypercube. Our underlying quadrature formulae (the U^i's) will determine the distribution of points throughout the cube, and will be a uniform distribution and certainly not adapted for electronic structure calculations where the region of interest is now $(-\infty, \infty)^d$ and most of our nodes should be placed near the nuclei. However, such a transformation can be constructed in the following way. For ease, we begin with the simplest one dimensional example. We define the coordinate transformation:

$$\Theta(x) = \frac{\int_{-\infty}^{x} w(x)dx}{\int_{-\infty}^{\infty} w(x)dx}. \tag{15}$$

Then,

$$\begin{aligned} \int_0^1 f(\Theta)d\Theta &= \int_{-\infty}^{\infty} f(x)\frac{d\Theta}{dx}dx \\ &= \int_{-\infty}^{\infty} f(x)w(x)dx. \end{aligned} \tag{16}$$

It is noted that the weight transformation function, $w(x)$, simply takes the points from $[0,1]$ and distributes them on $(-\infty, \infty)$ such that there are many points where $w(x)$ is large and few where it is small. This is readily extended to $d > 1$ [59]. For instance, for $d = 3$, Eqn. 15 becomes:

$$\begin{aligned} \Theta_1(x_1) &= \frac{\int_{-\infty}^{x_1}\int_{-\infty}^{\infty}\int_{-\infty}^{\infty} w(\mathbf{r})d\mathbf{r}}{\int_{-\infty}^{\infty}\int_{-\infty}^{\infty}\int_{-\infty}^{\infty} w(\mathbf{r})d\mathbf{r}} \\ \Theta_2(x_2) &= \frac{\int_{-\infty}^{x_2}\int_{-\infty}^{\infty} w(x_1, r_2, r_3)dr_2dr_3}{\int_{\infty}^{\infty}\int_{\infty}^{\infty} w(x_1, r_2, r_3)dr_2dr_3} \\ \Theta_3(x_3) &= \frac{\int_{-\infty}^{x_3} w(x_1, x_2, r_3)dr_3}{\int_{-\infty}^{\infty} w(x_1, r_2, r_3)dr_3}. \end{aligned} \tag{17}$$

For the purposes of a molecular or atomic electronic structure calculation, a suitable weight transformation might be something that looks like a molecular, or atomic, density. Consider such a 'molecular' density,

$$w(\mathbf{x}) = \rho_{\text{mol}}(\mathbf{x}) = \sum_{\alpha=1}^{N} \rho_\alpha(\mathbf{x} - \mathbf{R}_\alpha), \tag{18}$$

consisting of N atoms with the αth atom at $\mathbf{R}_\alpha$. To simplify this transformation we consider atomic densities that have been fit to s-type Gaussian functions

$$\rho_\alpha(\mathbf{x}) = \sum_i c_{\alpha i} \exp\left(-\beta_{\alpha i} (\mathbf{x} - \mathbf{R}_\alpha)^2\right). \tag{19}$$

We use the $c_{\alpha i}$ and $\beta_{\alpha i}$ as generated by Constans and Carbó (they have generated tables of coefficients and exponents fitted variationally using s-function Gaussians to atomic densities from H to Kr at the HF/6-311G level of theory [60, 61]). This weight function has the additional advantage that the Jacobian elements are easily computed, are lower diagonal and are trivially related to the weight.

The transformation we have introduced above is actually the reverse of the one we are interested in: it maps points such that $\mathbb{R}^d \rightarrow [0,1]^d$. To achieve the reverse, we couple the transformation with a non-linear equation solver and treat the problem as a system of equations. We solve consecutively from Θ_1 to Θ_d and as the finding of each x_i is a one dimensional problem, the bisection method is effective. To illustrate the adaptive properties of this transformation, in Fig. 1 we show the uniform 2 dimensional grid corresponding to the use of the rectangle rule within Eqn. 14. This grid has an effort (q) of 9 and is composed of 4097 points. We define a pseudomolecular 2 dimensional CO_2 molecule using Eqn. 19 and perform the transformation, and subsequently solve for x_1 and x_2. Within our weight function we have used a carbon atom at $(0,0)$ and two oxygen atoms at $(\pm 2, 0)$ respectively. This results in Fig. 2. It is clear that the points on the uniform grid have been moved towards the atomic nuclei, and the grid has been markedly changed. We have shown in an earlier publication that these grids are suitable for the kinds of cubature neccessary for electronic structure calculations [55]. In this chapter we extend the usefulness of this result to encompass the other fundamental numerical operations: interpolation, and differentiation.

5. Grid Curving: Interpolation and Differentiation

We begin by noting that any linear operator on a function can be written as:

$$\hat{L}[f(x)] \approx \sum_{i=0}^{n(q)-1} w_i f(x_i), \tag{20}$$

essentially the form shown in Eqn. 14. Here the effort (q), or grid spacing, is subsumed into the function $n(q) - 1$. Now for periodic functions the Fourier series is a well known function for interpolation:

$$f(x) \approx \frac{a_0}{2} + \sum_{j=1}^{M} a_j \cos(2\pi j x) + b_j \sin(2\pi j x), . \tag{21}$$

We wish to express this in the language of Eqn. 14 such that we can approximate our function as a sum of weights multiplied by function evaluations at x_i. To achieve this we will use a discrete Fourier transform to find the coefficients in terms of $f(x_i)$. In doing so

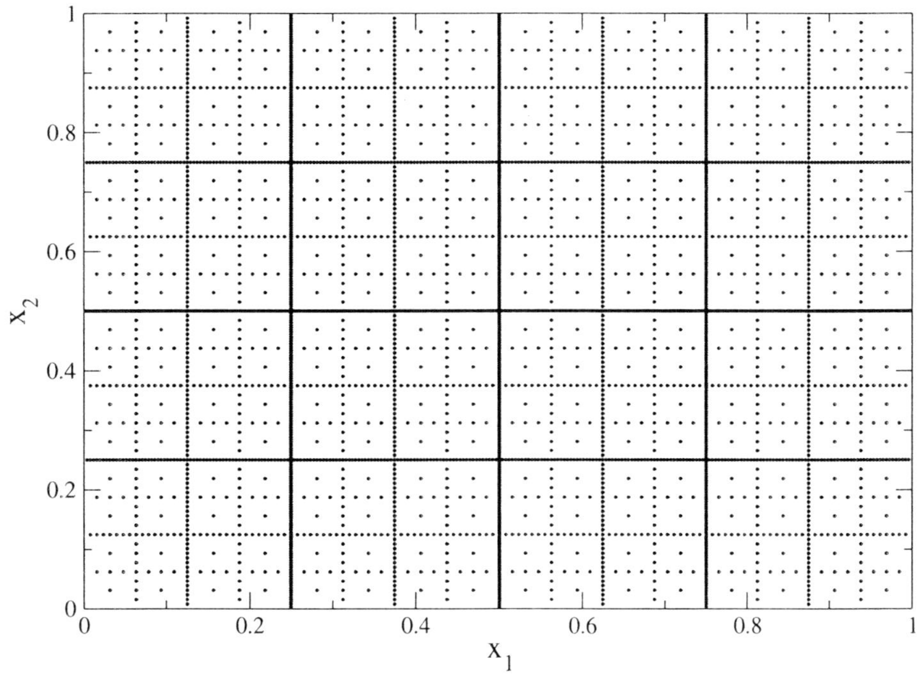

Figure 1. A 2 dimensional grid derived from the rectangle rule. The grid corresponds to an effort, q, of 9 and is composed of 4097 points.

we require that the number of points, x_i, is a simple power of two. Thus $n(q) - 1$ above becomes $2^q - 1$. We find that our Fourier coefficients are thus:

$$\begin{aligned} a_0 &= \frac{1}{n}\sum_{k=0}^{n-1} f(x_k) \\ a_{0<j<n/2} &= \frac{2}{n}\sum_{k=0}^{n-1} f(x_k)\cos\left(\frac{2\pi jk}{n}\right) \\ b_j &= \frac{2}{n}\sum_{k=0}^{n-1} f(x_k)\sin\left(\frac{2\pi jk}{n}\right) \\ a_{n/2} &= \frac{1}{n}\sum_{k=0}^{n-1} f(x_k)\cos(k\pi). \end{aligned} \tag{22}$$

Note that

$$b_0 = b_{n/2} = 0 \tag{23}$$

and

$$x_k = \frac{k}{n}. \tag{24}$$

We now have n points with n coefficients that can be solved for. Hence, M in the above expression becomes $n/2$. Substitution of Eqn. 23 into Eqn. 21 and applying the trigonometric

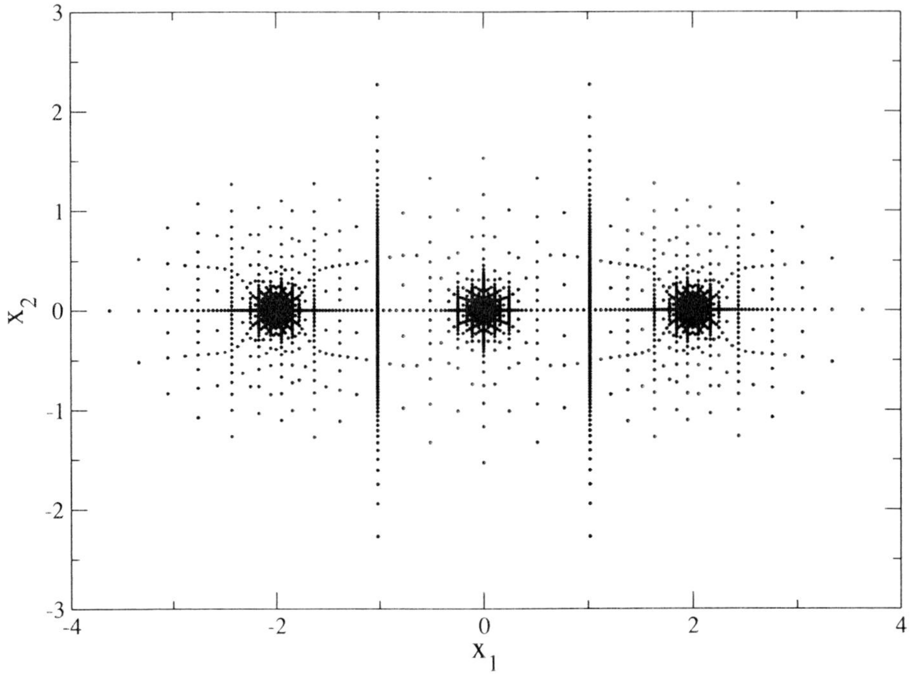

Figure 2. The grid in Fig. 1 transformed using a 2 dimensional CO_2 pseudo-molecular weight function. The carbon and oxygen atoms are at (0,0) and ($\pm$ 2,0) respectively. We observe a higher concentration of points where the weight transformation function is large corresponding to the positions of the carbon and oxygen nuclei.

identity

$$\cos(A - B) = \cos(A)\cos(B) - \sin(A)\sin(B) \tag{25}$$

one easily obtains the expression:

$$f(x) \approx \sum_{k=0}^{n-1} \left[\frac{1}{n} + \frac{2}{n} \sum_{j=1}^{n/2-1} \cos(2\pi j(x - x_k)) + \frac{1}{n} \cos(\pi(x - x_k)) \right] f(x_k). \tag{26}$$

The term in the square bracket is the weight for interpolation, w_k, and the portion outside is simply the value of the function at the point x_k. Having this result it is trivial to generate expressions for differentiation. Indeed, the mth derivative of the Fourier series converges pointwise to the mth derivative of the function if that function is at least m times differentiable.

$$f^{'}(x) \approx \sum_{k=0}^{n-1} \left[\frac{-4\pi}{n} \sum_{j=1}^{n/2-1} j \sin(2\pi j(x - x_k)) - \frac{\pi}{n} \sin(\pi(x - x_k)) \right] f(x_k) \tag{27}$$

and

$$f^{''}(x) \approx \sum_{k=0}^{n-1} \left[\frac{-8\pi^2}{n} \sum_{j=1}^{n/2-1} j^2 \cos(2\pi j(x - x_k)) - \frac{\pi^2}{n} \cos(\pi(x - x_k)) \right] f(x_k) \tag{28}$$

provide the weights of the first and second derivatives respectively. Integration of Eqn. 26 results in

$$\int_0^1 f(x)dx \approx \sum_{k=0}^{n-1} \frac{1}{n} f(x_k). \tag{29}$$

This is just the rectangle rule.

These results correspond to a function on $[0, 1]$, and in order to differentiate a function using our coordinate transformation we must apply the chain rule. In three dimensions, a derivative evaluated on our transformed grid becomes:

$$\frac{\partial f}{\partial x_i} = \sum_{j=1}^{3} \frac{\partial f}{\partial \Theta_j} \frac{\partial \Theta_j}{\partial x_i} \tag{30}$$

with the second derivative being

$$\frac{\partial^2 f}{\partial x_i^2} = \sum_{j=1}^{3} \left[\left[\sum_{k=1}^{3} \frac{\partial^2 f}{\partial \Theta_k \partial \Theta_j} \frac{\partial \Theta_k}{\partial x_i} \right] \frac{\partial \Theta_k}{\partial x_i} + \frac{\partial f}{\partial \Theta_j} \frac{\partial^2 \Theta_j}{\partial x_i^2} \right]. \tag{31}$$

The terms $\frac{\partial f}{\partial \Theta_j}$ and $\frac{\partial^2 f}{\partial \Theta_k \partial \Theta_j}$ are computed using the unit cube grid, and the terms $\frac{\partial \Theta_j}{\partial x_i}$ and $\frac{\partial \Theta_j^2}{\partial x_i^2}$ are determined analytically from the form of the transformation.

6. Results

To recap, our method takes the regular Smolyak grid of points, as defined through Eqn. 13, and transforms it using the d dimensional extension of Eqn. 15. The resulting grid points are now inhomogenously distributed throughout $\mathbb{R}^d$, with localization occurring where the transformation $w(\mathbf{x})$ is large. To illustrate the utility of this approach we have computed interpolants, and both 1st and 2nd derivatives, of a range of functions of varying dimension.

In Fig. 3 we see how the effort, q, of Eqn. 13 affects the log of the number of unique points for different dimensionalities. Fig. 4 illustrates the convergence of the 1st and 2nd derivatives as a function of the number of grid points used for the 6 dimensional Gaussian

$$f(\mathbf{x}) = \exp(-3x^2). \tag{32}$$

Here,

$$x^2 = \sum_{i=1}^{d} x_i^2 \tag{33}$$

represents the radius of a d dimensional hypersphere. In this example, the differentiation is performed on the untransformed Smolyak grid, with respect to x_1, at the point:

$$\mathbf{x}' = (0.4, 0.2, -0.3, 0.5, 0.01, -0.8). \tag{34}$$

Throughout this article, as a measure of our accuracy, we compute the logarithm of the absolute error:

$$\log(\text{Abs. Error}) = \log(|f^{(n)}_{Exact}(\mathbf{x}') - f^{(n)}_{Computed}(\mathbf{x}')|). \tag{35}$$

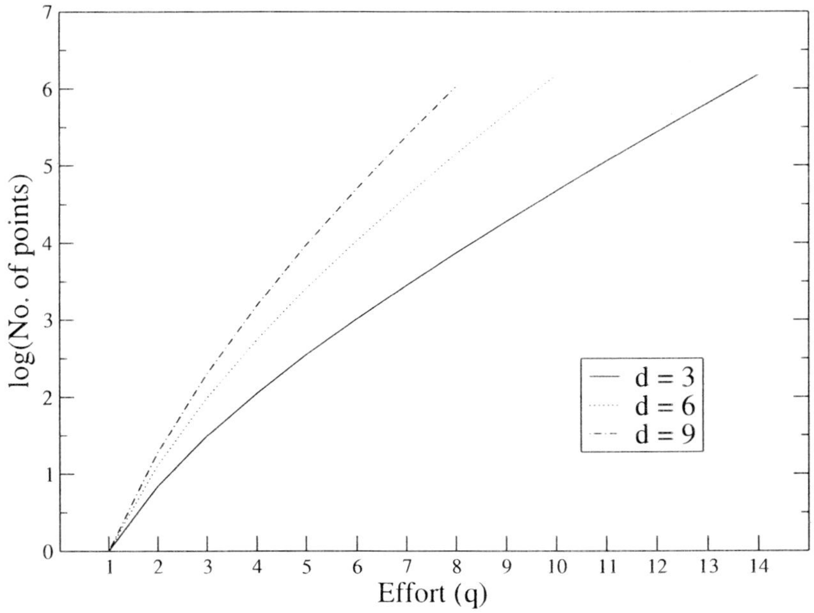

Figure 3. An illustration as to the rate of growth of the logarithm of the number of points as a function of effort, for a range of different dimensions.

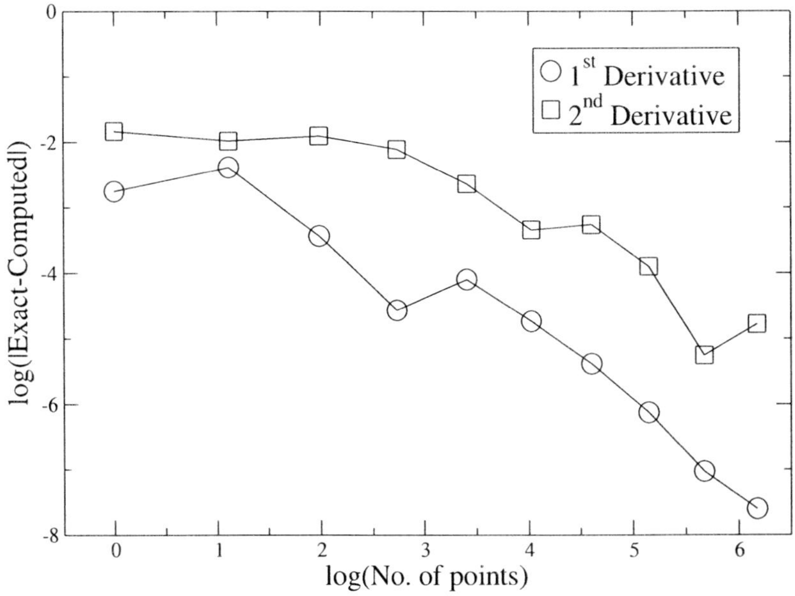

Figure 4. Convergence, with respect to the number of grid points, of the 1 st and 2nd derivatives of a 6 dimensional function, $f(\mathbf{x})$ [Eqn. 32], with respect to x_1. The derivatives are evaluated at $\mathbf{x}' = (0.4, 0.2, -0.3, 0.5, 0.01, -0.8)$ and are performed on an untransformed Smolyak grid.

For, $n = 0, 1$ or 2 coresponding to interpolation, 1 st, and 2nd derivatives respectively. We see a systematic improvement in the accuracy of both derivatives with increasing number of grid points. We also see that the 1 st derivative is computed to a higher accuracy than the 2nd, at all numbers of grid points.

How does our method do for a non-trivial function? Consider the 6 dimensional function:

$$g(\mathbf{x}) = \exp(-3x^2)\left[x_1 + 1 + \sin(x_2) - \cos(x_3) + (x_4 - 3)^2 + x_5^6\right] \tag{36}$$

and its 1 st and 2nd derivatives as computed at $\mathbf{x}$ on the untransformed Smolyak grid. Pleasingly, Fig. 5 shows much the same behaviour with respect to accuracy, as Fig. 4. Again, we observe a systematic increase in accuracy with respect to the number of grid points.

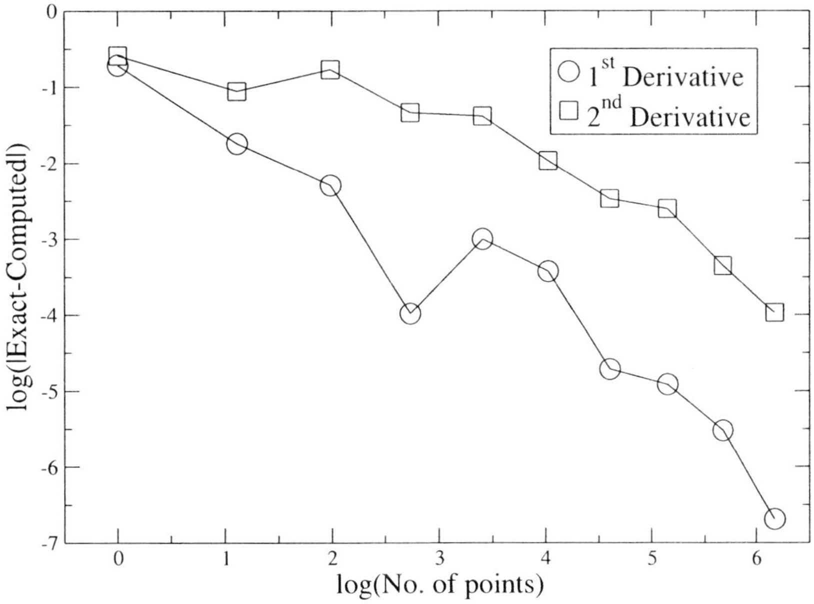

Figure 5. Convergence, with respect to the number of grid points, of the 1 st and 2nd derivatives of a 6 dimensional function, $g(\mathbf{x})$ [Eqn. 36], with respect to x_1. The derivatives are evaluated at $\mathbf{x}' = (0.4, 0.2, -0.3, 0.5, 0.01, -0.8)$ and are performed on an untransformed Smolyak grid.

In Fig. 6 we consider the 1st and 2nd derivatives, with respect of x_1, of a 9 dimensional function

$$h(\mathbf{x}) = \exp(-3x^2) \tag{37}$$

at

$$\mathbf{x}' = (0.1874, 0.003, 0.425, -0.68636, -0.23496, -0.095905, 0.198655, 0.40912, 0.6114) \tag{38}$$

on the untransformed Smolyak grid. In general, as the dimensionality increases, more grid points are required to achieve a given level of accuracy. Thus far we have shown that our underlying Smolyak grids are well suited to compute derivatives (both 1 st and 2nd) of a range of functions, and for very large dimensions. In all cases we have examined, the

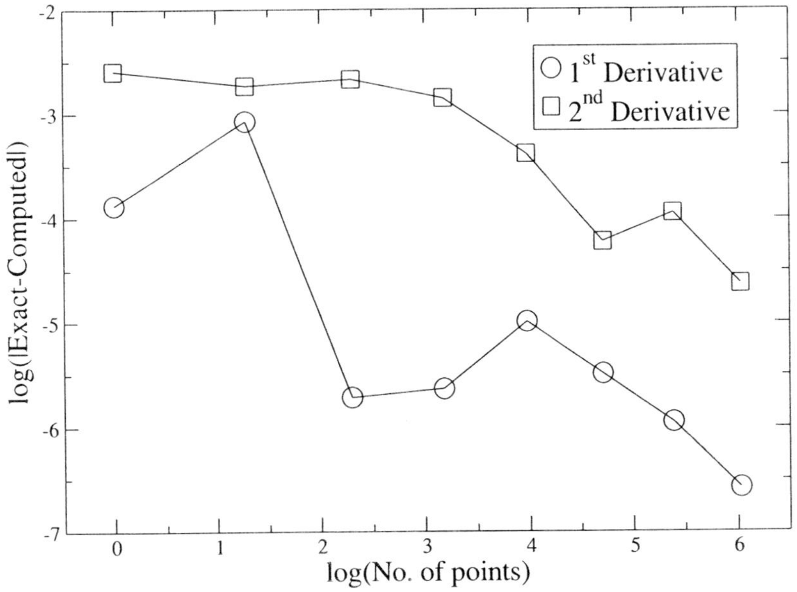

Figure 6. Convergence, with respect to the number of grid points, of the 1st and 2nd derivatives of a 9 dimensional function, $h(\mathbf{x})$ [Eqn. 37], with respect to x_1. The derivatives are evaluated at $\mathbf{x}' = (0.1874, 0.003, 0.425, -0.68636, -0.23496, -0.095905, 0.198655, 0.409012, 0.6114)$ and are performed on an untransformed grid.

1st derivative is computed to a greater accuracy than the second however, in both cases systematic improvements are seen upon increasing the effort, q.

We now investigate our method of computing interpolants and derivatives on a transformed grid in $\mathbb{R}^d$. Plotted in Fig. 7 are the logarithm of the absolute errors of the computed 1st and 2nd derivatives of the function

$$\begin{aligned} F(\mathbf{x}) &= \exp(-6[(x_1 - 0.01)^2 + (x_2 - 0.02)^2 + (x_3 - 0.03)^2]) \\ &+ 2\exp(-6[(x_1 + 0.01)^2 + (x_2 + 0.02)^2 + (x_3 - 0.03)^2]) \end{aligned} \quad (39)$$

at the point $\mathbf{x} = \mathbf{0}$, and with respect to x_1. This function is a sum of two 3 dimensional Gaussians offset slightly from the origin. The transformation used to map the grid from $[0, 1]^3 \rightarrow \mathbb{R}^3$ is a single Gaussian, with unit exponent, placed at the origin. The subsequent decrease in accuracy, with increasing effort, is attributed to the resulting expression involving differences between nearly equal numbers; this function presents a particularly challenging numerical example.

Finally, in Fig. 8, for the function $F(\mathbf{x})$, we compute the interpolant and both 1st and 2nd derivatives as a function of $\mathbf{x}'$ where $\mathbf{x}' = (x_1, 0.003, -0.523)$ and $x_1 = 0, \ldots, 1.5$ in steps of 0.1. As already seen previously, $\text{Error}_{\text{Interpolation}} < \text{Error}_{1^{st}\text{ derivative}} < \text{Error}_{2^{nd}\text{ derivative}}$ and may be a reflection of the cummulative errors arising from subsequent applications of the Fourier approximation.

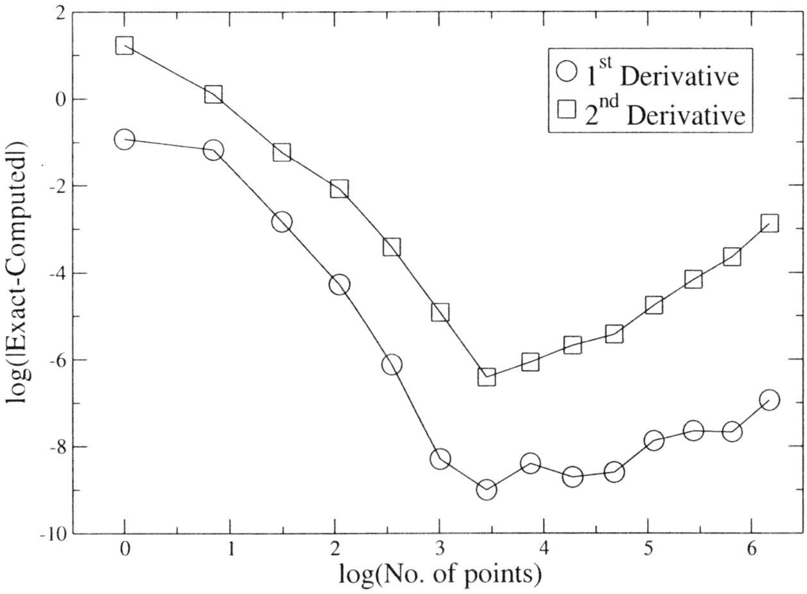

Figure 7. Convergence, with respect to the number of grid points, of the 1 st and 2nd derivatives of a 3 dimensional function, $F(\mathbf{x})$ [Eqn. 39], with respect to x_1. The derivatives are evaluated at $\mathbf{x} = \mathbf{0}$ and are performed on a transformed Smolyak grid.

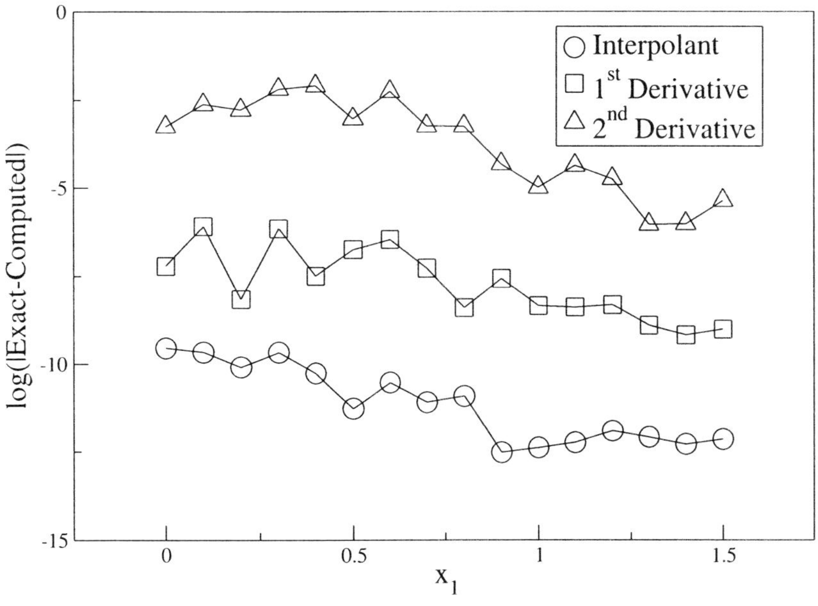

Figure 8. The interpolant, 1st, and 2nd derivatives of a 3 dimensional function, $F(\mathbf{x})$ [Eqn. 39, as a function of x_1. Here $\mathbf{x}' = (x_1, 0.003, -0.523)$ with $x_1 = 0, \ldots, 1.5$ in steps of 0.1 on a transformed Smolyak grid.

7. Conclusion

By coupling the Smolyak construction with a novel non-linear transformation of coordinates, we have shown that the resulting grid can be efficiently utilized for the purposes of integration, interpolation, and differentiation. The transformation of coordinates has been formalized such that points are moved to where the transforming weight function is largest and, one would hope, most relevant. This method joins the collection of adaptive coordinate real space methods; we have demonstrated that Smolyak derived grids are thus amply suited to solving problems within molecular electronic structure theory where the electronic density is a spatialy heterogeneous function.

The numerical results presented here are encouraging. We observe good convergence in the accuracy of computed interpolants and derivatives at points both on and off of the grid. This convergence is, in general, smooth and concomitant with an increase in 'effort'; this parameter is analogous to the wave vector in a plane wave basis, or the grid spacing in other real space methods. We have shown application of this method to a number of functions, of varying difficulty, and of varying dimensionality.

Work continues within our group to improve the utility of these algorithms; non-locality is a bane for numerical differentiation and efforts to improve the near sightedness of this method are ongoing.

References

[1] E. R. Davidson, *J. Math. Phys.* 10, 725 (1969).

[2] M. Taut, *Phys. Rev. A* 48, 3561 (1993).

[3] A. Alavi, *J. Chem. Phys.* 113, 7735 (2000).

[4] D. C. Thompson, and A. Alavi, *Phys. Rev. B* 66, 235118 (2002).

[5] J. Jung, and A. Avarellos, *J. Chem. Phys.* 118, 10825 (2003).

[6] D. C. Thompson, and A. Alavi, *J. Chem. Phys.* 122, 124107 (2005).

[7] P. Hohenberg, and W. Kohn, *Phys. Rev.* 136, B864 (1964).

[8] W. Kohn, and L. J. Sham, *Phys. Rev.* 140, A1133 (1965).

[9] Density Functional Repository as developed and distributed by the Quantum Chemistry Group, CCLRC Daresbury Laboratory, Daresbury, Cheshire, WA4 4AD United Kingdom.

[10] V. N. Staroverov, G. E. Scuseria, J. Tao, and J. P. Perdew, *J. Chem. Phys.* 119, 12129 (2003).

[11] A. D. Becke, *J. Chem. Phys.* 122, 064101 (2005).

[12] J. P. Perdew, A. Ruzsinszky, J. M. Tao, V. N. Staroverov, and G. E. Scuseria, *J. Chem. Phys.* 123, 062201 (2005).

[13] P. Mori-Sanchez, A. J. Cohen, and W. Yang, *J. Chem. Phys.* 124, 091102 (2006).

[14] R. M. Martin, *Electronic Structure, Basis Theory, and Practical Methods*, Cambridge University Press, Cambridge, (2004).

[15] P. W. Ayers, and W. Yang, in *Computational Medicinal Chemistry for Drug Discovery*, Marcel Dekker, New York, (2004).

[16] J. C. Dyson, and M. von Itzstein, *Aus. J. Chem.* 54, 663 (2001).

[17] H. Hövel, L. S. Johansson, and B. Reihl in *Metal clusters at Surfaces: Structure, Quantum Properties, Physical Chemistry*, Springer Verlag, Berlin, (2000).

[18] R. S. Berry, and H. Haberland, in *Clusters of Atoms and Molecules*, Springer Verlag, Berlin, (1994).

[19] I. L. Garzón, J. A. Reyes-Nava, J. I. Rodríguez-Hernández, I. Sigal, M. R. Beltrán, and K. Michaelian, *Phys. Rev. B* 66, 073403 (2002).

[20] I. L. Garzón, M. R. Beltrán, G. González, I. Gutíerrez-González, K. Michaelian, J. A. Reyes-Nava, and J. I. Rodríguez-Hernández, *Eur. Phys. J. D* 24, 105 (2003).

[21] R. G. Parr, and W. Yang, in *Density-Functional Theory of Atoms and Molecules*, Oxford University Press, Oxford, (1989).

[22] W. Yang, *Phys. Rev. Lett.* 66, 1438 (1991).

[23] A. P. Horsfield, A. M. Bratkovsky, M. Fearn, D. G. Pettifor, and M. Aoki, *Phys. Rev. B* 53, 12694 (1996).

[24] W. Yang, *Phys. Rev. Lett.* 56, 9294 (1997).

[25] J. N. Kim, F. Mauri, and G. Galli, *Phys. Rev. B* 52, 1640 (1995).

[26] L. Xi, W. Nunes, and D. Vanderbilt, *Phys. Rev. B* 47, 10891 (1993).

[27] D. Vanderbilt, *Phys. Rev. B,* 41, 7892 (1990).

[28] M. Payne, M. Teter, D. Allan, T. Arias, and J. Joannopoulos, *Rev. Mod. Phys.* 64, 1045 (1992).

[29] J. Hutter, P. Lüthi, and M. Parrinello, *Comput. Matter. Sci.* 2, 244 (1994).

[30] A. Szabo, and N. S. Ostlund, *Modern Quantum Chemistry*, McGraw-Hill, New York, (1989).

[31] A. J. Cohen, and N. C. Handy, *J. Chem. Phys.* 117, 1470 (2002).

[32] P. Pulay, *Chem. Phys. Lett.* 73, 393 (1980).

[33] P. Pulay, *J. Comput. Chem.* 3, 556 (1982).

[34] T. P. Hamilton, and P. Pulay, *J. Chem. Phys.* 84, 5728 (1986).

[35] G. Lippert, J. Hutter, and M. Parrinello, *Mol. Phys.* 92, 477 (1997).

[36] J. VandeVondele, M. Krack, F. Mohamed, M. Parrinello, T. Chassaing, and J. Hutter, *Comput. Phys. Commun.* 167, 103 (2005).

[37] M. Krack, and M. Parrinello, *Phys. Chem. Chem. Phys.* 2, 2105 (2000).

[38] G. Lippert, J. Hutter, and M. Parrinello, *Theor. Chem. Acc.* 103, 124 (1999).

[39] M. Challacombe, *Comput. Phys. Commun.* 128, 93 (2000).

[40] T. L. Beck, *Rev. Mod. Phys.* 72, 1041 (2000).

[41] K. Cho, T. A. Arias, J. D. Joannopoulos, and P. K. Lam, *Phys. Rev. B* 52, R5459 (1995).

[42] S. Q. Wei, and M. Y. Chou, *Phys. Rev. Lett.* 76, 2650 (1996).

[43] S. R. White, J. Wilkins, and M. P. Teter, *Phys. Rev. B* 39, 5819 (1989).

[44] E. Tsuchida, and M. Tsukada, *Solid State Commun.* 94, 5 (1995).

[45] J. Bernholc, J.-Y. Yi, and D. J. Sullivan, *Faraday Discuss.* 92, 217 (1991).

[46] E. J. Bylaska *et al.*, in *Proceedings of the 7th SIAM Conference on Parallel Processing for Scientific Computing*, SIAM, San Francisco, (1995).

[47] N. A. Modine, G. Zumbach, and E. Kaxiras, *Phys. Rev. B* 55, 10289 (1997).

[48] U. V. Waghmare, H. Kim, I. J. Park, N. Modine, P. Maragakis, and E. Kaxiras, *Comput. Phys. Commun.* 137, 341 (2001).

[49] F. Gygi, *Europhys. Lett.* 19, 617 (1992).

[50] F. Gygi, *Phys. Rev. B* 48, 11692, (1993).

[51] F. Gygi, *Phys. Rev. B* 51, 11190, (1995).

[52] D. R. Hamann, *Phys. Rev. B* 51, 7337 (1995).

[53] A. Devenyi, K. Cho, T. A. Arias, and J. D. Joannopoulos, *Phys. Rev. B* 49, 13373 (1994).

[54] F. Gygi, and G. Galli, *Phys. Rev. B* 52, R2229 (1995).

[55] David C. Thompson, Juan I. Rodriguez, James S. M. Anderson, Jordan Thomson, and Paul W. Ayers *In Preparation*.

[56] Y. Xu, *Common Zeros of Polynomials in Several Variables and Higher Dimensional Quadrature*, Wiley, New York, (1994).

[57] D. Rees, and G. G. Hall, *Int. J. Quantum. Chem.* 102, 19 (2005).

[58] S. A. Smolyak, *Soviet Math. Dokl.* 4, 240 (1963).

[59] D. C. Thompson, and P. W. Ayers, *Int. J. Quantum Chem.* 106, 787 (2006).

[60] P. Constans and R. Carbó, J. *Chem. Inf. Sci.* 35, 1046 (1995).

[61] *http://www.molspaces.com/*

In: Quantum Chemistry Research Trends
Editor: Mikas P. Kaisas, pp. 147-172
ISBN 978-1-60021-620-6

Chapter 4

A NOVEL APPROACH TO QUANTUM DOT STRUCTURE CALCULATION

J. B. Wang and R. D. Muhandiramge
School of Physics, The University of Western Australia, Perth 6907, Australia

Abstract

With the increased sophistication of semiconductor technology, it is now possible to create designer nanostructures named as quantum dots which display many of the characteristic properties normally associated with atoms and molecules. The ability to manipulate the size and shape of these quantum dots has opened up a wide range of possibilities and areas for examination. Transitions never observed in the spectra of natural atoms can be obtained from the artificial ones. Consequently, quantum dots may be used to build more efficient and precisely controlled lasers with otherwise inaccessible wavelengths, as well as vital components of nanoelectronic devices. It is also hoped that quantum dots may one day be able to help realize the dream of quantum computing.

Since these structures are made in the nano-meter regime, conventional classical or semi-classical theories and methods no longer apply. A pressing challenge to theoreticians is to provide an accurate description of the electronic properties of such nanostructures based on fully quantum mechanical theories. This chapter describes a newly developed method, the spin adapted configuration interaction method, for calculating multi-electron quantum dot structures. This method takes into account all perceivable interactions and full spin correlation effects, giving a level of accuracy not available in alternative approaches such as the density functional theory or the self-consistent Hartree-Fock method. It also has an advantage over the standard configuration interaction method using Slater determinants, in that a much smaller basis is needed since the spin eigenfunctions already have the required symmetry. The results presented in this paper, where the matrix elements are evaluated analytically and thus exactly, provide a benchmark against which one can test other approximate methodologies developed to study more complex systems.

PACS: 73.21.La, 71.15.Ap, 02.70.-c, 31.15.-p

1. Introduction

Quantum dots are artificially fabricated "atoms", in which charge carriers are confined in all three dimensions just like electrons in real atoms. Consequently, they exhibit similar properties normally associated with real atoms, such as quantised energy levels and shell structures, and are being investigated for possible uses in lasers, electronic appliances, and quantum computation. Their behaviours and properties are described by electrons' wavefunctions whose evolution is governed by the Schrödinger equation and the Pauli exclusion principle.

One of the first theoretical studies of quantum dots came from Pfannkuche et al. [1], who performed Hartree-Fock self-consistent calculations for a two-electron quantum dot (i.e. an artificial helium), and compared with results obtained from direct numerical digitalisation of the two-particle Hamiltonian. The confining potential was modelled as a circularly symmetric parabolic function. They obtained good agreement between these two calculations for the triplet state, but markedly different results for the singlet state. This indicates that important spin correlations were not properly treated in the Hartree-Fock model. Yannouleas et al. [2, 3] also studied circularly symmetric quantum dots using an unrestricted spin-space Hartree-Fock approach, however as the authors mentioned, it was not clear whether the symmetry breaking shown in their result was real or an artefact of the mean-field approximation. McCarthy, Wang and Abbott [4] developed a Hartree-Fock *Mathematica* package, which provided similar results as the earlier studies.

There are other theoretical approaches developed to study the electronic structure of quantum dots. For example, Macucci et al. [5, 6, 7] studied quantum dots with up to 24 electrons using a mean-field local-density-functional approach, in which the exchange and correlation potential was approximated by an empirical polynomial expression. Lee et al. [8] also studied an N-electron quantum dot using the density functional theory, where the generalized-gradient approximation was used for exchange-correlation potentials. Exchange interaction comes directly from the anti-symmetrisation of wavefunctions as required by the Pauli's exclusion principle. In the density function theory, this is a major problem since the mathematical object is electron density rather than electron wavefunction, making evaluation of the exchange interaction intrinsically difficult.

Ezaki et al. [9, 10], Eto [11, 12], and Reimann, et al. [13, 14] applied a brute force approach by numerically diagonalising the N-electron Hamiltonian, using Slater determinants composed of single-electron eigenstates as the basis functions. This approach, namely the configuration interaction (CI) method, takes into account the full interaction and correlation of the electrons in the system as long as the numerical results converge with increasing number of basis functions. However, such an approach involves the calculation of a very large number of interaction integrals and the inversion of large matrixes, which can be prohibitively expensive in terms of computer resources. Reimann et al. had employed matrices of dimensions up to 108,375 with 67,521,121 nonzero elements for a six-electron quantum dot. Calculations for any higher number of electrons in the system were not considered numerically possible using the conventional CI formalism [13, 14].

In this paper, we employ a spin-adapted basis to calculate the energy structure of an N-electron quantum dot. Similar approaches have been developed in quantum chemistry but not yet applied in mesoscopic Physics. An in-depth analysis of the methodology can be found in [15]. By isolating spin eigenstates of the system, which in turn restrict the possible form of the spatial wavefunctions, we are able to reduce significantly the size of the configuration interaction matrices in comparison with the standard CI methods. This has allowed us to investigate quantum dot systems with electron numbers greater than six using PCs and *Mathematica*.

In addition, we have developed a novel approach to the simplification of the interaction integrals that arise in the calculation. This approach transforms the two-electron interaction integrals of four variables into a sum of one-variable integrals, which can be evaluated algebraically. We also present a scheme in which the most computational intensive step of calculating the interaction matrix is carried out once only. The electronic structure of quantum dots with different well parameters and different magnetic field can be worked out by rescaling the interacting and non-interacting terms in the Hamiltonian matrix. This provides enormous saving in computing time and allows a systematic study of a wide range of experimental settings without having to recalculate the entire Hamiltonian.

Material units are used throughout this paper, in which the quantities $\hbar = m^* = e = a_0^* = 1$, where m^* is the effective mass of the electron, e is the unit of elementary charge, $\hbar$ is the fundamental unit of angular momentum, and a_0^* is the effective Bohr radius defined by the formula $a_0^* = \frac{4\pi\varepsilon\hbar^2}{m^*e^2}$. The value of calculated results in other units, for example the S.I. units, can be readily obtained by dimensional arguments at the completion of the calculations.

2. An Orthonormal Spin-Adapted Basis

2.1. Construction of Spin Eigenfunctions

The projection of the spin angular momentum operator for an N-electron system is given by

$$\hat{S}_z = \sum_{i=1}^{N} \hat{S}_z(i), \tag{1}$$

where $\hat{S}_z(i)$ is the projection of the spin operator for the ith electron. The eigenfunctions of $\hat{S}_z$ are products of the single electron eigenfunctions, namely

$$\theta_k = \sigma(1)\sigma(2)\ldots\sigma(N), \tag{2}$$

where $\sigma(i) = \alpha(i)$ or $\beta(i)$ represents spin up or spin down of the ith electron, respectively. We term θ_k the elementary spin eigenfunctions. The eigenvalue for a particular θ_k is simply the sum of the eigenvalues of the individual spin operators. If we define μ as the number of α and ν as the number of β in θ_k, then

$$\hat{S}_z\theta_k = \frac{(\mu - \nu)\hbar}{2}\theta_k, \tag{3}$$

where $\mu + \nu$ is equal to N, the number of electrons in the system.

For a given number of electrons, any linear combination of eigenfunctions θ_k with the same eigenvalues is also an eigenfunction of $\hat{S}_z$ with the same eigenvalue. It is easy to verify that a complete set of eigenfunctions for the particular eigenvalue $(\mu-\nu)\hbar/2$ is given by all possible permutations of the eigenfunction $\theta_1 = \alpha(1)\alpha(2)\ldots\alpha(\mu)\beta(\mu+1)\beta(\mu+2)\ldots\beta(\mu+\nu)$. There are $\binom{N}{\mu}$ of these functions, where $\binom{N}{\mu} = \frac{N!}{\mu!(1-\mu)!}$ is the binomial coefficient, and they form a complete orthonormal basis for eigenfunctions of $\hat{S}_z$ with that particular eigenvalue.

The elementary spin eigenfunctions are generally not eigenfunctions of $\hat{S}^2$, but linear combinations of elementary spin eigenfunctions can be constructed to be simultaneously eigenfunctions of $\hat{S}^2$. In the N-electron case, the $\hat{S}^2$ operator has the following form [15]:

$$\hat{S}^2 X = \sum_{i<j} P_{ij} X + \frac{N}{4}(4-N)X, \tag{4}$$

where P_{ij} is a permutation operator and $X = \sum_k c_k \theta_k$ are simultaneous eigenfunctions of both $\hat{S}^2$ and $\hat{S}_z$.

Let $X = \sum_k c_k \theta_k(N, M)$ be a linear combination of all elementary spin eigenfunctions of $\hat{S}_z$ for a given number of electrons N and the projection of total spin quantum number $M = (\mu-\nu)/2$. Let $\vec{v}$ be the corresponding vector of coefficients, i.e. $\vec{v} = (c_1, c_2, \ldots, c_{\binom{N}{\mu}})$. Define the matrix $\{\hat{S}^2\}$ by $\{\hat{S}^2\}_{ij} = \langle\theta_i|\hat{S}^2|\theta_j\rangle$, whose elements can be calculated using Eq. (4). Then the action of $\hat{S}^2$ on X can be represented by the matrix operation $\{\hat{S}^2\}.\vec{v}$. The eigenfunctions of $\hat{S}^2$ are then found by solving the matrix eigenvalue problem.

As an example, the calculation of the spin eigenfunctions for three electrons with $M = \frac{1}{2}$ is presented. In this case the possible elementary spin eigenfunctions are $\theta_1 = \alpha\alpha\beta$, $\theta_2 = \alpha\beta\alpha$ and $\theta_3 = \beta\alpha\alpha$.

$$\begin{aligned}
\hat{S}^2 |\theta_1\rangle &= \sum_{i<j} P_{ij} |\theta_1\rangle + \frac{N}{4}(4-N)|\theta_1\rangle \\
&= P_{12}|\alpha\alpha\beta\rangle + P_{13}|\alpha\alpha\beta\rangle + P_{23}|\alpha\alpha\beta\rangle + \frac{3}{4}(4-3)|\alpha\alpha\beta\rangle \\
&= |\alpha\alpha\beta\rangle + |\beta\alpha\alpha\rangle + |\alpha\beta\alpha\rangle + \frac{3}{4}|\alpha\alpha\beta\rangle \\
&= \frac{7}{4}|\theta_1\rangle + |\theta_2\rangle + |\theta_3\rangle.
\end{aligned} \tag{5}$$

Similar calculations lead to the matrix elements of $\{\hat{S}^2\}$:

$$\{\hat{S}^2\} = \begin{pmatrix} \frac{7}{4} & 1 & 1 \\ 1 & \frac{7}{4} & 1 \\ 1 & 1 & \frac{7}{4} \end{pmatrix}. \tag{6}$$

The eigenvalues of the above matrix are $\frac{15}{4} = \frac{3}{2}(\frac{3}{2}+1)$ and $\frac{3}{4} = \frac{1}{2}(\frac{1}{2}+1)$ which corresponds to the total angular momentum quantum numbers $S = \frac{3}{2}$ and $S = \frac{1}{2}$, respectively. The orthonormalised spin eigenfunctions are obtained by first solving for the eigenvectors and then applying the Gram-Schmidt orthonormalisation procedure. These are:

$$X_1(3, \frac{3}{2}, \frac{1}{2}) = \frac{1}{\sqrt{3}}(\alpha\alpha\beta + \alpha\beta\alpha + \beta\alpha\alpha), \tag{7}$$

$$X_1(3, \frac{1}{2}, \frac{1}{2}) = \frac{1}{\sqrt{2}}(-\alpha\alpha\beta + \beta\alpha\alpha), \tag{8}$$

$$X_2(3, \frac{1}{2}, \frac{1}{2}) = \frac{1}{\sqrt{6}}(-\alpha\alpha\beta + 2\alpha\beta\alpha - \beta\alpha\alpha). \tag{9}$$

Here $X_k(N, S, M)$ are termed as spin eigenfunctions as opposed to the elementary spin eigenfunctions and k is a positive integer denoting different eigenfunctions in a multidimensional spin eigenspace.

2.2. Spin Adapted Basis

Next we need to include the spatial wavefunctions to form an orthonormal and properly antisymmetrised basis. The basis elements have the form $C_\Phi A \Phi X_k(N, S, M)$, where C_Φ is a normalisation constant, $A = \frac{1}{\sqrt{N!}}(-1)^p P$ is the antisymmetriser, $X_k(N, S, M)$ is the spin eigenfunction, $\Phi = \phi_1\phi_2 \ldots \phi_N$ is a representative spatial wavefunction, and ϕ are single-electron orbitals (i.e. eigenfunctions of the single-electron Hamiltonian). We occasionally abbreviate $X_k(N, S, M)$ to X_k, when the values of N, S and M are fixed in a given calculation. Different elements of this basis have different spatial and spin combinations.

Note that both the spin and spatial variables are permuted in $C_\Phi A \Phi X_k$ as required by the antisymmetry principle. We also note that any product spatial wavefunction with three or more orbitals that are the same will vanish when multiplied by a spin eigenfunction and antisymmetrised. This implies the greatest number of orbitals that can be the same in a representative spatial wavefunction is two, which we call a doubly occupied orbital.

In the following, we demonstrate that by using the spin eigenfunctions that are also eigenfunctions of the permutation operators $P_{2i-1,2i}$ for $i = 1, 2, \ldots$ where $i \leq \frac{N}{2}$ and spatial wavefunctions that only have doubly occupied orbitals sequentially at positions $(2i-1, 2i)$ for $i = 1, 2, \ldots$ where $i \leq \frac{N}{2}$, we can construct a orthogonal basis in which vanishing elements can be easily identified and the normalisation constant is of a simple form. This is similar to the Serber construction [16]. Following the nomenclature of Salmon and Ruedenberg [17], as also adopted by Pauncz [15], we will call a pair of numbers of the form $(2i, 2i-1)$ a geminal pair. Correspondingly, we call the set of permutations mentioned in the previous paragraph the geminal permutations.

Since the S^2 operator commutes with any permutation and the geminal transposition operators pairwise commute, it is possible to construct a complete set of orthonormal spin eigenfunctions that commute with all of these operators. To compute this basis we first obtain the orthonormal spin eigenfunctions, $X^0 = \{X_1^0, \ldots, X_f^0\}$, as in the previous section. We then use these to compute the representation matrix $U(P_{12})$. The representation matrix for any permutation P is defined by

$$P\, X_k(N, S, M) = \sum_{i=1}^{f} U(P)_{ik} X_i(N, S, M). \tag{10}$$

It can be shown that for any two permutations R and P,

$$U(RP) = U(R)U(P), \tag{11}$$

satisfying the condition for a representation of the symmetric group. This is independent of the particular spin basis chosen.

Note that $X_k^0(N, S, M)$ are not, in general, eigenfunctions of P_{12}. However, the orthonormalised eigenvectors of $U(P_{12})$ will give the linear combination of vectors of X^0 that form a new orthonormal basis, $X^1 = \{X_1^1, \ldots, X_f^1\}$, for the spin space which are also eigenfunctions of P_{12}.

For example, we can start with the three electron spin eigenfunctions with $S = \frac{1}{2}$ and $M = \frac{1}{2}$ given by Eqs. (8) and (9), and obtain

$$\begin{aligned} P_{12}X_1^0(3, \frac{1}{2}, \frac{1}{2}) &= \frac{1}{2}X_1^0(3, \frac{1}{2}, \frac{1}{2}) + \frac{\sqrt{3}}{2}X_2^0(3, \frac{1}{2}, \frac{1}{2}), \\ P_{12}X_2^0(3, \frac{1}{2}, \frac{1}{2}) &= \frac{\sqrt{3}}{2}X_1^0(3, \frac{1}{2}, \frac{1}{2}) - \frac{1}{2}X_2^0(3, \frac{1}{2}, \frac{1}{2}). \end{aligned} \tag{12}$$

Therefore

$$U(P_{12}) = \begin{pmatrix} \frac{1}{2} & \frac{\sqrt{3}}{2} \\ \frac{\sqrt{3}}{2} & \frac{-1}{2} \end{pmatrix}. \tag{13}$$

The eigenvalues of the above matrix are ± 1 corresponding to the eigenvectors $\left(\frac{\sqrt{3}}{2}, \frac{1}{2}\right)$ and $\left(\frac{-1}{2}, \frac{\sqrt{3}}{2}\right)$. This means that the new spin eigenfunction basis is:

$$\begin{aligned} X_1^1(3, \frac{1}{2}, \frac{1}{2}) &= \frac{\sqrt{3}}{2}X_1^0(3, \frac{1}{2}, \frac{1}{2}) + \frac{1}{2}X_2^0(3, \frac{1}{2}, \frac{1}{2}) \\ &= \frac{1}{\sqrt{6}}\left(-2\alpha\alpha\beta + \alpha\beta\alpha + \beta\alpha\alpha\right), \end{aligned} \tag{14}$$

$$\begin{aligned} X_2^1(3, \frac{1}{2}, \frac{1}{2}) &= \frac{-1}{2}X_1^0(3, \frac{1}{2}, \frac{1}{2}) + \frac{\sqrt{3}}{2}X_2^0(3, \frac{1}{2}, \frac{1}{2}) \\ &= \frac{1}{\sqrt{2}}(\alpha\beta\alpha - \beta\alpha\alpha). \end{aligned} \tag{15}$$

The basis elements are now eigenfunctions of P_{12}. Note that the first spin eigenfunction is symmetric in the first two electrons, while the second is antisymmetric. This means that the first will vanish if we multiply it by a spatial wavefunction in which the first two electrons are in the same orbital and then antisymmetrise. This will not be the case for the second spin eigenfunction, thus we are allowed to combine it with spatial wavefunctions in which the first two orbitals are the same.

The symmetry properties of the spin eigenfunctions dictate with which spatial orbitals it can be combined. This is a general feature that is developed in the next section. Specifically we will see that the choice of spin eigenfunctions that are simultaneous eigenfunctions of geminal transpositions allows us to restrict the spatial representative wavefunctions we need to consider to those with the doubly occupied orbitals in the geminal positions.

If we have more than three electrons, we can use X^1 to calculate the representation matrix $U(P_{34})$, which creates a new basis X^2. This process continues till we use all possible geminal transposition operators. The pairwise commutative property of the geminal transposition operators guarantees that subsequent bases will still be eigenfunctions of earlier geminal transpositions after each transformation. After this process, we will have produced a spin adapted basis, $X = \{X_1, \ldots, X_f\}$, where the spin eigenfunctions are simultaneous eigenfunctions of all the geminal transposition operators.

There are other construction methods for, such as the Serber construction [16], which produce bases with the same properties. In the Serber construction, spin eigenfunctions are constructed by adding two electron singlet and triplet spin eigenfunctions using angular momentum addition formula and the Clebsch-Gordan coefficients. Degenerate eigenfunctions can be traced using the Serber path symbol [18, 15]. For an odd number of electrons, the last electron can be added using the one electron addition formulas. This produces a orthonormal spin eigenspace with the spin eigenfunctions also being eigenfunctions of the geminal transposition operators.

As the calculation of the spin eigenfunctions in our studies takes only a small proportion of the computational effort compared to the calculation of the Hamiltonian matrix, the above matrix eigenvalue method was chosen for simplicity of implementation. The Serber method may prove useful for systems with larger numbers of electrons.

3. Properties of the Spin Adapted Basis

In this section, we discuss the general properties of the spin adapted basis $X = \{X_1, \ldots, X_f\}$. Each X_k is a simultaneous eigenfunction of $\hat{S}^2$ and the geminal transposition operators. We also derive the formula for the normalisation constant C_Φ.

We are restricting the position of any doubly occupied orbitals in the spatial wavefunctions to be sequentially in the geminal positions. Also, two representative wavefuctions for different basis elements are never non-invariant permutations of each other. With this restriction, we show that we can construct an orthonormal basis. The following two theorems are used in proving the properties of the spin adapted basis.

Theorem 3.1. *Let $\Phi = \phi_1\phi_2\ldots\phi_n$ with $\phi_{2i-1} = \phi_{2i}$, that is there is a doubly occupied orbital. Then if $A\Phi X_k \neq 0$ we have $P_{2i-1,2i}X_k = -X_k$ and $U(P_{2i-1,2i})_{kk} = -1$.*

Proof. By definition we have:

$$\begin{aligned} A\Phi X_k &= \frac{1}{\sqrt{N!}}\sum_P (-1)^p P^r \Phi \; P^\sigma X_k, \\ &= \frac{1}{\sqrt{N!}}\sum_Q (-1)^{qp_{2i-1,2i}} Q^r P^r_{2i-1,2i}\Phi \; Q^\sigma P^\sigma_{2i-1,2i} X_k. \end{aligned}$$

We also know $P_{2i-1,2i}\Phi = \Phi$ as the orbitals at $2i-1$ and $2i$ are the same. Therefore,

$$\begin{aligned} A\Phi X_k &= \frac{1}{\sqrt{N!}}\sum_Q (-1)^q(-1)^{p_{2i-1,2i}} Q^r \Phi Q^\sigma (P^\sigma_{2i-1,2i} X_k), \\ &= \frac{1}{\sqrt{N!}}\sum_Q -(-1)^q Q^r \Phi\, Q^\sigma \sum_{m=1}^{f} U(P_{2i-1,2i})_{mk} X_m. \end{aligned}$$

Rearranging over the summation gives us:

$$\begin{aligned} A\Phi X_k &= \sum_{m=1}^{f} -U(P_{2i-1,2i})_{mk} \frac{1}{\sqrt{N!}}\sum_Q (-1)^q Q^r \Phi\, Q^\sigma X_m, \\ &= \sum_{m=1}^{f} -U(P_{2i-1,2i})_{mk} A\Phi X_m. \end{aligned}$$

Since X_m are eigenfunctions of $P_{2i-1,2i}$, we get:

$$\begin{aligned} A\Phi X_k &= \sum_{m=1}^{f} - \langle X_m | P_{2i-1,2i} | X_k \rangle A\Phi X_m, \\ &= \sum_{m=1}^{f} - \langle X_m | \lambda_m | X_k \rangle A\Phi X_m, \\ &= \sum_{m=1}^{f} -\lambda_m \delta_{m,k} A\Phi X_m, \\ &= -\lambda_k A\Phi X_k. \end{aligned}$$

As $A\Phi X_k \neq 0$ we must have $\lambda_k = -1$ and therefore $P_{2i-1,2i}X_k = -X_k$.

□

The above theorem is simply a result of the antisymmetry principle. As the spatial wavefunction is symmetric under $P_{2i-1,2i}$, the spin eigenfunction must be antisymmetric to compensate. The condition $A\Phi X_k \neq 0$ is included because we can not have zero elements in the basis if we want a well constructed matrix eigenvalue problem.

If the spatial wavefunction Φ has d pairs of doubly occupied orbitals then the invariance group S_Φ is given by all products of transpositions in the set $\{P_{1,2}, P_{3,4}, \dots P_{2d-1,2d}\}$. As each of these elements is of order 2, as $P^2_{i,j}$ is the identity permutation, this set has 2^d elements given by $P = P^{n_1}_{1,2} P^{n_2}_{3,4} \dots P^{n_d}_{2d-1,2d}$ where $n_i = 0$ or 1 , $i = 1, .., d$.

Theorem 3.2. *Let $\Phi = \phi_1\phi_2\dots\phi_n$ with $\phi_1 = \phi_2, \dots, \phi_{2d-1} = \phi_{2d}$ where $d = \Pi(\Phi)$ is the number of doubly occupied orbitals in Φ. Then if $P \in S_\Phi = \{P \in S_N : P\Phi = \Phi\}$, with $A\Phi X_k \neq 0$ we have $U(P)_{kk} = (-1)^p$ where $(-1)^p$ is the parity of the permutation P.*

Proof. Now if $P \in S_\Phi$ then $P = P_{1,2}^{n_1} P_{3,4}^{n_2} \ldots P_{2d-1,2d}^{n_d}$ where $n_i = 0$ or 1 , $i = 1, .., d$. Thus using the multiplicative property of the representation matrices:

$$\begin{aligned} U(P) &= U(P_{1,2}^{n_1}) U(P_{3,4}^{n_2}) \ldots U(P_{2d-1,2d}^{n_d}) \\ &= U(P_{1,2})^{n_1} U(P_{3,4})^{n_2} \ldots U(P_{2d-1,2d})^{n_d}, \end{aligned}$$

where $U(P)^0$ is the identity matrix. The $U(P_{2i-1,2i})$ are diagonal, so

$$\begin{aligned} U(P)_{kk} &= U(P_{1,2})_{kk}^{n_1} U(P_{3,4})_{kk}^{n_2} \ldots U(P_{2d-1,2d})_{kk}^{n_d} \\ &= (-1)^{n_1} (-1)^{n_2} \ldots (-1)^{n_d} \\ &= (-1)^{n_1+n_2+\ldots+n_d}, \end{aligned} \tag{16}$$

where we have used $U(P_{2i-1,2i})_{kk} = -1$ [Theorem 3.1]. The parity of P is given by $(-1)^p = (-1)^{n_1+n_2+\ldots+n_d}$, so $U(P)_{kk} = (-1)^p$ which is the result we require. □

Theorem 3.3. *The basis functions $A\Phi X_j$ and $A\Psi X_k$ are orthogonal if $\Phi \neq P\Psi$ for all permutations P.*

Proof. As the Antisymmetriser is Hermitian and proportional to its square[15], we can simplify the inner product of basis elements

$$\begin{aligned} \langle A\Phi X_j | A\Psi X_k \rangle &= \langle \Phi X_j | A A^\dagger \Psi X_k \rangle \\ &= \langle \Phi X_j | A^2 \Psi X_k \rangle \\ &= \langle \Phi X_j | \sqrt{N} A \Psi X_k \rangle \\ &= \sum_P (-1)^p \langle \Phi | P\Psi \rangle \langle X_j | P X_k \rangle \\ &= \sum_P (-1)^p \langle \Phi | P\Psi \rangle U(P)_{jk} \\ &= 0, \end{aligned} \tag{17}$$

since $\langle \Phi | P\Psi \rangle = 0 \, \forall P \in S_N$. □

This means that the basis elements are orthogonal for different spatial wavefunctions. Note that the basis does not include spatial wavefunctions that are non-invariant permutations of each other. The only case where two basis elements would have the same spatial wavefunction is when it is multiplied by a different spin eigenfunction. This case is dealt with by the next theorem.

Theorem 3.4. *The basis elements $A\Phi X_j$ and $A\Phi X_k$ are orthogonal if $A\Phi X_j \neq 0$ and $A\Phi X_k \neq 0$. That is, for $j \neq k$, their inner product is 0. Furthermore for $j = k$ their inner product is 2^d, where d is the number of pairs of doubly occupied orbitals in Φ.*

Proof. Since

$$\langle A\Phi X_j | A\Phi X_k \rangle = \sum_P (-1)^p \langle \Phi | P\Phi \rangle U(P)_{jk} \tag{18}$$

and

$$\langle \Phi | P\Phi \rangle = \begin{cases} 1, & P \in S_\Phi \\ 0, & \text{otherwise} \end{cases} \tag{19}$$

Eq. (18) transforms to

$$\langle A\Phi X_j | A\Phi X_k \rangle = \sum_{P \in S_\Phi} (-1)^p U(P)_{jk}. \tag{20}$$

If $P \in S_\Phi$ then $P = P_{1,2}^{n_1} P_{3,4}^{n_2} \ldots P_{2d-1,2d}^{n_d}$ where $n_i = 0$ or 1 , $i = 1, .., d$ and $d = \Pi(\Phi)$ is the number of doubly occupied orbitals in Φ.

If $j \neq k$ then

$$\langle A\Phi X_j | A\Phi X_k \rangle = 0 \tag{21}$$

as $U(P)_{jk}$ is diagonal for $P \in S_\Phi$.

If $j = k$ then

$$\begin{aligned} \langle A\Phi X_k | A\Phi X_k \rangle &= \sum_{P \in S_\Phi} (-1)^p U(P)_{kk} \\ &= \sum_{P \in S_\Phi} (-1)^p (-1)^p \\ &= |S_\Phi| \\ &= 2^d, \end{aligned} \tag{22}$$

where we have used theorem 3.2, which states $U(P)_{kk} = (-1)^p$ when $A\Phi X_k \neq 0$ and $P \in S_\Phi$. Thus $C_\Phi A\Phi X_k$ is a properly normalised basis functions with $C_\Phi = \frac{1}{\sqrt{2^d}}$. □

Theorem 3.5. *The basis function $A(P^r\Phi)X_k$ is a linear combination of the basis functions $A\Phi X_i$ for $i = 1, \ldots, f$ where f is the dimension of the spin eigenfunction space.*

Proof. Let P^r be the part of permutation P that acts on the spatial component of the basis function and P^σ be the part that acts on the spin component.

$$\begin{aligned}
A(P^r\Phi)X_k &= \frac{1}{\sqrt{N!}}\sum_Q(-1)^q Q^r P^r\Phi\, Q^\sigma X_k \\
&= \frac{1}{\sqrt{N!}}\sum_T(-1)^q T^r\Phi\, T^\sigma (P^{-1})^\sigma X_k \\
&= \frac{1}{\sqrt{N!}}\sum_T(-1)^q T^r\Phi\, T^\sigma \sum_{m=1}^{f} U(P^{-1})_{mk} X_m \\
&= \sum_{m=1}^{f} U(P^{-1})_{mk}\frac{1}{\sqrt{N!}}\sum_T(-1)^t(-1)^{p^{-1}} T^r\Phi\, T^\sigma X_m \\
&= \sum_{m=1}^{f} (-1)^p U(P^{-1})_{mk}\frac{1}{\sqrt{N!}}\sum_T(-1)^t T^r\Phi\, T^\sigma X_m \\
&= \sum_{m=1}^{f} (-1)^p U(P^{-1})_{mk} A\Phi X_m, \qquad (23)
\end{aligned}$$

noting that $T = QP$. Thus $A(P\Phi)X_k$ is a linear combination of the basis functions $A\Phi X_m$ with the coefficients given by $(-1)^p U(P^{-1})_{mk}$. □

This means that the spatial wavefunctions from two different basis elements are never non-invariant permutations of each other. Therefore if representative spatial wavefunctions have doubly occupied orbitals, we can choose to put the doubly occupied orbitals sequentially in the geminal positions. Thus if Φ has d doubly occupied orbitals they will be in the positions $(1,2), (3,4)\ldots(2d-1, 2d)$. For example, we would use $\phi_1\phi_1\phi_2\phi_2\phi_3$ but not $\phi_1\phi_2\phi_1\phi_3\phi_2$.

To construct a basis for the matrix eigenvalue problem for N electron, we first calculate the spin adapted basis. We then choose a set of representative spatial wavefunctions that are products of the single electron energy eigenfunctions, making sure that any doubly occupied orbitals appear first and that none is a non-invariant permutation of another in the set. We then combine the spatial wavefunctions with each spin eigenfunction and antisymmetrise, eliminating any basis elements that are zero. Lastly, we multiply each basis element by the appropriate normalization constant. The accuracy of the final calculation will depend on which spatial wavefunctions are chosen but will improve by increasing basis size.

4. Simplifying the Hamiltonian Elements

We wish to use the basis described in the previous section to calculate the Hamiltonian matrix elements of the form $\langle C_\Phi A\Phi X_j|\hat{H}|C_\Psi A\Psi X_k\rangle$, where $\hat{H} = \hat{H}_0 + \hat{H}_{int}$, $\hat{H}_0 = \sum_{i=1}^{N}\hat{H}_{0i}$ is the single electron component, and $\hat{H}_{int} = \sum_{i<j}^{N}\hat{H}(i,j)$ is the interaction component that acts pairwise. This is the general form of a spin-free Hamiltonian in which spin-orbit and spin-spin interactions are neglected. This gives us,

$$\begin{aligned}
\langle C_\Phi A\Phi X_j|\hat{H}|C_\Psi A\Psi X_k\rangle &= \langle C_\Phi A\Phi X_j|\hat{H}_0+\hat{H}_{int}|C_\Psi A\Psi X_k\rangle \\
&= \langle C_\Phi A\Phi X_j|\hat{H}_0|C_\Psi A\Psi X_k\rangle + \langle C_\Phi A\Phi X_j|\hat{H}_{int}|C_\Psi A\Psi X_k\rangle \qquad (24)
\end{aligned}$$

4.1. Single Electron Integral

Calculation of the single electron integral is straight forward noting that the single electron orbitals are the eigenfunctions of $\hat{H}_{0i}$, that is $\hat{H}_{0i}\Phi(\vec{r_i}) = E_i\Phi(\vec{r_i})$ where E_i is the energy eigenvalue. Therefore,

$$\begin{aligned} \langle C_\Phi A\Phi X_j|\hat{H}_0|C_\Psi A\Psi X_k\rangle &= C_\Phi C_\Psi \sum_P (-1)^p \langle\Phi|\hat{H}_0|P\Psi\rangle\, U(P)_{jk} \\ &= C_\Phi C_\Psi \sum_{i=1}^{N}\sum_P (-1)^p E_{P(i)} \langle\Phi|P\Psi\rangle\, U(P)_{jk}, \quad (25) \end{aligned}$$

where $\hat{H}_{0i}(P\Psi) = E_{P(i)}P\Psi$ and $E_{P(i)}$ is the single electron energy eigenvalue of the ith orbital in $P\Psi$. However, $\langle\Phi|P\Psi\rangle = 0\ \forall P$ unless $\Phi = P\Psi$ for some $P \in S_n$. In the basis construction we only need to use representative wavefunctions that are not permutations of each other. Thus if $\Phi = P\Psi$ for some P then $P \in S_\Phi$, i.e.

$$\langle C_\Phi A\Phi X_j|\hat{H}_0|C_\Psi A\Psi X_k\rangle = 0 \text{ for } \Phi \neq \Psi. \quad (26)$$

Otherwise,

$$\begin{aligned} \langle C_\Phi A\Phi X_j|\hat{H}_0|C_\Phi A\Phi X_k\rangle &= C_\Phi^2 \sum_{i=1}^{N}\sum_{P\in S_\Phi} (-1)^p E_{P(i)} \langle\Phi|\Phi\rangle\, U(P)_{jk} \\ &= C_\Phi^2 \sum_{i=1}^{N} 2^{\Pi(\Phi)} E_i \delta_{jk} \\ &= \delta_{jk}\sum_{i=1}^{N} E_i, \quad (27) \end{aligned}$$

where δ_{jk} is the Kronecker delta. In other words, if the basis functions $A\Phi X_j$ and $A\Psi X_k$ are the same in both the spin and spatial components, i.e. $\Phi = \Psi$ and $j = k$, then $\langle C_\Phi A\Phi X_j|\hat{H}_0|C_\Phi A\Phi X_j\rangle$ is the sum of single electron eigenvalues $\sum_{i=1}^{N} E_i$. Otherwise $\langle C_\Phi A\Phi X_j|\hat{H}_0|C_\Psi A\Psi X_k\rangle = 0$.

4.2. Reduction of the Sum over Permutations in the Interaction Integral

Since the interaction Hamiltonian acts pairwise, i.e. $H_{int} = \sum_{i<j} H(i,j)$ where $H(i,j)$ is the interaction term between the ith and j^{th} electron,

$$\begin{aligned} \langle\Phi|\hat{H}(i,j)|\Psi\rangle &= \langle\phi_1\phi_2\ldots\phi_N|\, \hat{H}(i,j)\, |\psi_1\psi_2\ldots\psi_N\rangle \\ &= \langle\phi_i\phi_j|\, \hat{H}(i,j)\, |\psi_i\psi_j\rangle\, \Pi_{k\neq i,j}^{N} \langle\phi_k|\psi_k\rangle\,. \quad (28) \end{aligned}$$

Thus $\langle\Phi|\hat{H}(i,j)|\Psi\rangle = 0$ unless $\Phi_k = \Psi_k\ \forall k \neq i,j$.

Consequently,

$$
\begin{aligned}
\langle C_\Phi A\Phi X_k|\hat{H}_{int}|C_\Psi A\Psi X_l\rangle &= C_\Phi C_\Psi \sum_P (-1)^p \langle \Phi|\hat{H}_{int}|P\Psi\rangle\, U(P)_{kl} \\
&= C_\Phi C_\Psi \sum_{i<j}\sum_P (-1)^p \langle \Phi|\hat{H}(i,j)|P\Psi\rangle\, U(P)_{kl} \\
&= C_\Phi C_\Psi \sum_{i<j}\sum_P \left((-1)^p \langle \phi_i\phi_j|\hat{H}(i,j)|(P\Psi)_i(P\Psi)_j\rangle \prod_{m\neq i,j} \langle \phi_m|(P\Psi)_m\rangle\, U(P)_{kl} \right),
\end{aligned}
\tag{29}
$$

where we have used the notation that $(P\Psi)_i$ is the ith orbital in $P\Psi$, for example, if $\Psi = \psi_1\psi_2\psi_3$ and $P\Psi = \psi_2\psi_3\psi_1$, then $(P\Psi)_2 = \psi_3$ as ψ_3 is the second orbital in $P\Psi$.

Let $L_{i,j}$ be a permutation operator that aligns Φ and Ψ such that $\phi_k = (L_{i,j}\Psi)_k\ \forall k \neq i,j$. If this permutation does not exist for a particular Φ, Ψ, i and j, then define L_{ij} as 0 and $U[L_{ij}]_{kl} = 0$. For this permutation $\prod_{k\neq i,j} \langle \Phi_k|(L_{i,j}\Psi)_k\rangle = 1$.

This permutation will not be unique as the set of permutations $L_{i,j}Q$ and $P_{i,j}L_{i,j}Q$ for $Q \in S_\Psi$ will also be line-up permutations, where $P_{i,j}$ is the transposition of the electron i and j. These sets are not necessarily distinct as $L_{i,j}Q$ and $P_{i,j}L_{i,j}Q$ will produce the same integral if $(P\Psi)_i = (P\Psi)_j$.

Define ω_i as $(L_{i,j}\Psi)_i$ and ω_j as $(L_{i,j}\Psi)_j$. For all permutations not in $L_{i,j}Q$ or $P_{i,j}L_{i,j}Q$, we have $\prod_{k\neq i,j} \langle \Phi_k|(P\Psi)_k\rangle = 0$. Therefore,

$$
\begin{aligned}
\langle C_\Phi A\Phi X_j|\hat{H}_{int}|C_\Psi A\Psi X_k\rangle &= C_\Phi C_\Psi \sum_{i<j}\sum_{Q\in S_\Psi} 2^{-\delta(\omega_i,\omega_j)} \\
&\quad ((-1)^{l_{i,j}q} U(L_{i,j}Q)_{kl} \langle \phi_i\phi_j|\hat{H}(i,j)|\omega_i\omega_j\rangle + (-1)^{p_{i,j}l_{i,j}q} U(P_{i,j}L_{i,j}Q)_{kl} \langle \phi_i\phi_j|\hat{H}(i,j)|\omega_j\omega_i\rangle) \\
&= C_\Phi C_\Psi \sum_{i<j}\sum_{Q\in S_\Psi} 2^{-\delta(\omega_i,\omega_j)}((-1)^{l_{i,j}}(-1)^q U(L_{i,j})_{km} U(Q)_{ml} \langle \phi_i\phi_j|\hat{H}(i,j)|\omega_i\omega_j\rangle \\
&\quad -(-1)^{l_{i,j}}(-1)^q U(P_{i,j}L_{i,j})_{km} U(Q)_{ml} \langle \phi_i\phi_j|\hat{H}(i,j)|\omega_j\omega_i\rangle) \\
&= C_\Phi C_\Psi \sum_{i<j}\sum_{Q\in S_\Psi} 2^{-\delta(\omega_i,\omega_j)}((-1)^{l_{i,j}}(-1)^q U(L_{i,j})_{kl}(-1)^q \langle \phi_i\phi_j|\hat{H}(i,j)|\omega_i\omega_j\rangle \\
&\quad -(-1)^{l_{i,j}}(-1)^q U(P_{i,j}L_{i,j})_{kl}(-1)^q \langle \phi_i\phi_j|\hat{H}(i,j)|\omega_j\omega_i\rangle) \\
&= C_\Phi C_\Psi \sum_{i<j} |S_\Psi|(-1)^{l_{i,j}} 2^{-\delta(\omega_i,\omega_j)}(U(L_{i,j})_{kl} \langle \phi_i\phi_j|\hat{H}(i,j)|\omega_i\omega_j\rangle \\
&\quad -U(P_{i,j}L_{i,j})_{kl} \langle \phi_i\phi_j|\hat{H}(i,j)|\omega_j\omega_i\rangle) \\
&= \frac{C_\Phi}{C_\Psi} \sum_{i<j} (-1)^{l_{i,j}} 2^{-\delta(\omega_i,\omega_j)}(U(L_{i,j})_{kl} \langle \phi_i\phi_j|\hat{H}(i,j)|\omega_i\omega_j\rangle - U(P_{i,j}L_{i,j})_{kl} \langle \phi_i\phi_j|\hat{H}(i,j)|\omega_j\omega_i\rangle),
\end{aligned}
\tag{30}
$$

which is an reduction of Eq. (29) using the line-up permutation. In particular the $2^{-\delta(\omega_i,\omega_j)}$ factor makes sure integrals are not counted twice. The sum over all permutations is removed as permutations produce vanishing integrals due to the orthogonality of the orbitals.

4.3. Hamiltonian Elements in Special Cases

Let us first define the orbital difference between two spatial wavefunctions Φ and Ψ as the number of orbitals that appear in Ψ but do not appear in the corresponding number of times in Φ. For example, the orbital difference between $\Phi = \phi_1\phi_1\phi_2\phi_3\phi_4$ and $\Psi = \phi_1\phi_2\phi_3\phi_4\phi_5$ is 1, because the orbital ϕ_5 appears in Ψ but not in Φ, whereas the other four orbitals are in both functions.

We can now simplify Eq. (30) further based on the orbital difference of the two spatial functions Φ and Ψ in $\langle C_\Phi A\Phi X_j|\hat{H}_{int}|C_\Psi A\Psi X_k|\rangle$. The following formulas are general in that they will work for any set of orthonormal single electron orbitals. They are similar to the Slater-Condon rules for matrix elements between Slater determinates.

4.3.1. Orbital Difference Equals Zero

With the basis we have chosen, if the orbital difference of two spatial wavefunctions is 0 the orbitals must be equal. This is because we are not including spatial wavefunctions that are non-invariant permutations of each other in the basis. This means the line-up permutation $L_{i,j}$ will be the same for any pair (i, j), namely the identity permutation P_0, whose corresponding matrix is the identity matrix. In this case, we have

$$\langle C_\Phi A\Phi X_j|\hat{H}_{int}|C_\Psi A\Psi X_k\rangle = \frac{C_\Phi}{C_\Psi}\sum_{i<j}(-1)^{l_{i,j}}2^{-\delta(\omega_i,\omega_j)}(U(L_{i,j})_{kl}\langle\phi_i\phi_j|\hat{H}(i,j)|\omega_i\omega_j\rangle$$
$$-U(P_{i,j}L_{i,j})_{kl}\langle\phi_i\phi_j|\hat{H}(i,j)|\omega_j\omega_i\rangle)$$

$$= \frac{C_\Phi}{C_\Psi}\sum_{i<j}2^{-\delta(\omega_i,\omega_j)}\left(\delta_{kl}\langle\phi_i\phi_j|\,\hat{H}(i,j)\,|\omega_i\omega_j\rangle - U(P_{i,j})_{kl}\langle\phi_i\phi_j|\,\hat{H}(i,j)\,|\omega_j\omega_i\rangle\right) \quad (31)$$

4.3.2. Orbital Difference Equal to One

In this case there is one orbital in Φ that does not appear in Ψ. We label this orbital ϕ_{dif}. The only pairs (i, j) in Eq. (30) that give non-zero results are those containing the orbital ϕ_{dif}. If ϕ_{dif} appears once in position i, then these pairs are:

$$(1, i), (2, i), \ldots, (i-1, i), (i, i+1), \ldots, (i, N).$$

If ϕ_{dif} appears twice in position i and $i+1$, remembering that doubly occupied orbitals are next to each other in the basis, then these pairs are:

$$(1, i), (2, i), \ldots, (i-1, i), (i, i+1), \ldots, (i, N) \text{ and}$$

$$(1, i+1), (2, i+1), \ldots, (i-1, i+1), (i+1, i+2), \ldots, (i+1, N).$$

It is important to note that $\phi_i = \phi_{i+1} = \phi_{dif}$, so the integrals in the second case are exactly the same as the integrals in the first case except the pair $(i, i+1)$ only appears once. We also note that if $L_{i,j}$ is the line-up permutation for the pair (i, j) then $P_{i,i+1}L_{i,j}$ is the line-up permutation for the pair $(i+1, j)$. Using this information and Eq. (30) we obtain:

$$\begin{aligned}
&\langle C_\Phi A\Phi X_k|\hat{H}_{int}|C_\Psi A\Psi X_l\rangle = \\
&\quad \frac{C_\Phi}{C_\Psi}\sum\nolimits_{j=1,j\neq i}^{N} 2^{-\delta(\omega_i,\omega_j)}(-1)^{l_{i,j}}\Big(U(L_{i,j})_{kl}\,\langle\phi_i\phi_j|\hat{H}(i,j)|\omega_i\omega_j\rangle - U(P_{i,j}L_{i,j})_{kl}\,\langle\phi_i\phi_j|\hat{H}(i,j)|\omega_j\omega_i\rangle\Big) \\
&\quad +(n_{dif}-1)\frac{C_\Phi}{C_\Psi}\Big(\sum\nolimits_{j=1,j\neq i+1}^{N}\Big(2^{-\delta(\omega_{i+1},\omega_j)}(-1)^{p_{i,i+1}l_{i,j}}\Big(U(P_{i,i+1}L_{i,j})_{kl}\,\langle\phi_{i+1}\phi_j|\hat{H}(i+1,j)|\omega_{i+1}\omega_j\rangle \\
&\quad -U(P_{i,i+1}P_{i,j}L_{i,j})_{kl}\,\langle\phi_{i+1}\phi_j|\hat{H}(i+1,j)|\omega_j\omega_{i+1}\rangle\Big)\Big) - 2^{-\delta(\omega_{i+1},\omega_j)}(-1)^{l_{i,i+1}} \\
&\quad \times\Big(U(L_{i,i+1})_{kl}\,\langle\phi_i\phi_{i+1}|\hat{H}(i,i+1)|\omega_i\omega_i+1\rangle - U(P_{i,i+1}L_{i,i+1})_{kl}\,\langle\phi_i\phi_{i+1}|\hat{H}(i,i+1)|\omega_i+1\omega_i\rangle\Big)\Big),
\end{aligned} \tag{32}$$

where n_{dif} is the number of times the orbital ϕ_{dif} appears in Φ. The $(n_{dif}-1)$ term accounts for the cases where ϕ_{dif} occurs twice in Φ.

Using the fact that $\phi_i = \phi_{i+1}$ and

$$U(P_{i,i+1}R)_{kl} = U(P_{i,i+1})_{kk}U(R)_{kl} = (-1)^{p_{i,i+1}}U(R)_{kl} = -U(R)_{kl},$$

we obtain

$$\begin{aligned}
&\langle C_\Phi A\Phi X_k|\hat{H}_{int}|C_\Psi A\Psi X_l\rangle = \frac{C_\Phi}{C_\Psi}\sum\nolimits_{j=1,j\neq i}^{N} 2^{-\delta(\omega_i,\omega_j)}(-1)^{l_{i,j}}\Big(U(L_{i,j})_{kl} \\
&\quad \times\langle\phi_i\phi_j|\hat{H}(i,j)|\omega_i\omega_j\rangle - U(P_{i,j}L_{i,j})_{kl}\,\langle\phi_i\phi_j|\hat{H}(i,j)|\omega_j\omega_i\rangle\Big) + (n_{dif}-1) \\
&\quad \times\frac{C_\Phi}{C_\Psi}\Big(\sum\nolimits_{j=1,j\neq i}^{N} 2^{-\delta(\omega_i,\omega_j)}(-1)(-1)^{l_{i,j}}\Big((-1)U(L_{i,j})_{kl}\,\langle\phi_i\phi_j|\hat{H}(i,j)|\omega_i\omega_j\rangle - (-1)U(P_{i,j}L_{i,j})_{kl} \\
&\quad \times\langle\phi_i\phi_j|\hat{H}(i,j)|\omega_j\omega_i\rangle\Big) - 2^{-\delta(\omega_{i+1},\omega_j)}(-1)^{l_{i,i+1}}\Big(U(L_{i,i+1})_{kl} \\
&\quad \times\langle\phi_i\phi_{i+1}|\hat{H}(i,i+1)|\omega_i\omega_{i+1}\rangle - U(P_{i,i+1}L_{i,i+1})_{kl}\,\langle\phi_i\phi_{i+1}|\hat{H}(i,i+1)|\omega_{i+1}\omega_i\rangle\Big)\Big).
\end{aligned} \tag{33}$$

Grouping terms and simplifying further we get,

$$\begin{aligned}
&\langle C_\Phi A\Phi X_k|\hat{H}_{int}|C_\Psi A\Psi X_l\rangle = \frac{C_\Phi}{C_\Psi}n_{dif}\sum\nolimits_{j=1,j\neq i}^{N} 2^{-\delta(\omega_i,\omega_j)}(-1)^{l_{i,j}}(U(L_{i,j})_{kl} \\
&\quad \times\langle\phi_i\phi_j|\,\hat{H}(i,j)\,|\omega_i\omega_j\rangle - U(P_{i,j}L_{i,j})_{kl}\,\langle\phi_i\phi_j|\hat{H}(i,j)|\omega_j\omega_i\rangle) \\
&\quad -\frac{C_\Phi}{C_\Psi}(n_{diff}-1)(2^{-\delta(\omega_{i+1},\omega_j)}(-1)^{l_{i,i+1}}(U(L_{i,i+1})_{kl}\,\langle\phi_i\phi_{i+1}|\hat{H}(i,i+1)|\omega_i\omega_{i+1}\rangle \\
&\quad -U(P_{i,i+1}L_{i,i+1})_{kl}\,\langle\phi_i\phi_{i+1}|\hat{H}(i,i+1)|\omega_{i+1}\omega_i\rangle)).
\end{aligned} \tag{34}$$

Note that we have used many of the mathematical properties developed earlier to obtain the above formula.

4.3.3. Orbital Difference Equal to Two

If the orbital difference of Φ and Ψ is 2, then only one selection of orbitals ϕ_i and ϕ_j will give a valid line-up permutation, i.e. the orbitals that appear in Φ but do not appear in Ψ. Although there is only one choice for the orbitals, this could correspond to more than one pair of i, j, as each of ϕ_i and ϕ_j could appear up to twice in Φ.

Let v_i and v_j denote the orbitals in Φ that do not appear in Ψ. Let $n(v, \Phi)$ be the number of times the orbital v appears in the function Φ.

$$\begin{aligned} \langle C_\Phi A\Phi X_j|\hat{H}_{int}|C_\Psi A\Psi X_k\rangle &= \frac{C_\Phi}{C_\Psi}2^{-\delta(\omega_i,\omega_j)}2^{s(v_i,v_j,\Phi)}(-1)^{l_{i,j}}(U(L_{i,j})_{kl}\langle v_i v_j|\hat{H}(i,j)|\omega_i\omega_j\rangle \\ &\quad -U(P_{i,j}L_{i,j})_{kl}\langle v_i v_j|\hat{H}(i,j)|\omega_j\omega_i\rangle), \end{aligned} \tag{35}$$

where $s(v_i, v_j, \Phi)$ is define by

$$s(v_i, v_j, \Phi) = \begin{cases} 0, & v_i = v_j; \\ n(v_i, \Phi) + n(v_j, \Phi) - 2, & v_i \neq v_j. \end{cases}$$

The justification for this formula is similar to that for orbital difference of 1.

4.3.4. Orbital Difference Greater than or Equal to Three

In this case a line-up permutation, L_{ij}, does not exist for any i and j no matter what permutation you choose. Consequently,

$$\begin{aligned} \langle C_\Phi A\Phi X_j|\hat{H}_{int}|C_\Psi A\Psi X_k|\rangle &= \frac{C_\Phi}{C_\Psi}\sum_{i<j}(-1)^{l_{i,j}}2^{-\delta(\omega_i,\omega_j)(U(L_{i,j})_{kl}}\langle\phi_i\phi_j|\hat{H}(i,j)|\omega_i\omega_j\rangle \\ &\quad -U(P_{i,j}L_{i,j})_{kl}\langle\phi_i\phi_j|\hat{H}(i,j)|\omega_j\omega_i\rangle). \\ &= 0, \end{aligned} \tag{36}$$

where the last step follows as we define $U(L_{ij})$ to be the zero matrix if L_{ij} does not exist.

Note that in each of these cases only one line-up permutation needs to be calculated. We also know how to calculate the special spin basis in which the spin eigenfunctions are also eigenfunctions of the transpositions $P_{1,2}$, $P_{3,4}$, $P_{5,6}$, etc and elements of the representation matrices $U(P)$. Thus all we need now is to solve the single electron Schrödinger equation for the set of orthonormal orbitals and calculate the two electron orbital interactions $\langle\phi_i\phi_j|\hat{H}(i,j)|\psi_i\psi_j\rangle$ for arbitrary orbitals $\phi_i, \phi_j, \psi_i, \psi_j$.

5. Circularly Symmetric Harmonic Well

5.1. Choice of an Appropriate Multi-electron Basis

Quantum dot structures are often made from layers of semiconductor materials, where the electron motion is effectively confined in a two-dimensional plane. Metallic gates are then applied to form quantum dots with various geometric shapes. For illustration purpose, we will consider circularly symmetric quantum dots of the form $V(\vec{r}) = \frac{1}{2}m^*\omega_0^2 r^2$, where m^* is the effective mass and ω_0 relates to the steepness of the harmonic well, and any applied magnetic field is perpendicular to the plane of the dot.

In this case, the single electron Schrödinger equation can be solved analytically, which was first done by Fock [19] and later independently by Darwin [20]. The single electron Hamiltonian is

$$\hat{H}_0 = \frac{1}{2m^*}(\hat{p} - \frac{e}{c}\vec{A})^2 + \frac{1}{2}m^*\omega_0^2 r^2, \tag{37}$$

where $\vec{A} = (-\frac{1}{2}B\vec{y}, \frac{1}{2}B\vec{x}, 0)$ is the magnetic vector potential, c is the speed of light, and $\hat{p}$ is the momentum operator. The eigenenergies and wavefunctions for this system are known as

$$E_{nm} = (2n + |m| + 1)\hbar\Omega - \frac{1}{2}m\hbar\omega_c, \tag{38}$$

and

$$\psi_{nm}(r, \theta) = \left[2k^{|m|+1}\frac{n!}{(n + |m|)!}\right]^{\frac{1}{2}} r^{|m|}e^{-\frac{k}{2}r^2}L_n^{|m|}(kr^2)\frac{1}{\sqrt{2\pi}}e^{im\theta}, \tag{39}$$

where $\omega_c = \frac{eB}{m^*c}$ is the cyclotron frequency, $\Omega = \sqrt{\frac{1}{4}\omega_c^2 + \omega_0^2}$, $k = \frac{m^*\Omega}{\hbar}$ is the effective harmonic well constant, and $L_n^{|m|}(x)$ is the generalised Laguerre polynomials. From Eq. (38) we can deduce that the nth energy level is n-fold degenerate at zero magnetic field. This degeneracy is lifted when an external magnetic field is applied.

The multi-electron basis functions have the form $A\Phi X_k$ where the spatial wavefunction $\Phi = \phi(\vec{r_1})\phi(\vec{r_2})\ldots\phi(\vec{r_N})$ with each $\phi(\vec{r_i})$ chosen to be one of the solutions to the single electron Schrödinger equation, as given by Eq. (39). A complete set of basis functions would include an infinite number of product spatial wavefunctions. In practice, we need to select the most appropriate finite basis to approximate the true wavefunction. Note that each of the single electron wavefunctions has an energy given by Eq. (38). In the non interaction case, the multi-electron wavefunction would have energy $\sum_{i=1}^{N}(2n_i + |m_i| + 1)\hbar\omega_0$ at zero magnetic field. In this work, we choose to include all basis functions which have non-interacting energy below a certain value, and then increase this value until numerical results converge.

5.2. The Interaction Integral

The Hamiltonian describing the interactions of many electrons in a quantum dot is

$$\hat{H} = \sum_{i=1}^{N}\hat{H}_{0i} + \sum_{j=2}^{N}\sum_{i=1}^{j}\frac{1}{|\vec{r_i} - \vec{r_j}|}, \tag{40}$$

where $\hat{H}_{0i}$ is the single electron Hamiltonian acting on the ith electron. The second term describes the Coulomb interaction between each pair of electrons.

To construct the Hamiltonian matrix we need to calculate inner products $\langle\psi_{mn}(r,\theta)\psi_{pq}(s,\phi)|\hat{H}_{int}|\psi_{ab}(r,\theta)\psi_{cd}(s,\phi)\rangle$, which has the explicit form:

$$\int_0^\infty\int_0^\infty\int_0^{2\pi}\int_0^{2\pi} rs\frac{\psi_{nm}^*(r,\theta)\psi_{pq}^*(s,\phi)\psi_{ab}(r,\theta)\psi_{cd}(s,\phi)}{\sqrt{r^2 + s^2 - 2rs\cos(\theta - \phi)}}d\theta d\phi dr ds$$

$$= \int_0^\infty\int_0^\infty\int_0^{2\pi}\int_0^{2\pi} rs\frac{R_{nm}(r)R_{pq}(s)R_{ab}(r)R_{cd}(s)e^{i(b-m)\theta+i(d-q)\phi}}{4\pi^2\sqrt{r^2 + s^2 - 2rs\cos(\theta - \phi)}}d\theta d\phi dr ds \tag{41}$$

where $R_{nm}(r) = \sqrt{2k^{|m|+1}\frac{n!}{(n+|m|)!}}r^{|m|}e^{-\frac{k}{2}r^2}L_n^{|m|}(kr^2)$.

Noting that the integrand is periodic in both θ and ϕ with period 2π, we can apply the transformation $\phi \to \theta+\delta$ or equivalently $\theta \to \phi+\delta$ with delta ranging from 0 to 2π. These respectively give:

$$\frac{1}{4\pi^2}\int_0^{2\pi} e^{i(b-m+d-q)\theta}d\theta \int_0^{\infty}\int_0^{\infty}\int_0^{2\pi} rs\frac{R_{nm}(r)R_{pq}(s)R_{ab}(r)R_{cd}(s)e^{i(d-q)\delta}}{\sqrt{r^2+s^2-2rs\cos(\delta)}}d\delta dr ds, \quad (42)$$

and

$$\frac{1}{4\pi^2}\int_0^{2\pi} e^{i(b-m+d-q)\phi}d\phi \int_0^{\infty}\int_0^{\infty}\int_0^{2\pi} rs\frac{R_{nm}(r)R_{pq}(s)R_{ab}(r)R_{cd}(s)e^{-i(m-b)\delta}}{\sqrt{r^2+s^2-2rs\cos(-\delta)}}d\delta dr ds. \quad (43)$$

Evaluating the single integral gives us

$$\int_0^{2\pi} e^{i(b-m+d-q)\theta}d\theta = \begin{cases} 0, & b-m+d-q \neq 0; \\ 2\pi, & b-m+d-q = 0. \end{cases} \quad (44)$$

This means the interaction integral is zero if the sum of the angular momentum quantum numbers is different for the two orbitals, i.e. $b+d \neq m+q$.

We can also see that if $b-m+d-q=0$ then $d-q=-(m-b)$. Using the fact that $\cos(-x)=cos(x)\ \forall x \in \mathbb{R}$ and equating the two integrals (42) and (43), the complex part of the integral vanishes and we can replace $e^{i(d-q)\delta}$ with $\cos(|(d-q)|\delta)$. If we define a new constant $\alpha=|b-m|=|d-q|$, the integral (43) becomes:

$$\frac{1}{2\pi}\int_0^{\infty}\int_0^{\infty}\int_0^{2\pi} rs\frac{R_{nm}(r)R_{pq}(s)R_{ab}(r)R_{cd}(s)\cos(\alpha\delta)}{\sqrt{r^2+s^2-2rs\cos(\delta)}}d\delta dr ds. \quad (45)$$

We can break the integral down further by noting that

$$rsR_{nm}(r)R_{pq}(s)R_{ab}(r)R_{cd}(s) = 4\sqrt{k}^{|b|+|d|+|m|+|q|+4}e^{-k(r^2+s^2)}r^{|m|+|b|+1}s^{|q|+|d|+1}\frac{1}{|b|!|d|!|m|!|q|!}$$
$$\times\sqrt{\frac{(a+|b|)!}{a!}}\sqrt{\frac{(c+|d|)!}{c!}}\sqrt{\frac{(n+|m|)!}{n!}}\sqrt{\frac{(p+|q|)!}{p!}}\,{}_1F_1(-a,1+|b|,kr^2)\,{}_1F_1(-c,1+|d|,ks^2)$$
$$\times\,{}_1F_1(-n,1+|m|,kr^2)\,{}_1F_1(-p,1+|q|,ks^2),$$

where

$${}_1F_1(-n,1+|m|,kr^2) = \sum_{i=0}^{\infty}\frac{(-n)_i}{(1+|m|)_i i!}(kr^2)^i$$

and $(a)_i = a(a+1)(a+2)..(a+i-1)$ is the Pochhammer symbol or rising factorial. We can see from the definition that the Pochhammer symbol will be zero for $i \geq n+1$. This gives us a finite polynomial in $\sqrt{k}r$ of degree $2n$ with only even powers of $\sqrt{k}r$.

It is clear that if we can do integrals of the form :

$$\int_0^{\infty}\int_0^{\infty}\int_0^{2\pi}\frac{\cos(\alpha\delta)e^{-k(r^2+s^2)}k(\sqrt{k})^{a+b}r^a s^b}{\sqrt{r^2+s^2-2rs\cos(\delta)}}dr ds d\delta, \quad (46)$$

we can do the original integral as it is just a linear combination of integral of the above form. A point worth mentioning that will come up later is that the difference of a and b is even. This can be seen by the following reasoning.

Any power of r and s coming from the expansion of the hypergeometric function will be even, so we only have to worry about the powers of r and s coming from $r^{|m|+|b|+1}s^{|q|+|d|+1}$. Define the function $\sigma(n)$ by:

$$\sigma(n) = \begin{cases} o, & \text{n odd;} \\ e, & \text{n even.} \end{cases} \tag{47}$$

This function has the property that $\sigma(n+m) = \sigma(\pm n \pm m)$. Remembering that $b - m + d - q = 0$, then $\sigma(b - m + d - q) = e \Rightarrow \sigma(|b| + |m| - (|d| + |q|)) = e$, so $(|b| + |m| + 1) - (|d| + |q| + 1)$ is even. This means $a - b$ will always be even.

We apply another transformation $r \rightarrow \rho\sin(\lambda)$ and $s \rightarrow \rho\cos(\lambda)$ with $\rho \in [0, \infty]$ and $\lambda \in [0, 2\pi]$. This gives us:

$$\int_0^\infty \int_0^{\frac{\pi}{2}} \int_0^{2\pi} \rho \frac{\cos(\alpha\delta)e^{-k\rho^2}k(\sqrt{k})^{a+b}\rho^{a+b}\sin^a(\lambda)\cos^b(\lambda)}{\sqrt{\rho^2\sin^2(\lambda) + \rho^2\cos^2(\lambda) - 2\rho^2\sin(\lambda)\cos(\lambda)\cos(\delta)}} d\delta d\rho d\lambda$$

$$= \int_0^\infty e^{-k\rho^2}k(\sqrt{k})^{a+b}\rho^{a+b}d\rho \int_0^{\frac{\pi}{2}} \int_0^{2\pi} \frac{\cos(\alpha\delta)\sin^a(\lambda)\cos^b(\lambda)}{\sqrt{1 - \sin(2\lambda)\cos(\delta)}} d\delta d\lambda. \tag{48}$$

Now $\int_0^\infty e^{-k\rho^2}k(\sqrt{k})^{a+b}\rho^{a+b}d\rho = \frac{1}{2}\sqrt{k}\Gamma(\frac{1+a+b}{2})$, where Γ is the Euler gamma function. Thus our integral becomes:

$$\frac{1}{2}\sqrt{k}\Gamma(\frac{1+a+b}{2}) \int_0^{\frac{\pi}{2}} \int_0^{2\pi} \frac{\cos(\alpha\delta)\sin^a(\lambda)\cos^b(\lambda)}{\sqrt{1 - \sin(2\lambda)\cos(\delta)}} d\delta d\lambda. \tag{49}$$

The important point to note here is that the integral is proportional to $\sqrt{k}$ for all the interaction terms, where k is the effective harmonic well constant. This allows us to calculate the interaction integrals for $k = 1$ and then use these for any other value of k by multiplying by an appropriate factor. This integral has a logarithmic singularity at $\lambda = \frac{\pi}{4}$, which is resulted from the Coulomb interaction diverging when the two electrons are in the same location. However, this integral can be done in closed form, noting that the integrand is symmetric in a and b and furthermore a and b can only differ by an even number.

If we let Z_0 represent the part of the integrand that does not involve $\sin^a(\lambda)\cos^b(\lambda)$ and assume, without loss of generality, that $b \geq a$ and let $b - a = 2n$, we get for $n = 1$

$$\begin{aligned} \int_0^{\frac{\pi}{2}} Z_0 \sin^a(\lambda)\cos^{a+2}(\lambda)d\lambda &= \int_0^{\frac{\pi}{2}} Z_0 \sin^a(\lambda)\cos^a(\lambda)(1 - \sin^2(\lambda))d\lambda \\ &= \int_0^{\frac{\pi}{2}} Z_0 \sin^a(\lambda)\cos^a(\lambda)d\lambda - \int_0^{\frac{\pi}{2}} Z_0 \sin^{a+2}(\lambda)\cos^a(\lambda)d\lambda. \end{aligned} \tag{50}$$

Since $\int_0^{\frac{\pi}{2}} Z_0 \sin^a(\lambda)\cos^{a+2}(\lambda)d\lambda = \int_0^{\frac{\pi}{2}} Z_0 \sin^{a+2}(\lambda)\cos^a(\lambda)d\lambda$, we have

$$\int_0^{\frac{\pi}{2}} Z_0 \sin^a(\lambda)\cos^{a+2}(\lambda)d\lambda = \frac{1}{2}\int_0^{\frac{\pi}{2}} Z_0 \sin^a(\lambda)\cos^a(\lambda)d\lambda. \tag{51}$$

Let $Z_n(a) = \int_0^{\frac{\pi}{2}} Z_0 \sin^a(\lambda)\cos^{a+2n}(\lambda)d\lambda$, we can use the same working as before to show that for $n \geq 2$

$$Z_n(a) = Z_{n-1}(a) - Z_{n-2}(a+2). \tag{52}$$

It can also be easily verified that

$$Z_0(a) = \frac{\pi^{\frac{3}{2}}\Gamma(\frac{1+\alpha+a}{2}))}{2^{3\alpha-a}\Gamma(\frac{2+\alpha+a}{2})}\binom{2\alpha}{\alpha}{}_3F_2\left(\{\frac{1+2\alpha}{4},\frac{3+2\alpha}{4},\frac{1+\alpha+a}{2}\},\{1+\alpha,\frac{2+\alpha+a}{2}\},1\right) \quad (53)$$

and

$$Z_1(a) = \frac{1}{2}Z_0(a). \quad (54)$$

The above recursion relations (52, 53, 54) are sufficient to calculate the integral in terms of a sum of hypergeometric functions. *Mathematica* evaluates the hypergeometric functions exactly which allows the exact calculation of interaction elements. This eliminates numerical problems that may arise in evaluating these integrals.

6. Results and Discussion

6.1. Energy Levels and Wavefunctions

The main advantage of the SACI approach is that it allows the calculation of energy levels and system wavefunctions to arbitrarily high precision, while using less computational resources than the conventional CI methods. The results presented in this paper can then act as benchmarks for other approximate methodologies developed to study more complex systems. Fig. 1 shows the convergence of the ground state energies for multi-electron quantum dots at zero magnetic field with increasing basis size. Here we take the harmonic well parameter $\hbar\omega_0$ to be $10meV$ and the values of $m^* = 0.067$ and $\varepsilon = 13.1$ to correspond to the experiments using GaAs dots [21]. From the convergence calculation, we are confident in asserting that the ground state energies obtained from our calculations for 2, 3, 4, 5, 6, and 7 electrons are accurate to within 0.03%, 0.024%, 0.045%, 0.05%, 0.1%, 0.15%, and 0.2%, respectively.

From the same calculation, we also obtained accurate information on the excited states of these systems. Fig. 2 shows the energy level structure for a two-electron quantum dot. In the triplet states the two electrons have the same spin and are thus forbidden from being in the same orbital by Pauli's exclusion principle, while in the singlet states the two electrons have different spin and thus can be in the same orbital. As a result, the basis functions used in the singlet wavefunction expansion include both doubly and singly occupied orbitals, while the basis for the triplet system can only include singly occupied orbitals. This means the $1/r$ Coulomb interaction is less significant for the triplet states than the singlet states and thus there is more degeneracy in the triplet energy levels of the system.

The charge densities for two and three electrons at zero magnetic field are shown in Figs. 3 and 4. The most notable aspect of these plots is the spatial extent of the electron density is ranges from 50-100nm. Thus the electron density at the edge of a dot of diameter 500nm is negligible and therefore edge effects should not be important in the modelling. The fact that the electron density is concentrated close to the centre of the dot also means the harmonic well approximation should be quite reasonable as any circularly symmetric potential well can be approximated as a harmonic well near its centre.

The quantity measured by the experiments of Tarucha et al. [21] is the addition energy, which is the energy required to add an extra electron to the system. The chemical potential

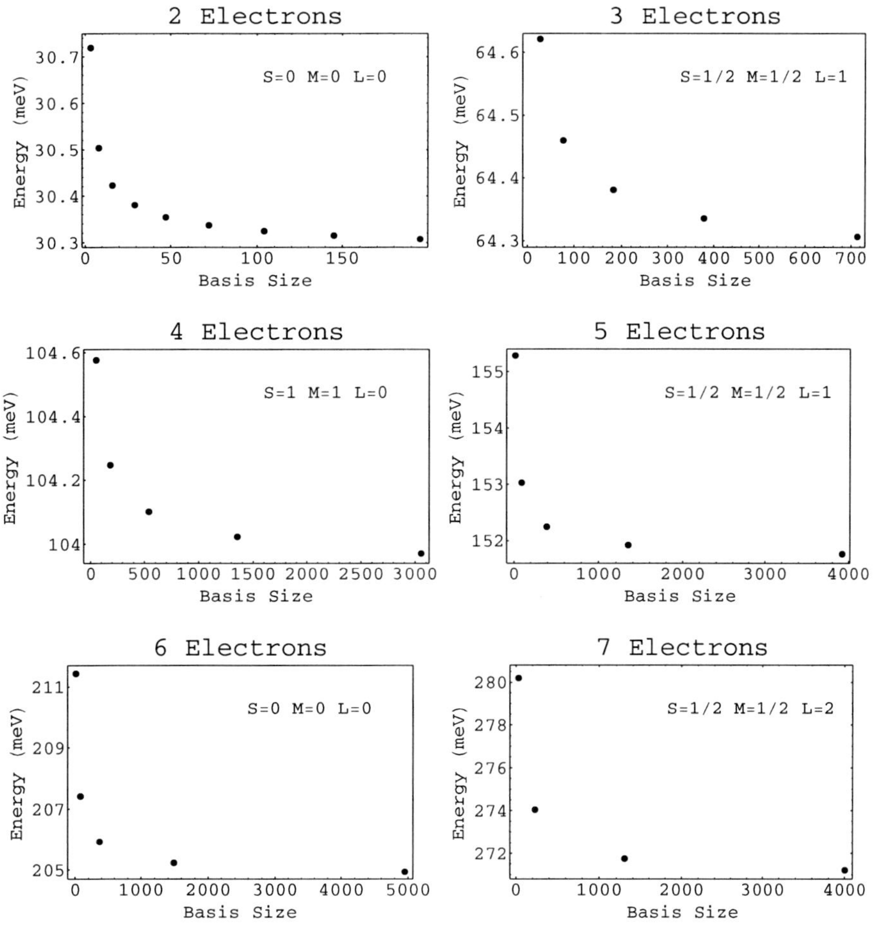

Figure 1. Convergence of ground state energies with increasing basis size. The harmonic well parameter $\hbar\omega_0 = 10meV$.

$\mu(N)$ is defined as

$$\mu(N) = E(N) - E(N-1) \tag{55}$$

with $E(N)$ being the ground state energy for the N-electron system. The addition energy is the difference between the chemical potential of N and $N+1$ electrons, i.e.

$$\Delta\mu(N) = \mu(N+1) - \mu(N) = E(N+1) - 2E(N) + E(N-1). \tag{56}$$

Fig. 5 gives the calculated addition energy at zero magnetic field at various values of $\hbar\omega_0$. The plot shows large maxima at 2 and 6 electrons, where the shells are filled and a relative maxima at 4 electrons due to the shell being half filled. From the graphs, we see that the peaks are more pronounced for larger values of $\hbar\omega_0$, which is a more tightly confined well. This is in agreement with the measurements of Tarucha et al. [21].

The wavefunctions of the electrons in quantum dots are spread over a much larger area than those in natural atoms and the excited state energies are much closer to the ground

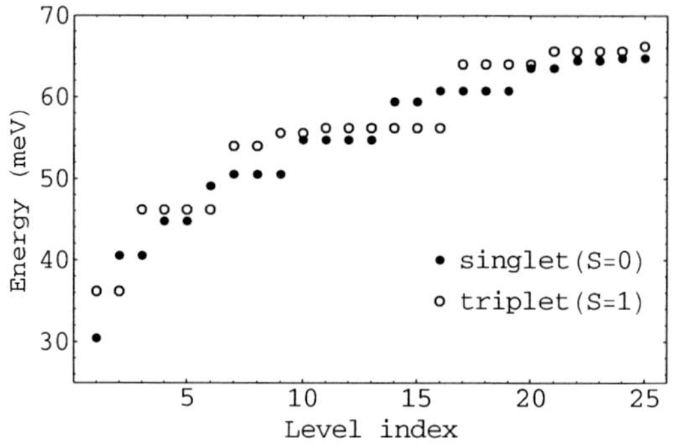

Figure 2. The first 50 energy levels of a two-electron quantum dot with $\hbar\omega_0 = 10meV$.

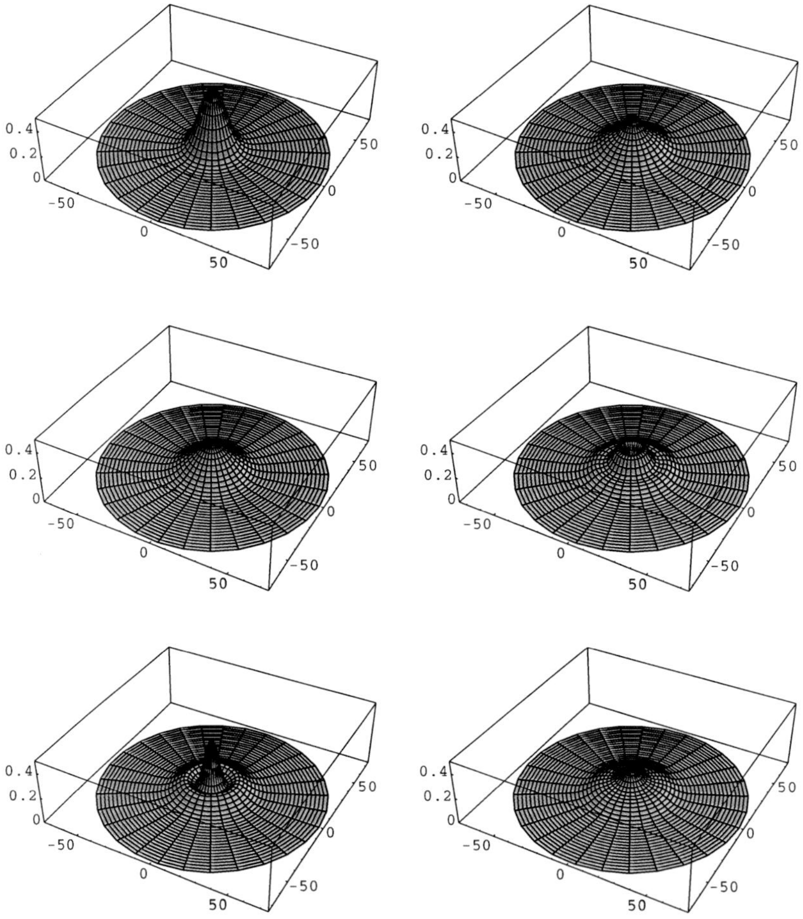

Figure 3. Charge densities of the lowest 6 states with $L = 0$ for a two-electron quantum dot (from top left to bottom right).

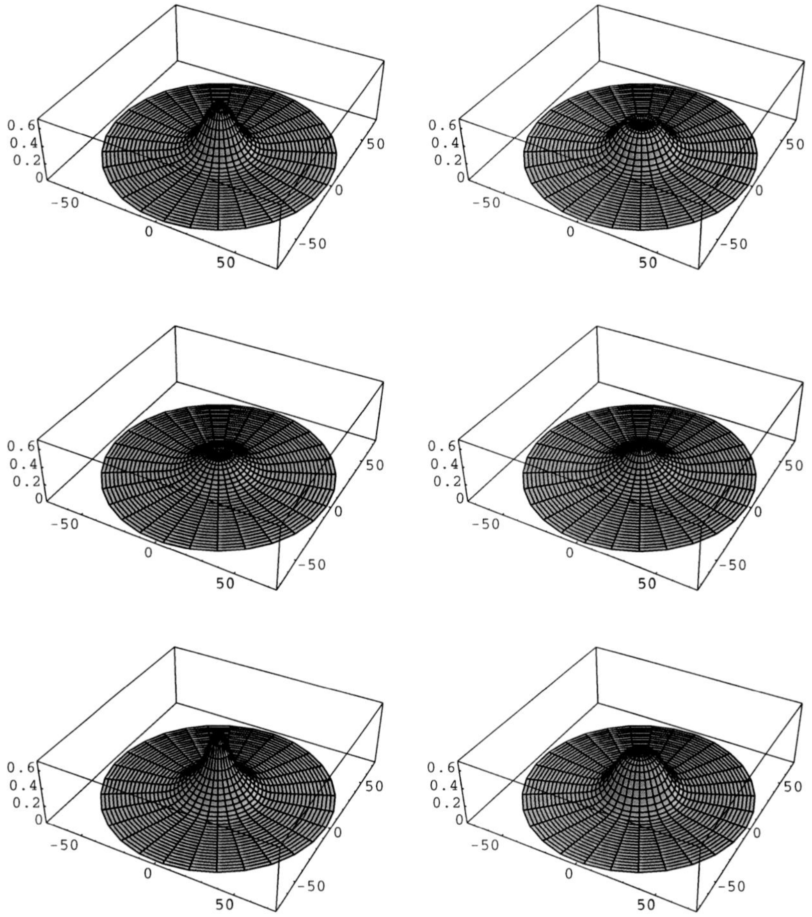

Figure 4. Charge densities of the lowest 6 states with $L = 1$ for a two-electron quantum dot (from top left to bottom right).

state energies. This makes the energy levels in the quantum dot more sensitive to changes in magnetic field than in natural atoms. As an example, Fig. 6 shows the energies of various states as a function of magnetic field for a quantum dot with 2 electrons. In this case, the harmonic well parameter $\hbar\omega_0$ is set to be $5.5meV$ in order to compare with the experimental results of Tarucha et al. [21]. The most noticeable feature of this figure is the singlet to triplet transition in the ground state that occurs at around $4T$, which is in agreement with the experimental data.

7. Conclusions

In this paper we have demonstrated how to use antisymmetrised products of spatial wavefunction and spin eigenfunctions as basis functions in a configurational interaction approach to solving multi-electron Schrödinger's equations. A spin adapted CI basis was formed

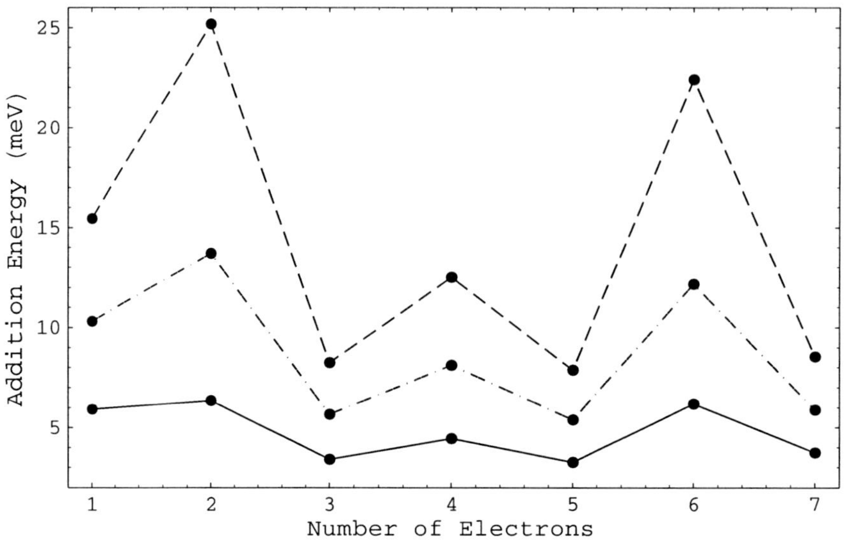

Figure 5. Addition energies obtained using the SACI approach, where the energies of 8-electron dots were also required and evaluated with accuracy within 0.2%.

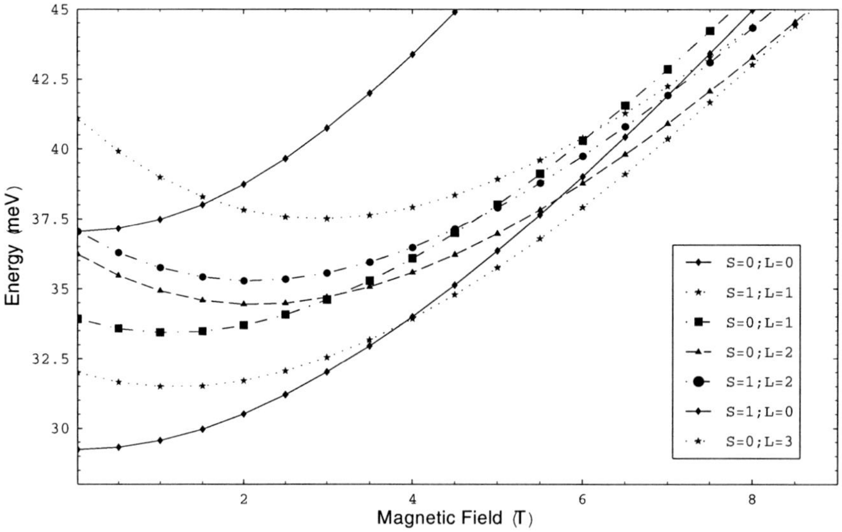

Figure 6. The evolution of the energies of a 2-electron quantum dot with external magnetic field. The harmonic well parameter $\hbar\omega_0 = 5.5meV$.

by selecting spin eigenfunctions that are also eigenfunctions of the geminal transposition operators and representative spatial wavefunctions that have doubly occupied orbitals in the sequential geminal positions. We have also showed that any two representative spatial wavefunctions in the basis are not non-invariant permutations of each other. The orthogonality of this basis was established and formulas for the normalisation constant and the non-zero elements were developed.

This theory was then employed to model electrons confined in a circularly symmetric quantum dot. The electric potential of the dot was approximate as a two-dimensional harmonic well and an external magnetic field is applied perpendicularly to the two-dimensional plane. The elements of the Hamiltonian matrix can be evaluated analytically in this case. The formula for determining Hamiltonian matrix elements are quite general and can be applied to other forms of confinement potentials. This framework allows computations to be done faster and more accurately than the conventional CI approach, providing benchmark results against which others approximate methods can be tested.

The calculations presented in this paper were carried out using *Mathematica* on a PC with the Pentium IV 2.4GHz processor. This work can be readily extended by switching to a more efficient programming language such as Fortran or C^{++} and using a more powerful computer to study quantum dots containing more electrons or having more complex geometry.

References

[1] D. Pfannkuche, V. Gudmundsson, and P. Maksym, *Physics Review B* 47, 2244 (1993).

[2] C. Yannouleas and U. Landman, *Physics Review Letters* 82, 5325 (1999).

[3] C. Yannouleas and U. Landman, *Physics Review Letters* 85, 1726 (2000).

[4] S. McCarthy, J. Wang, and P. Abbott, *Computer Physics Communications* 141, 175 (2001).

[5] M. Macucci, K. Hess, and G. Jafrate, *Phys. Rev. B* 48, 17354 (1993).

[6] M. Macucci, K. Hess, and G. Jafrate, *Journal of Applied Physics* 77, 3267 (1995).

[7] M. Macucci, K. Hess, and G. Jafrate, *Physics Review B* 55, R4879 (1997).

[8] I. Lee, V. Rao, R. Martin, and J. Leburton, *Physics Review B* 57, 9035 (1998).

[9] T. Ezaki, N. Mori, and C. Hamagushi, *Physics Review B* 56, 6428 (1997).

[10] T. Ezaki, Y. Sugimoto, N. Mori, and C. Hamaguchi, *Semiconductor Science and Tecnology* 13, A1 (1998).

[11] M. Eto, *Journal of Applied Physics* 36, 3924 (1997).

[12] M. Eto, *Journal of Applied Physics* 38, 376 (1999).

[13] S. Reimann, M. Koskinen, and M. Manninen, *Physics Review B* 62, 8108 (2000).

[14] S. Reimann and M. Manninen, *Reviews of Modern Physics* 74, 1283 (2002).

[15] R. Pauncz, *The Construction of Spin Eigenfunctions An Exercise Book* (Kluwer Academic/Plenum Publishers, New York, 2000).

[16] R. Serber, *Phys. Rev* 45, 461 (1934).

[17] W. Salmon and K. Ruedenberg, *Journal of Chemical Physics* 57, 2776 (1972).

[18] R. Pauncz, *International Journal of Quantum Chemistry* 12, 369 (1977).

[19] V. Fock, *Zeitschrift für Physik* 47, 446 (1928).

[20] C. Darwin, in *Proc. Cambridge Philos. Soc.* (1930), vol. 27, pp. 86–90.

[21] S. Tarucha, T. Honda, D. Austing, Y. Tokura, K. Muraki, T. Oosterkamp, J. Janssen, and L. Kouwenhoven, *Physica E* 3, 112 (1998).

In: Quantum Chemistry Research Trends
Editor: Mikas P. Kaisas, pp. 173-204
ISBN: 978-1-60021-620-6

Chapter 5

NEW APPROACH TO RESEARCHES IN RELATIVISTIC QUANTUM CHEMISTRY USING HAMILTON-JACOBI EQUATION

A. L. A. Fonseca and D. L. Nascimento
Institute of Physics, University of Brasília, P.O. Box: 04455
70919-970, Brasília-DF, Brazil

Abstract

The aim of this work is to show that it is possible to obtain a new approach of relativistic quantum mechanics based on relativistic classical field theory to study atomic and molecular systems. Recently, relativistic effects have been shown to be important to study several chemical systems. In this work, starting from the basic principles of Special Relativity Theory (SRT), without using the usual concepts of Quantum Mechanics (QM), we obtain the equation of motion for the studied systems. The interaction between two particles is introduced through their respective metric functions. This method is accomplished by taking the interaction among particles moving with the standing wave radiated by them in the four dimensional space-time. According to the classical theory of fields, the geometrical properties of the solutions are governed by a classical-like equation of motion. The eigenvalue equation is derived from an improvement of the relativistic variational method introduced by Schrödinger in 1926. As this work is developed in a classical formalism, we use real action functions and thus we can obtain well-defined trajectories, which can be useful to study chemistry reactions. As examples, this formalism is applied to one, two and three-body problems, yielding solutions of the equation of motion by integrals of action or by solving the associated differential eigenvalue equation. Both set of solutions reproduce the correct structure of the energy levels.

1. Introduction

As is well known relativistic effects play a crucial role in the description of the electronic structure of heavy elements. Nowadays, the importance of relativistic quantum mechanics in the study of atomic and molecular systems is met even in calculations with light elements.

For example, in very precise energy calculations with H_2^+ and H_2, relativistic contributions may not be neglected. In spite of this, the relevance of relativistic effects in heavy-elements chemistry was only recognized in the early 1970 decade.

The structure, spectrum and chemistry of molecules and solids containing heavy atoms are strongly modified by relativistic effects; depending on the accuracy, they become relevant even for a middle mass element, such as the Cu atom. Considering the periodic table of elements, the relativistic effects become as important as the electronic correlation effects out of the 5th row (Rb-Xe) and 6th row (Cs-Rn) elements of the table. The size of the relativistic effects becomes greater than the size of electronic correlation effects particularly for Au and its neighbors. For example, for the AuH molecule, the relativistic effects are responsible for a reduction of 15% in the molecule bond length and for an increase of 60% in the molecule binding energy. Furthermore, relativistic effects are essential for the actinides elements.

The quantum mechanical description of many electron atoms and molecules is knowingly very difficult to be accomplished even in the framework of the standard non-relativistic Schrödinger equation. Further, such a description becomes prohibitively difficult in the Dirac equation relativistic framework because its one-electron spinorial solution is a two-vector whose components are scalar wave functions.

Unfortunately, Dirac's theory itself is incomplete. Therefore, a full relativistic treatment of the problem of the atomic and molecular electronic structure demands the application of methods of the Quantum Electrodynamics (QED), which is up to now considered as the best established quantum theory of fields. As a consequence, the difficult of implementing a rigorously relativistic computational treatment of a many electron system becomes even lesser attainable.

After the above brief review of the current situation in the relativistic many body calculations (see refs. [1-3] and references therein) we have enough motivation to state that the aim of this new approach is to search for a less complicated alternative method for computational calculus in atomic and molecular systems.

We must stress here that we are neither questioning the validity of QED nor even proposing a new theory, but simply re-interpreting already existing ideas in the realm of QM with the purpose of trying to obtain a less complicated computational framework. The net profit of the new approach would be a worthwhile reduction of the time required to perform calculations in atomic and molecular systems already known, which can be seen as a prototype to successful calculations in many body systems of current interest that are still not partially or completely understandable.

Our line of thought may be also seen as a consequence of the free space in theoretical physical thought opened with the so called Einstein-Bohr debate on the foundations of Quantum mechanics [4], particularly on the role of probabilities in the process of measurement and the possibility of sharp separation among individual components which are interacting in a many body system [5-7].

In the current interpretation of QM, N.ewmann and Birkoff [8] have shown that there are only three possibilities of calculus: Real Hilbert Space (RHS), Complex Hilbert Space (CHS) and Quarternion Hilbert Space (QHS). The Octonion Hilbert Space (OHS) was further formulated in the 60's [9]. These are the four known division algebras in which it is possible to built standard calculus of QM up today (more about this subject in ref. [10]).

Stueckelberg [11] has shown that the RHS scheme faces a logical contradiction when it tries to express the uncertainty principle in its formalism, so that QM is incomplete in RHS scheme.

The CHS scheme is able to represent ordinary QM, both in Schrödinger's non-relativistic framework as well as in Dirac's relativistic one. CHS contains also several not very well succeeded interpretations of QM, as for example, the de Broglie-Bohm mechanics, where a particle moves along a real trajectory which is generated by a quantum potential associated with the non-relativistic Schrödinger's equation [12].

The authors believe that the failure with the RHS scheme is more a consequence of the standard interpretation of QM than of the physical reality itself, so that we believe it is still possible to build a real number model of atomic and molecular systems. Of course, such a new formulation cannot have the same structure of the usual QM. For instance, in agreement with Bohm [13-15], the so called intrinsic or spin degrees of freedom are already taken into account in the orbital angular momentum, and particles have well-defined effective position and momentum.

Our approach follows the historical fact that Lorentz invariance leads naturally to the transformation of the classical equations of motion, the Hamilton-Jacobi equation in particular, to the corresponding quantum eigenvalue-equation, which is Klein-Gordon-like [16] but, as is well known, the Klein-Gordon equation is not appropriate to study the atomic and molecular system. Therefore, we are proposing an alternative new quantum equation that may be directly derived from the fundamental space-time metrical function through a variational formulation or by means of the de Broglie-Einstein relations through an operator formalism similar to that of usual QM. Thus, there is no direct relationship either with Dirac's theory or with Quantum Electrodynamics.

The new model is however not a new version of a hidden variables theory [4], since it incorporates the usual wave-particle duality principle in such a way that the average or effective values of the physical variables are well determined, while it does not matter for us the status of the actual values of them, if they really exist.

In fact, we have started from the basic principles of the SRT for a two-particle system, which can interact through standing wave fields, from which we reach the following two equivalents formulations, depending on which of the particle or wave feature is focused. On the one hand, a one body effective Hamilton-Jacobi equation (HJE) is obtained, whose momenta coordinates must obey Sommerfeld-like quantization conditions, which are due to the periodicity of the possible modes of vibration of the interacting standing waves [17].

On the other hand, an improvement of the relativistic variational method introduced by Schrödinger in 1926 [18], interpreted as furnishing the conditions for the stability of motion under the action of interacting wave-fields, leads to a Klein-Gordon-like eigenvalue equation. Therefore, the Sommerfeld-integral-like and the eigenvalue-differential-like quantization conditions become perfectly equivalent in the new approach.

After testing the new equations with one and two body standard problems [19-24], we have made a first attempt of generalizing the model to taken into account general many body problems, whose first results are reviewed in this book.

2. Equations of Motion and Stability Conditions

The whole mathematical formalism to be developed will be started out of the following fundamental hypotheses (natural units, $c = \hbar = 1$, are used):

(*i*) The existence of a time-space continuum manifold whose geometrical properties are described by the *(t,x,y,z)* coordinates of a Cartesian coordinate system associated to a reference frame O_R, which is to be considered at least approximately as an inertial frame, so that the usual formulations of the SRT may be employed. However, the universal methods of the General Relativity Theory (GRT) are to be intended to be behind and in some particular situations may be taken explicitly into account. The metric function is given by

$$ds^2 = -dx_\mu dx^\mu = -g_{\mu\nu} dx^\mu dx^\nu = dt^2 - dx^2 - dy^2 - dz^2 \quad (1)$$

The connection between the contravariant and the covariant coordinates [25] appearing in Eq. (1) is given by $g_{\mu\nu}$, which in this paper it is assumed to represent the Minkowski flat space metric tensor.

(*ii*) Any interaction function introduced in the model must agree with the covariance and the invariance requirements of at least the SRT [25].

(*iii*) The interaction functions in (*ii*) must satisfy the field version of a least action principle.

It is important to remark that since no macroscopic or empirical principle of motion is considered here, the interaction between two particles can only be introduced through their respective metric functions.

We begin by considering that the coordinates of Eq. (1) represent the geometry of the one body free-motion problem consisting of a point particle of mass m and mechanical four-momentum [25],

$$p^\nu = m\frac{dx^\nu}{ds} = \gamma m\left(1, \vec{v}\right), \quad (2)$$

where Eq.(1) has been rewritten as $ds = \gamma^{-1} dt$, the time dilatation factor is $\gamma = \left(1 - v^2\right)^{-\frac{1}{2}}$, the velocity four-vector of the particle is given by $\left(1, \vec{v}\right) \equiv (1, v_x, v_y, v_z)$ and $\vec{v} = d\vec{r} / dt$ is the usual three-dimensional velocity of the particle. The particle position vector, $\vec{r} = (x, y, z)$, and the time t are measured in the O_R frame of reference.

Eq. (1) can be rewritten as

$$p_\nu p^\nu + m^2 = 0, \quad (3)$$

which in this free particle context describes a purely inertial motion parallel to the geodesic of the continuum determined by the tensor $g_{\mu\nu}$, that is, a straight line[25].

2.1. The N-body Problem

Let us now consider the existence of N particles with masses m_a, with $a = 1, 2, ..., N$, respectively. They can interact only through the waves that they produce in the continuum time-space by their motions. Therefore, although there are mutual interactions among the particles, the motion of the system as a whole must obey some kind of inertial law. This means that we may consider an effective moving point particle of mass

$$M = m_1 + ... + m_N, \tag{4}$$

as representing the N particle system as a whole. We shall define this effective moving point particle as the center of momentum (CM) of the system and the respective rest frame as O_{CM}.

The equations of motion for each particle in the O_R frame of reference, whose space coordinates $\mathbf{r}_a = (x_a, y_a, z_a)$ are measured in a time t, are written, from Eq.(3), as

$$p_{a\nu} p_a^\nu + m_a^2 = 0, \tag{5}$$

$$p_a^\nu = \gamma_a m_a (1, \mathbf{v}_a), \tag{6}$$

where $\gamma_a = \left(1 - v_a^2\right)^{-\frac{1}{2}}$ and $\mathbf{v}_a = d\mathbf{r}_a / dt$. (We shall use throughout this work the convention that Greek letters run into the range 0,1,2,3, while Latin letters run into the range 1,2,3,...,N, or any smaller subset.) In this frame, the total momentum four-vector,

$$P^\mu = \gamma_{CM} M (1, \mathbf{V}) = \sum_{a=1}^{N} p_a^\mu, \tag{7}$$

describes the motion of the effective mass M, whose velocity is $\mathbf{V} = d\mathbf{R}/dt$, where $\mathbf{R} = (X, Y, Z)$ is the position vector of the center of momentum, determined at time t, in the frame O_R, and $\gamma_{CM} = \left(1 - \mathbf{V}^2\right)^{-\frac{1}{2}}$. In three dimensional notation Eq. (7) becomes the following system of equations

$$M \gamma_{CM} = \sum_{a=1}^{N} \gamma_a m_a \tag{8a}$$

$$M\mathbf{V}\gamma_{CM} = \sum_{a=1}^{N}\mathbf{p}_a = \sum_{a=1}^{N}\gamma_a m_a \mathbf{v}_a \,. \tag{8b}$$

Eqs. (8) may be immediately seen to reduce, in the low velocity limit, to the momentum conservation form characteristic of the non-relativistic Newtonian physics, in which case (8b) may be directly integrated to give the classical center of mass equation

$$M\mathbf{R} = \sum_{a=1}^{N} m_a \mathbf{r}_a \,. \tag{8c}$$

In order to taken into account the mutual interactions between any pair of different particles of the n-body system, we must introduce field potential functions $A_{ab}^{\mu} = (\mathbf{A}_{ab}, \phi_{ab})$, where $a \neq b = 1,2,...,N$, and a system characteristic action function S, so that

$$p_a^{\mu} = \partial_a^{\mu} S - \sum_{b\neq a=1}^{N} A_{ab}^{\mu} \,, \tag{9}$$

where $\partial_a^{\mu} = (\nabla_a, \alpha_a \partial / \partial t)$ is a N-body differential operator with $\sum_{a=1}^{N}\alpha_a = 1$, the α_a coefficient being proportional to the mass of each particle. From Eq.(9) the Eq.(5) is rewritten in its four and three dimensional forms respectively as

$$\left(\partial_{a\mu} S - \sum_{b\neq a=1}^{N} A_{ab\mu}\right)\left(\partial_a^{\mu} S - \sum_{b\neq a=1}^{N} A_{ab}^{\mu}\right) + m_a^2 = 0 \,, \tag{10a}$$

$$\left(\nabla_a S - \sum_{b\neq a=1}^{N} \mathbf{A}_{ab}\right)^2 - \left(\alpha_a \frac{\partial S}{\partial t} - \sum_{b\neq a=1}^{N} \phi_{ab}\right)^2 + m_a^2 = 0 \,, \tag{10b}$$

which in any case is a system of N equations of motion to be simultaneously solved. The time derivative $H = \partial S / \partial t$ is called the Hamiltonian or energy function of the system. Eq.(10b) is usually referred as the Hamilton-Jacobi equation for the many body system. In a conservative system we must have [16]

$$H = E = \sum_{a=1}^{N} E_a = const., \tag{11a}$$

where the partial energies E_a and the ratio $\alpha_a = \frac{E_a}{E}$ between the energy of the particle a and the system energy E are introduced for the sake of allowing a formal separation of the Eqs.(10b) in each component.

When we pass to the rest frame of reference O_{CM}, in which by definition $\mathbf{V} = \mathbf{0}$, Eq.(8b) allows us to express the momentum four vector of the N-th particle as a function of the momentum of the remaining $N-1$ particles,

$$\mathbf{p}_N = -\sum_{a=1}^{N-1} \mathbf{p}_a , \tag{11b}$$

which reduces the original N-body simultaneous problem to a single effective ($N-1$)-body problem for the N-th particle. Thus, considering Eq.(9), the Eq.(10b) becomes

$$\sum_{a,b=1}^{N-1}\left(\nabla_a S - \sum_{c=1}^{N-1}\mathbf{A}_{ac}\right)\left(\nabla_b S - \sum_{d=1}^{N-1}\mathbf{A}_{bd}\right) - \left(E_N - \sum_{b=1}^{N-1}\phi_{Nb}\right)^2 + m_N^2 = 0 , \tag{12a}$$

which is written only in the three-dimensional form, because the four dimensional form is of no valuation. After determining $S(\mathbf{r}_1, \mathbf{r}_2, ..., \mathbf{r}_N, t)$ and E_N from Eq.(12), the other energies E_a, for $a = 1, 2, ..., N-1$ may be immediately determined by using Eq.(10b).

However, the problem has been solved in O_{CM}, so that we should obtain the corresponding solutions in the O_{lab} laboratory frame, where the observer is in fact, through a Lorentz transformation, if O_{CM} is considered, at least approximately, as an inertial frame with respect to O_{lab}, or by the tensor transformation methods of the GRT otherwise.

2.2. Stability Conditions of Motion

Stability against radiation requires the system be strictly periodic in each space coordinate and time. The periodicity is expressed by the prescription that each conjugated pair of coordinates and momenta describes a closed path in configuration space [16]. This means that the path integral along a closed path of the action function, that is, its finite difference, must be zero or vary proportionally to an integer number n

$$\oint_C \left(p_a^{\mu} + \sum_{b \neq a=1}^{N} A_{ab}^{\mu} \right) dx_{a\mu} = \oint_C \partial_a^{\mu} S dx_{a\mu} = \Delta S = 2\pi n . \tag{12b}$$

The condition expressed in Eq.(12b) is a generalization of the Sommerfeld quantization condition of the old quantum theory [17].

The condition given by Eq.(12b) is exact, however is effectively impossible to operate it in a many body system. Thus we have considered the alternative Eq.(12a), in the rest frame O_{CM}, to get an approximate way of expressing the stability condition, Eq.(12b), by introducing an auxiliary action function ψ through the transformation

$$S = \ln \psi \,. \tag{13}$$

The function ψ is an always real function which here plays the role of the "wave function" in ordinary Quantum Mechanics. Using Eq.(9) we obtain the following Lagrangean density

$$\Lambda_0 = \left\{ \sum_{a,b=1}^{N-1} \left(\nabla_a \psi - \psi \sum_{c=1}^{N-1} \mathbf{A}_{ac} \right)\left(\nabla_b \psi - \psi \sum_{d=1}^{N-1} \mathbf{A}_{bd} \right) + \left[m_N^2 - \left(E_N - \sum_{b=1}^{N-1} \phi_{Nb} \right)^2 \right] \psi^2 \right\} \rho \,, \tag{14}$$

where $\rho = \rho(x_a, y_a, z_a, \ldots)$ is a weight function to be defined, and then to consider instead solving Eq.(12a), to solve the 3N-dimensional variational problem

$$\delta \int d^{3N} x \Lambda_0 = 0 \,, \tag{15a}$$

which can be directly executed by partial integration or through the field Lagrange equations

$$\sum_{a=1}^{N-1} \nabla_a \frac{\partial \Lambda_0}{\partial (\nabla_a \psi)} - \frac{\partial \Lambda_0}{\partial \psi} = 0 \,, \tag{15b}$$

from which we get

$$\sum_{a=1}^{N-1} \Delta_a \psi + \frac{1}{\rho} \nabla_a \rho \cdot \nabla_a \psi + \left[\left(E_N - \sum_{b=1}^{N-1} \phi_{Nb} \right)^2 - \sum_{a,b,c,d=1}^{N-1} \mathbf{A}_{ac} \cdot \mathbf{A}_{bd} - m_N^2 \right] \psi = 0 \,, \tag{16}$$

which is a second order partial differential equation, whose solutions are easier to find than the solutions of Eq.(12a).

However, as p_a^μ must satisfy the HJE exactly at least at the orbital fixed points, it follows that the solutions coming from Eq.(17) must additionally satisfy the requirements

$$\delta p_a^\mu \big| \mathbf{r}_{a0} = \partial^\mu \psi \big| \mathbf{r}_{a0} = 0 \,, \tag{17}$$

where $\mathbf{r}_{a0}$ is a fixed orbital point for the particle a. Therefore $\rho = \rho(x_a, y_a, z_a, \ldots)$ must be chosen so that Eq. (17) can effectively be satisfied.

In particular, for a two body problem, as the hydrogen atom to be considered below, Eq.(17) becomes

$$\Delta_2\psi + \frac{1}{\rho}\vec{\nabla}_2\rho\cdot\vec{\nabla}_2\psi + \left[\left(E_1-\phi_{12}\right)^2 - \mathbf{A}_{12}^2 - m_1^2\right]\psi = 0, \tag{18}$$

where the Laplacian is taken in the proton coordinates $\mathbf{r}_2$, and E_1 is the electron energy. We must remark that Eq. (18) is an approximation in the O_{CM} frame.

2.3. Equations of the Electromagnetic Field

Now we consider the determination of the actual electromagnetic field, from which it comes the effective values for the field potentials. The simplest invariant Lagrangean density that can be suggested from the potentials and its partial derivates is [25].

$$\Lambda_1 = \sum_{a=1}^{N}\sum_{b\neq a=1}^{N}\left(\varsigma_a A_{ba\mu}A_{ab}^{\mu} - \frac{1}{2}F_{ab\mu\nu}F_{ab}^{\mu\nu}\right), \tag{19}$$

where $\zeta A_\mu A^\mu$ is the gauge-fixing term and ζ is a constant to be determined. The Eq.(19) is written by considering the only possible contractions that produce quadratic forms in A^μ and $F^{\mu\nu}$. This assures that linear differential equations will be obtained from the variation of Λ_1, as required by the superposition principle. The electromagnetic tensor is [25]

$$F_{ab\mu\nu} = \partial_{a\mu}A_{ba\nu} - \partial_{a\nu}A_{ba\mu}. \tag{20}$$

We consider now the following variational problem

$$\delta\int d^4x\Lambda_1 = 0, \tag{21}$$

whose execution yields the field potentials equations:

$$\partial_{a\alpha}\partial_a^{\alpha}A_{ba}^{\beta} + \varsigma_a A_{ba}^{\beta} = 0. \tag{22}$$

This equation may be seen to be invariant under a gauge transformation of the kind

$$A'^{\mu} = A^{\mu} + \partial^{\mu}\chi, \tag{23}$$

if and only if $\partial^{\mu}\chi$ also satisfies Eq.(22). Hence, the sum of any pair of solutions of Eq.(22) becomes a new accepted solution, therefore there is not a unique solution for equation Eq.(22). In order to univocally fix the field potentials, use may be made of the Lorentz gauge

$$\partial_{a\alpha}A^{\alpha}_{ba}=0\,, \tag{24}$$

which also holds for $\partial^{\mu}\chi$. In three-dimensional notation Eq.(22) and Eq.(24) become

$$\Delta_a\phi_{ab}-\frac{\partial^2\phi_{ab}}{\partial t^2}+\varsigma_a\phi_{ab}=0 \tag{25a}$$

$$\Delta_a\mathbf{A}_{ab}-\frac{1}{c^2}\frac{\partial^2\mathbf{A}_{ab}}{\partial t^2}+\varsigma_a\mathbf{A}_{ab}=\mathbf{0}\,, \tag{25b}$$

$$\frac{1}{c}\frac{\partial\phi_{ab}}{\partial t}+\nabla_a\cdot\mathbf{A}_{ab}=0\,. \tag{25c}$$

Eqs.(25) are proposed to be the set of potential equations corresponding to the microscopic version of Maxwell's Equations.

For any particular pair of indices in Eqs.(25), the associated solutions are spherical waves which can be obtained by fixing a spherical coordinate system $(r_{ab},\theta_{ab},\varphi_{ab})$ in the rest frame of one of the particles, where r_{ab} is the distance between the particles a and b. The general form of these waves is given by [26]:

$$A_{r\pm}=\phi_{\pm}=\frac{\sin\left(\pm\varphi+k\left(r\mp t\right)\right)}{r\sin\theta}\,,\quad A_{\theta\pm}=\frac{\sin\left(k\left(r\mp t\right)\right)}{r\sin\theta}\,,\quad A_{\varphi\pm}=\frac{\cos\left(\pm\varphi+k\left(r\mp t\right)\right)}{r}\,. \tag{26}$$

Here, the upper signs refer to progressive, the lower to regressive waves respectively and the subscripts a and b are omitted for simplicity. These waves form a standing pattern in the space between any pair of particles, whose effective envoltory, defined for $\theta=\frac{\pi}{2}$(plane motion) and $kr=\frac{\pi}{2}$, is given by (see Fig.1)

$$\phi_{ab}=\sqrt{\langle\phi^2\rangle}\Big|_{\theta=\pi/2}=\frac{|\sin kr|}{r}\approx\frac{1}{r}\,. \tag{27}$$

The 1/r approximation allows to get relatively simple solutions for Eq.(7), for large r, however we must stress the fact that ϕ_{ab} is finite at the origin for the radial coordinate (this discards the need of the renormalization process of the QED). The determination of the constant k in Eqs. (15) is done by observing that, in the absence of external interactions, $E_a = 0$ in the rest frame O_a of the particle a, that is, $\phi_{ab}\left(0, \frac{\pi}{2}\right) = m_a$.

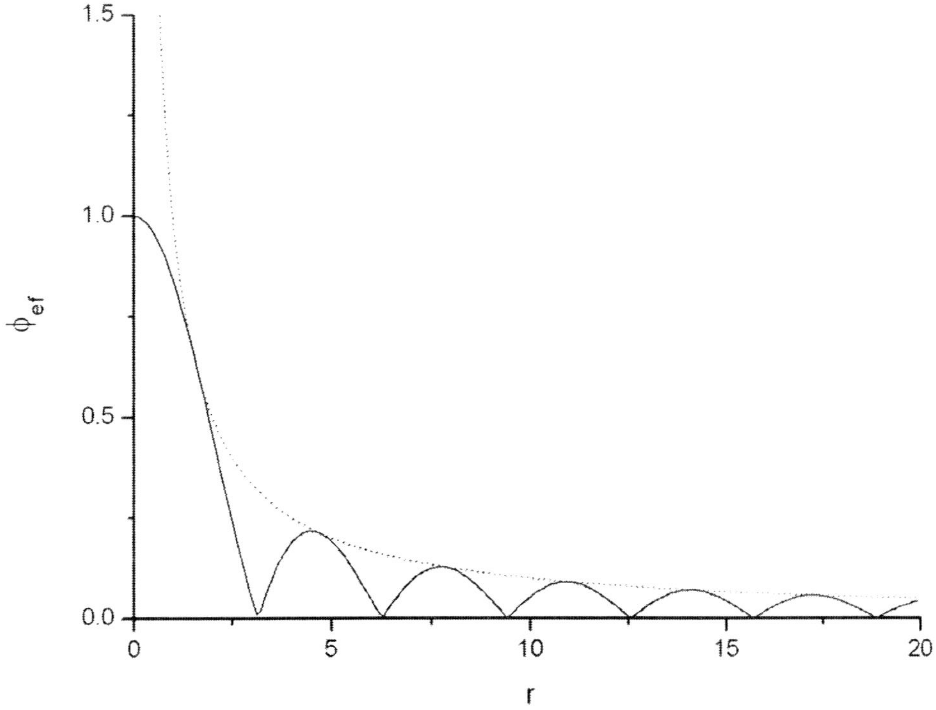

Figure 1. The effective scalar potential $\frac{|\sin kr|}{r}$, for $k = 1$, is represented by the full line while the envoltory $\frac{1}{r}$ is represented by the dotted line.

2.4. Field Stationary Conditions

Considering the Lagrangean density, Eq.(19), which satisfies Eq.(21), it is possible to obtain the following conservation equation

$$\partial_\nu T_\mu^\nu = 0. \tag{28}$$

The momentum energy tensor T_μ^ν is defined as

$$T_\mu^\nu = \Lambda_1 \delta_\mu^\nu - A_{,\mu}^\nu \frac{\partial \Lambda_1}{\partial A_{,\nu}^\beta}. \tag{29}$$

Eq.(28) expresses the invariance of Λ_1 under time-space translations. T_μ^ν is not, in general, unique since it is possible to add an arbitrary term to it, whose divergence is null. To eliminate this ambiguity, it is necessary to observe that the system must be also invariant under a four-rotation in the time-space, whose generator is the system angular tensor $M^{\mu\nu}$. In this way, we must have the additional conservation condition

$$\partial_\mu M^{\mu\nu} = 0, \tag{30}$$

where

$$M^{\mu\nu} = \int \left(x^\mu T_\alpha^\nu - x^\nu T_\alpha^\mu \right) dS^\alpha, \tag{31}$$

and the integration is performed across a hyper surface whose projections are dS^α. The condition given by Eq.(31) can only be satisfied if $T_\mu^\nu = T_\nu^\mu$, i.e., if the energy moment tensor is symmetric. This tensor is, according to Eq.(19), given by

$$T_\mu^\nu = \Lambda_1 \delta_\mu^\nu - 2 \sum_{a,b\neq a=1}^{N} F_{ab}{}^{\nu\beta} A_{ab\beta,\mu}, \tag{32}$$

which is obviously not symmetric. In order to make T_μ^ν symmetric, one subtracts from it the term $\sum_{a=1}^{N} \sum_{b\neq a=1}^{N} \left(\varsigma_a A_{ba\mu} A_{ab}^\nu + A_{ab\mu,\beta} F_{ab}^{\nu\beta} \right)$, whose divergence vanishes due to Eq.(22). The resulting tensor with correct symmetry is

$$T_\mu^\nu = \sum_{a,b\neq a=1}^{N} \left[2\left(F_{ab\mu\beta} F_{ab}{}^{\nu\beta} - \tfrac{1}{4} F_{ab\alpha\beta} F_{ab}{}^{\alpha\beta} \right) - k_{1a} \left(A_{ab\mu} A_a{}^\nu + \delta_\mu^\nu A_{ab\alpha} A_{ab}{}^\alpha \right) \right]. \tag{33}$$

The time component of the T_μ^ν divergence can be explicitly written in three-dimensional notation as

$$\frac{\partial W}{\partial t} + \nabla . \mathbf{S} = 0, \tag{34}$$

where W is the field energy density:

$$W = -T_0^0 = \sum_{a=1}^{N} \left[\mathbf{E}_a{}^2 + \mathbf{B}_a{}^2 + 2\varsigma_a \left(\phi_a{}^2 - \mathbf{A}_a{}^2 \right) \right], \tag{35}$$

$\mathbf{S}$ is the field transported momentum or energy flow vector:

$$\vec{S}\Big|_i = -T_0^i = \sum_{a,b\neq a=1}^{N}\left[2F_{ab}^{\ \ \nu\beta}F_{ab0\beta} - k_1 A_{ab0}A_{ab}^{\ \ i}\right] = \sum_{a,b\neq a=1}^{N}\left[\mathbf{E}_a \times \mathbf{B}_a - k_{1a}\phi_{ab}\mathbf{A}_{ab}\right]. \quad (36)$$

with i = 1,2,3, both of which are defined through the associated electric $\mathbf{E}$ and magnetic $\mathbf{B}$ fields:

$$\mathbf{E}\Big|_{ai} = \sum_{b\neq a=1}^{N} F_{ab0i} = -\sum_{b\neq a=1}^{N}\left(\frac{\partial \mathbf{A}_{ab}}{\partial t} + \nabla\phi_{ab}\right)\Bigg|_i, \quad (37a)$$

$$\mathbf{B}\Big|_{ai} = \tfrac{1}{2}\varepsilon_{ijk}\sum_{b\neq a=1}^{N} F_{ab}^{\ \ jk} = \sum_{b\neq a=1}^{N}\nabla_a \times \mathbf{A}_{ab}\Big|_i, \quad (37b)$$

These are obtained from the usual contractions of the electromagnetic tensor $F^{\mu\nu}$.

It must be remarked that the energy contained in the electromagnetic field will remain stationary in the volume containing the particles, if at least the average time variation of W is zero. Therefore, W can vary, but only in a periodic way, so that the direction of $\mathbf{S}$ changes periodically. This corresponds to the energy flow that periodically points outward or inward the space region containing the particles. To make W a constant it suffices to require that $\mathbf{S} = 0$ so that the stationary condition may be trivially satisfied. This may be accomplished, according to Eqs.(37), simply by imposing that $\mathbf{A} = 0$, which is the usual assumption in standard MQ calculations. Therefore, the most general conditions to assure that the system energy will remain stationary are

$$\left\langle \frac{\partial W}{\partial t}\right\rangle\Bigg|_{ciclo} = 0, \ \langle \nabla.\mathbf{S}\rangle\Big|_{ciclo} = 0. \quad (38)$$

These demand simply that $\langle \mathbf{A}\rangle\big|_{ciclo} = \mathbf{0}$ and, as a consequence, allows the possibility of stationary oscillations along the orbital dimensions, just like those waves postulated by de Broglie.

3. One, Two and Three Body Calculations

In this section, some applications will be done using this formalism to one, two and three-body problems. It yields solutions to the equation of motion by integrals of action or by solving the associated differential eigenvalue equation.

For one body problem we consider a classical particle moving in the magnetic field, and we show that the HJE solution combines the effects of the classical motion (trajectory) and the quantum effect (Landau levels) [25].

As an application of our equation for two-body problem, we obtain a correct solution for the hydrogen atom, including the energy levels fine structure using both methods: the action integral solution and a second order differential equation associated eigenvalue equation [23].

In the three-body calculations we apply an adiabatic non-relativistic approximation to our approach of the HJE in order to obtain the same results found by Hylleraas for the ion H_2^+ [27]. For the helium atom, we use a non-relativistic approximation in our approach obtaining exactly the same solution obtained by Hylleraas [28] for the ground state energy. As the Hylleraas method does not give any information about electron coordinates, we use the effective charge approximation in our approach to obtain solution for the helium atom dependent of the electrons coordinates. We found that the energy is in agreement with that obtained by literature, but we show that for the stable solution the equilibrium distance of the electrons are not the same. Then, these explain the Bohr-Sommerfeld model collapse.

3.1. Energy Levels for an Electron Moving in a Magnetic Field Through an Action Integral

As it is well known, in classical physics, an ordinary negative charged particle moving in the magnetic field produces the helicoidally trajectory along of the magnetic field. In this work, its motion is obtained by solving the Hamilton-Jacob equation with a vector potential $\mathbf{A}$, which represents the action of a strong and uniform magnetic field oriented in the direction of an arbitrary z-axis. The negative charged particle here is an electron that is moving in an external strong, uniform, magnetic field oriented in the positive z direction of a cylindrical coordinate system, has an associated vector potential given by $\mathbf{A} = -\frac{1}{2} rB\mathbf{e}_\varphi$, where $\mathbf{e}_\varphi$ is the unitary azimuthal vector. We neglect the effects of any scalar potential, i.e., we take $\phi = 0$. The Lorentz gauge $\frac{\partial \phi}{\partial t} + \nabla \cdot \mathbf{A} = 0$ is automatically satisfied with this choice for the potential.

We can write the conservative HJE, i.e., Eq.(10b) for $a = 1$ and $H = E$, as:

$$m^2 + p_r^2 + p_z^2 + \frac{J_z^2}{r^2} + eJB + \frac{e^2 B^2 r^2}{4} - E^2 = 0, \tag{39}$$

where particle subscripts are omitted and remembering that natural units ($\hbar = c = 1$) are used. Since the Hamiltonian function in Eq,(39) is separable, the associated system action function must be of the form:

$$S(\vec{r}, t) = -E\,t + \int J_z d\varphi + \int p_r dr + \int p_z dz\,. \tag{40}$$

By using the action integral quantization condition, Eq.(12b), in each coordinate, the electron angular momentum conjugated to the azimuthal coordinate is

$$\oint J_z d\varphi = 2\pi j\,, \tag{41}$$

from what we get $J = j$, or in conventional units, $J = j\hbar$.

Since there is no periodic condition in the z coordinate, the corresponding action integral cannot be evaluated and the z-axis linear momentum may have any constant value:

$$p_z = \text{constant} \tag{42}$$

so that the particle can move freely parallel to the magnetic field direction.

Now, the radial momentum can be obtained from Eq.(39) yielding

$$p_r = \pm\sqrt{\sigma_j^2 - \gamma^2 r^2 - \frac{j^2}{r^2}}, \tag{43}$$

where the constants are $\gamma = \frac{eB}{2}$, $\sigma_j^2 = \sigma^2 - p_z^2 - 2j\gamma$ and $\sigma^2 = E^2 - m^2$. The coordinates $r_{o\pm}$ for which $p_r = 0$ are given by

$$r_{o\pm} = r_o\sqrt{\left(1 \pm \sqrt{1 - \frac{2j^2}{\sigma_j^2 r_o^2}}\right)} \tag{44a}$$

$$r_o = \frac{\sigma_j}{\gamma\sqrt{2}}. \tag{44b}$$

The radial action integral can be obtained by integrating Eq. (43) along a closed path between r_{o-} to r_{o+}, the result is [16]

$$k = \frac{1}{2\pi}\oint p_r dr = \frac{\sigma^2}{2\gamma} - j. \tag{45}$$

Now, solving Eq. (45) for E, we get the electron energy levels:

$$E_{k,j,p_z} = \sqrt{m^2 + p_z^2 + neB}, \tag{46}$$

where the energy quantum numbers are:

$$n = 2j + k = 0, 1, 2, \ldots \tag{47a}$$

$$\text{k} = 0, \pm 1, \pm 2, \ldots \tag{47b}$$

$$j = 0, 1, 2, \ldots \tag{47c}$$

Eq. (46) gives the same Landau levels that were obtained by using Dirac's equation in the usual framework of relativistic quantum mechanics. Here, $k = j = 0$ corresponds to the state in which the electron is moving with constant linear momentum p_z parallel to the magnetic field. When $j > 0$ and $k = 0$ the orbits are circular with a rotation radius obtained from (44a)

$$r_o = \sqrt{\frac{j}{\gamma}} = \sqrt{\frac{2j}{eB}}, \tag{48}$$

and when $j > 0$ and $k \neq 0$ the orbits are elliptical.

From the relation between the canonical and mechanical momentum for the azimuthal coordinate calculated for the effective radius we have:

$$p_\varphi\Big|_{r_0} = \left(\frac{1}{r\sin\theta\psi}\frac{\partial\psi}{\partial\varphi} + eA_\varphi\Big|\right)_{r_0,\frac{\pi}{2}} = \frac{j}{r_0}\left(1 + \frac{er_0^2 B}{2j}\right). \tag{49}$$

And from the definition of the electron azimuthal component of the linear momentum $p_\varphi = \gamma m r_0 \omega_0$ we get the effective cyclotron frequency expression:

$$\omega_0 = \frac{eB}{\gamma_0 m}, \tag{50}$$

in which the relativistic γ_0 factor is given by:

$$\gamma_0 = \gamma(r_0) = \sqrt{1 + \frac{p_z^2 + neB}{m^2}}. \tag{51}$$

The classical trajectory may be easily seen, from Eqs.(42), (45), and (48) to be an elliptical helices.

3.2. Determination of the Energy Levels of the Hydrogen Atom

3.2.1. Determination Through Action Integrals

In spherical coordinates in the frame $O_{\mathbf{j}}$, which has its z axis parallel to $\mathbf{j} = \mathbf{r} \times \mathbf{p}$, the HJE, Eq. (10b), becomes

$$p_r^2 + \frac{j^2}{r^2} + m^2 - \left(E + \frac{\alpha}{r}\right) = 0. \tag{52}$$

where, due to the orbital periodicity, we must have $j = |\mathbf{j}| = 1, 2, ...$ From Eq.(52) above we obtain

$$p_r = \pm\sqrt{-\sigma^2 + \frac{\lambda}{r} - \frac{j'^2}{r^2}}, \tag{53}$$

where $\sigma^2 = m^2 - E^2$, $\lambda = 2\alpha E$, $j' = \sqrt{j^2 - \alpha^2}$.

The quantization of the radial momentum is given by the Sommerfeld condition [17,25]:

$$k_r = \frac{1}{2\pi}\oint p_r dr = \frac{\lambda}{2\sigma} - j' = 0, 1, 2, ... \tag{54}$$

From the algebraic inversion of Eq.(54) the electron energy eigenvalue is obtained

$$E = m\left(1 + \frac{\alpha^2}{(k_r + j')^2}\right)^{-1/2} \cong m\left\{1 - \frac{\alpha^2}{2n^2}\left[1 + \frac{\alpha^2}{n^2}\left(\frac{n}{j} - \frac{3}{4}\right)\right]\right\}, \tag{55}$$

in which $1 \leq j \leq n = k_r + j = 1, 2, 3...$ Eq.(55) was originally obtained by Sommerfeld [17], and later by Dirac [29]. For $k_r = 0$ we have the Bohr circular orbits de and for $k_r = 1, 2, ...$ the Sommerfeld ellipses:

$$r = \frac{r_{0j}}{1 + \varepsilon_{nj}\cos\varphi}, \; r_{0j} = j^2 a_B, \; \varepsilon_{nj} \cong \sqrt{1 - \frac{j^2}{n^2}}. \tag{56}$$

In order to pass from the non inertial frame $O_{\mathbf{j}}$ for the inertial O_{lab}, we consider a rotation of $O_{\mathbf{j}}$ about O_{lab}, which is due to the Thomas Precession [21,25], whose frequency is

$$\boldsymbol{\omega} = \tfrac{1}{2}\mathbf{a} \times \mathbf{v} \cong -\frac{v^2}{2mr_{0j}^2}\mathbf{j}. \tag{57}$$

From this follows that $\mathbf{j}$ has now components $\mathbf{j}_z$ and $\mathbf{j}_\theta$, it being $j = j_z + j_\theta$. As $j = 1, 2, ...$, we should have j_z , $j_\theta = \frac{1}{2}, \frac{3}{2}, \frac{5}{2}, ...$ Thus the electron "spin" appears naturally here as a consequence of the precession of the vector $\mathbf{j}$ about O_{lab}, making with its z-axis an angle $\theta_0 = \cos^{-1}\left(\frac{j_z}{j}\right)$ [21].

3.2.2. Determination Through Eigenvalues Equation

As discussed above in sect.2, the two-body problem is reduced to a problem in the particle 1 coordinates. The use of a weight function of the form $\rho = r^{-2\beta}$ in Eq.(18) results in the following electron eigenvalue effective relativistic differential equation:

$$\Delta\psi - \frac{2\beta}{r}\frac{\partial\psi}{\partial r} + \left[\left(E_1 + \frac{\alpha}{r}\right)^2 - m_1^2\right]\psi = 0\,. \tag{58}$$

The wave function which is the solution for Eq.(58) may be expressed as a product of independent functions of each variable

$$\psi(\vec{r},t) = Y(\theta)r^{\beta}R(r)\cos(m_\ell\varphi)\,. \tag{59}$$

Substituting Eq.(59) in Eq.(58), we get an eigenvalue polar equation:

$$\frac{1}{\sin\theta}\frac{d}{d\theta}\left(\sin\theta\frac{dY}{d\theta}\right) + \left[\ell(\ell+1) - \frac{m_\ell^2}{\sin^2\theta}\right]Y = 0\,, \tag{60}$$

whose solutions are the associated Legendre functions Y_{ℓ,m_ℓ} [17,18].

We also obtain a radial equation

$$\frac{1}{r^2}\frac{d}{dr}\left(r^2\frac{dR}{dr}\right) + \left[\frac{\lambda}{r} - \sigma^2 - \frac{s(s+1)}{r^2}\right]R = 0\,, \tag{61}$$

where $s(s+1) = \ell(\ell+1) + \beta(\beta-1) - \alpha^2$, from which $s_\pm \cong -\frac{1}{2} \pm \left(\ell + \frac{1}{2}\right) \pm \frac{\beta_\pm(\beta_\pm - 1) - \alpha^2}{2\ell+1}$.

In any case, the variations coming from the solutions of Eqs. (17a) satisfy

$$\delta p_r\big|_{r_{0\pm}} = \frac{1}{R}\frac{dR}{dr}\bigg|_{r_{0\pm}} = 0\,,\ \delta p_\theta\big|_{\frac{\pi}{2}} = \frac{1}{Y}\frac{\partial Y}{\partial\theta}\bigg|_{\frac{\pi}{2}} = 0\,. \tag{62}$$

The general solutions of the radial equation, which are everywhere finite, are

$$R(r)_{\pm} = r^{s_{\pm}} e^{\mp\sigma r} {}_1F_1\left[s_{\pm} + 1 - \zeta_{\pm};\ 2(s_{\pm} + 1);\ \pm 2\sigma r\right], \tag{63a}$$

where ${}_1F_1$ are the one pole hypergeometrical functions [26]. Furthermore $\zeta_{\pm}$ is given by:

$$\zeta_{\pm} = n_{0\pm} + s_{\pm} + 1 \cong n\left[1 \pm \frac{\beta_{\pm}(\beta_{\pm} - 1) - \alpha^2}{n(2\ell + 1)}\right]. \tag{63b}$$

The radial solution physically allowed may be obtained as

$$R(r) = r^{\beta+} R_{+} + b r^{\beta-} R_{-}\ , \tag{64}$$

The eigenvalues are

$$E_{1\pm} = m_1\left(1 + \frac{\alpha^2}{\zeta_{\pm}^2}\right)^{-1/2}, \tag{65a}$$

From the equivalency between the EHJ and the eigenvalue equation, we must have $\zeta_{\pm} = \zeta$, from what $n_{0+} = k_r$ and $s_{+} + 1 = j'$. Thus $j = \ell + 1$ and therefore

$$\beta_{+} = 1 + \alpha^2 / 2j\,,\ \beta_{-} = 1 + (2 - 1/2j)\alpha^2, \tag{65b}$$

so that Eq.(65a) along with Eq.(65b) gives the same result of Eq.(55) for the electron energy eigenvalue.

The orbital stability condition $R'(r_0) = 0$ is used to find b from the ${}_1F_1$ function. That becomes

$$b = -\frac{R'_{a+}(r_0)}{R'_{a-}(r_0)}, \tag{65c}$$

where we can see that $b = 0$ for $1 \le j \le n \le 2$ and $b \ne 0$ for $n > 2$.

3.3. Non-relativistic HJE Solution for the H_2^+ Molecule

The relativistic Hamiltonian Eq.(10b) for the electron is given by:

$$H_{rel} = \sqrt{p_1^2 + m_1^2} - \alpha\left(\frac{1}{r_a} + \frac{1}{r_b}\right), \tag{66}$$

where $\alpha = e^2$ and, as usual, $\hbar = c = 1$. The non-relativistic version of the Hamiltonian in Eq. (66) for the electron and the system Hamiltonian in the adiabatic approximation are respectively given by:

$$H_1 = \frac{p_1^2}{2m_1} - \alpha\left(\frac{1}{r_a} + \frac{1}{r_b}\right), \tag{67a}$$

$$H \cong H_1 + \frac{\alpha}{R}. \tag{67b}$$

Here r_a e r_b are the distances from protons a and b to the electron and R is the distance between the protons. The electron problem can be separated only in the confocal elliptic coordinate system [26] defined by

$$\xi = \frac{r_a + r_b}{R},\ \eta = \frac{r_a - r_b}{R}, \tag{68}$$

where the η, ζ ranges are $-1 \le \eta \le +1$ and $-1 \le \xi < \infty$.

The Hamiltonian H_1 in this new system of coordinates is written as

$$H_1 = \frac{H_\xi + H_\eta}{\xi^2 - \eta^2}, \tag{69a}$$

where H_ξ and H_η are given by:

$$H_\xi = \frac{1}{2m_1}\left[\left(\xi^2 - 1\right)p_\xi^2 + \frac{k_\varphi^2}{\xi^2 - 1}\right] - 2\alpha\xi\ ,\ H_\eta = \frac{1}{2m_1}\left[\left(1 - \eta^2\right)p_\eta^2 + \frac{k_\varphi^2}{1 - \eta^2}\right] \tag{69b}$$

In this work, instead of the direct solution of the HJE, Eq.(10b), with $H_1 = E_1$, we consider the corresponding eigenvalue differential equation, Eq.(17):

$$\begin{aligned}&\frac{\partial}{\partial\xi}\left[\left(\xi^2 - 1\right)\frac{\partial\psi}{\partial\xi}\right] + \frac{\partial}{\partial\eta}\left[\left(1 - \eta^2\right)\frac{\partial\psi}{\partial\eta}\right] + \\ &+\left(\frac{1}{\xi^2 - 1} + \frac{1}{1 - \eta^2}\right)\frac{\partial^2\psi}{\partial\xi^2} + \left[\tfrac{1}{2}R^2\left(\xi^2 - \eta^2\right)E_1 + 2\alpha R\xi\right]\psi = 0,\end{aligned} \tag{70}$$

whose solution $\psi(\xi,\eta,\varphi)$ is connected with the old momenta through the relations

$$p_\xi = \frac{1}{\psi}\frac{\partial \psi}{\partial \xi}\,,\quad p_\eta = \frac{1}{\psi}\frac{\partial \psi}{\partial \eta}. \tag{71}$$

This solution is expected to be separable as a product of functions in each variable [30] :

$$\psi(\xi,\eta,\varphi) = e^{ik_\varphi \varphi} Y(\eta) Q(\xi), \tag{72}$$

yielding the following set of ordinary differential equations:

$$\frac{d}{d\xi}\left[\left(\xi^2-1\right)\frac{dQ}{d\xi}\right]+\left(\lambda+2\alpha R\xi+\tfrac{1}{2}R^2\xi^2E_1-\frac{k_\varphi^2}{\xi^2-1}\right)Q=0 \tag{73a}$$

$$\frac{d}{d\eta}\left[\left(1-\eta^2\right)\frac{dY}{d\eta}\right]-\left(\lambda+\tfrac{1}{2}R^2\eta^2E_1+\frac{k_\varphi^2}{1-\eta^2}\right)Y=0. \tag{73b}$$

A solution for Eq.(10) can immediately be found by the Hylleraas differential method [27], starting from the expansion in the $P_\ell^{k_\varphi}(\eta)$ Legendre polynomials:

$$Y(\eta) = \sum_{\ell=k}^{\infty} c_\ell P_\ell^{k_\varphi}(\eta), \tag{74}$$

For k_φ=0,1,2,3... we will have molecular orbitals σ,π,δ,γ,.... On the other hand for $\ell = 0,2,4,...$ we obtain levels s,d,g,j,l... (even atomic orbitals) and for $\ell = 1,3,5,...$ levels p,f,i,k... (odd atomic orbitals).

The recurrence rule is determined by substituting Eq.(75) into Legendre's equation

$$\frac{d}{d\eta}\left[\left(1-\eta^2\right)\frac{dY}{d\eta}\right]+\left(n(n+1)-\frac{k_\varphi^2}{1-\eta^2}\right)Y=0, \tag{75}$$

from which we obtain the following recurrence relation for coefficients c_ℓ:

$$A_{\ell-2}c_{\ell-2}+B_\ell c_\ell+C_{\ell+2}c_{\ell+2}=0, \tag{76}$$

where,

$$A_{\ell-2} = \frac{(\ell - k_\varphi)(\ell - 1 - k_\varphi)b^2}{(2\ell-3)(2\ell-1)}, \quad (77)$$

$$B_\ell = \frac{(\ell + 1 - k_\varphi)(\ell + 1 + k_\varphi)b^2}{(2\ell+3)(2\ell+1)} + \frac{(\ell - k_\varphi)(l + k_\varphi)b^2}{(2\ell+1)(2\ell-1)} - \lambda - \ell(\ell+1), \quad (78)$$

$$C_{\ell+2} = \frac{(\ell + 2 + k_\varphi)(\ell + 1 + k_\varphi)b^2}{(2\ell+3)(2\ell+5)}, \quad (79)$$

and $b^2 = \frac{1}{2}R^2E_1$. Equations (77-79) can be considered as an infinite set of linear homogeneous equations, requiring the determinant of matrix coefficient to vanish:

$$\begin{vmatrix} \frac{1}{3}(1-k)(1+k)b^2 + k^2b^2 - \lambda & \frac{1}{15}(2+k)(1+k)b^2 & 0 & 0 & \dots \\ \frac{1}{3}(2-k)(1-k)b^2 & \frac{11}{21}b^2 - \frac{2}{21}b^2k^2 - \lambda - 6 & \frac{1}{63}(4+k)(3+k)b^2 & 0 & \dots \\ 0 & \frac{1}{35}(4-k)(3-k)b^2 & \frac{39}{77}b^2 + \frac{2}{77}b^2k^2 - \lambda - 20 & \dots & \dots \\ 0 & 0 & 0 & \dots & \dots \\ \dots & \dots & \dots & \dots & \dots \end{vmatrix}. \quad (80)$$

Equation (73a) on the other hand has been solved by the Jaffe's method [31] starting from the trial solution

$$Q(\xi) = (\xi^2 + 1)^{k/2}(\xi + 1)^\sigma y\left[\frac{(\xi - 1)}{(\xi + 1)}\right]e^{-b\xi}, \quad (81)$$

where

$$\sigma = \frac{R}{b} - 1 - k \quad (82)$$

$$y = \sum_{r=0}^{\infty} g_r \left(\frac{\xi - 1}{\xi + 1}\right)^r, \quad (83)$$

whose recurrence relations are given by

$$A_{r,r-1}g_{r-1} + B_{r,r}g_r + C_{r,r+1}g_{r+1} = 0, \tag{84}$$

$$A_{r,r-1} = (r-1-\sigma)(r-1-\sigma-k), \tag{85}$$

$$B_{r,r} = \lambda - b^2 + 2b\sigma + (\sigma+k)(k+1) + 2r(\sigma-2b) - 2r^2, \tag{86}$$

$$C_{r,r+1} = (r+1)(r+k+1). \tag{87}$$

The infinite system represented by Eq.(84) has then been solved by the same method as the one used for Eq.(76), i.e., by solving the determinant:

$$\begin{vmatrix} B_{00} & C_{01} & 0 & 0 & 0 & \dots \\ A_{10} & B_{11} & C_{12} & 0 & 0 & \dots \\ 0 & A_{21} & B_{22} & C_{23} & 0 & \dots \\ 0 & 0 & A_{32} & B_{33} & C_{34} & \dots \\ 0 & 0 & 0 & A_{43} & B_{44} & \dots \\ \dots & \dots & \dots & \dots & \dots & \dots \end{vmatrix} = 0. \tag{88}$$

However, up to this time only a numerical solution is possible [32]. We can use, for example, the Newton-Raphson method to solve the equations (84) and (76) together, allowing us to find the electronic energy $E = E_1 + 1/R$ and the parameter λ. With this method, it is possible to find the equilibrium distances R and electronic energies for all states with a reasonably good agreement with experimental results of ions H_2^+.

Proceeding further, by using Eqs.(74) and (81) into Eq. (72) we find the wave function according to:

$$\Psi(\eta,\xi,\phi) = Ne^{ik_\varphi\varphi}\sum_{\ell=k}^{\infty} c_\ell P_\ell^{k_\varphi}(\eta)\left(\xi^2+1\right)^{k/2}(\xi+1)^\sigma \sum_{r=0}^{\infty} g_r\left(\frac{\xi-1}{\xi+1}\right)^r e^{-b\xi}, \tag{89}$$

where N is the normalization constant which is to be determined by the condition

$$\int_{-1}^{\infty}\int_{-1}^{1}\int_{0}^{2\pi} N^2\left|\Psi(\eta,\xi,\phi)\right|^2 dV = 1. \tag{90}$$

The elliptic confocal coordinates (η,ξ,ϕ) relate to Cartesian coordinates (x,y,z) through transformation equations given by

$$\begin{aligned} x &= \frac{R}{2}\sqrt{(1-\eta^2)(\xi^2+1)}\cos(\phi) \\ y &= \frac{R}{2}\sqrt{(1-\eta^2)(\xi^2+1)}\sin(\phi) \\ z &= \frac{R}{2}\eta\xi \\ dV &= \frac{R^3}{8}(\xi^2-\eta^2)d\xi d\eta d\phi \end{aligned} \tag{91}$$

The expansion coefficients are determined from Eqs. (84) and (76) by the continuous fraction method:

$$\begin{aligned} &\frac{c_\ell}{c_{\ell-2}} = \frac{-A_{\ell-2}}{B_\ell + C_{\ell+2}\dfrac{c_{\ell+2}}{c_\ell}} \\ &\frac{g_r}{g_{r-1}} = \frac{-A_{r,r-1}}{B_{r,r} + C_{r,r+1}\dfrac{g_{r+1}}{g_r}} \\ &\ell \to \infty \quad \frac{c_{\ell+2}}{c_\ell} = 0 \\ &r \to \infty \quad \frac{g_{r+1}}{g_r} = 0 \end{aligned} \tag{92}$$

3.4. Non-relativistic HJE Approach to Study the He Atom

3.4.1. Exact Solution of the Helium Atom Using a Variational Method

The non-relativistic Hamiltonian function Eq.(12a), of the system is written in terms of the radial and relative momenta $\vec{p}_{r1}$, $\vec{p}_{r2}, \vec{p}_{r12}$ respectively and the electrons angular momenta $\mathbf{J}_1, \mathbf{J}_2$, that is

$$\begin{aligned} 2H = {} & p_{r1}^2 + p_{r2}^2 + p_{r12}^2 + \frac{r_1^2 - r_2^2 + r_{12}^2}{r_1 r_{12}} p_{r1} p_{r12} + \frac{r_2^2 - r_1^2 + r_{12}^2}{r_2 r_{12}} p_{r2} p_{r12} - \frac{2}{r_1} - \frac{2}{r_2} + \frac{1}{r_{12}} + \\ & + \Big[2(\vec{p}_{r1} + \vec{p}_{r12}) \cdot (a_{11}\vec{J}_1 + a_{12}\vec{J}_2) + (a_{11}\vec{J}_1 + a_{12}\vec{J}_2)^2 + \\ & + 2(\vec{p}_{r2} - \vec{p}_{r12}) \cdot (a_{21}\vec{J}_1 + a_{22}\vec{J}_2) + (a_{21}\vec{J}_1 + a_{22}\vec{J}_2)^2 \Big] \end{aligned} \tag{93}$$

Where the original momenta are related to the transformed momenta through the relations

$$\vec{p}_1 = \vec{\nabla}_1 S = \vec{p}_{r1} + \vec{p}_{r12} + a_{11}\vec{J}_1 + a_{12}\vec{J}_2 \quad (94a)$$

$$\vec{p}_2 = \vec{\nabla}_2 S = \vec{p}_{r2} - \vec{p}_{r12} + a_{21}\vec{J}_1 + a_{22}\vec{J}_2, \quad (94b)$$

with the coefficients

$$a_{12} = \frac{\hat{r}_1 \cdot \hat{r}_{12} p_{r12}}{\hat{r}_1 \cdot \vec{J}_2},\ a_{11} = \frac{\hat{r}_{12} \cdot \left(\vec{p}_1 - \vec{p}_{r1} - a_{12}\vec{J}_2\right) - p_{r12}}{\hat{r}_{12} \cdot \vec{J}_1}, \quad (95a)$$

$$a_{21} = \frac{\hat{r}_2 \cdot \vec{p}_{r12}}{\hat{r}_2 \cdot \vec{J}_1},\ a_{22} = \frac{\hat{r}_{12} \cdot \left(\vec{p}_2 - \vec{p}_{r2} - a_{21}\vec{J}_1\right) + p_{r12}}{\hat{r}_{12} \cdot \vec{J}_2}. \quad (95b)$$

The motion problem given by the constraint $H = E = \text{constant}$, correspond to the stable trajectory against the small variations of energy and the momentum obtained by HJE. So, it can be solved using the auxiliary function $S = \ln\psi\left(r_1, r_2, r_{12}, \vec{J}_1, \vec{J}_2\right)$ that represents the unique relevant variables of the system:

$$p_{r1} = \frac{\partial \ln\psi}{\partial r_1},\ p_{r2} = \frac{\partial \ln\psi}{\partial r_2},\ p_{r12} = \frac{\partial \ln\psi}{\partial r_{12}}. \quad (96)$$

We consider here explictly the variational problem

$$\delta \int (H - E)\psi^2 r_1^a r_2^b r_{12}^c dr_1 dr_2 dr_{12} = 0, \quad (97)$$

that is

$$\delta \int dr_1 dr_2 dr_{12} \left\{ \frac{r_1^a r_2^b r_{12}^c}{2m}\left[\left(\frac{\partial\psi}{\partial r_1}\right)^2 + \left(\frac{\partial\psi}{\partial r_2}\right)^2 + 2\left(\frac{\partial\psi}{\partial r_{12}}\right)^2\right] + \frac{r_1^{a-1} r_2^b r_{12}^{c-1}}{2m}\left(r_1^2 - r_2^2 + r_{12}^2\right)\frac{\partial\psi}{\partial r_1}\frac{\partial\psi}{\partial r_{12}} + \right.$$
$$\left. -e^2\left[r_1^a r_2^b r_{12}^c E + 2r_{12}^c\left(r_1^a r_2^{b-1} + r_1^{a-1} r_2^b\right) - r_1^a r_2^b r_{12}^{c-1}\right]\psi^2 \right\} = 0 \quad (98)$$

The execution of the variation yields

$$\frac{\partial^2\psi}{\partial r_1^2}+\left[\frac{a+1}{r_1}+\frac{c-1}{2r_1r_{12}^2}\left(r_1^2-r_2^2+r_{12}^2\right)\right]\frac{\partial\psi}{\partial r_1}+\frac{\partial^2\psi}{\partial r_2^2}+\left[\frac{b+1}{r_2}+\frac{c-1}{2r_2r_{12}^2}\left(r_2^2-r_1^2+r_{12}^2\right)\right]\frac{\partial\psi}{\partial r_2}+$$
$$2\frac{\partial^2\psi}{\partial r_{12}^2}+\left[\frac{2(c+1)}{r_{12}}+\frac{a-1}{2r_{12}r_1^2}\left(r_1^2-r_2^2+r_{12}^2\right)+\frac{b-1}{2r_{12}r_2^2}\left(r_2^2-r_1^2+r_{12}^2\right)\right]\frac{\partial\psi}{\partial r_{12}}+$$
$$\frac{r_1^2-r_2^2+r_{12}^2}{r_1r_{12}}\frac{\partial^2\psi}{\partial r_1\partial r_{12}}+\frac{r_2^2-r_1^2+r_{12}^2}{r_2r_{12}}\frac{\partial^2\psi}{\partial r_2\partial r_{12}}-j^2\left[\frac{4r_2^2}{\left(r_2^2+r_1^2-r_{12}^2\right)^2}+\frac{1}{r_2^2}\right]\psi+ \tag{99}$$
$$+2m\left[E+2e^2\left(\frac{1}{r_1}+\frac{1}{r_2}\right)-\frac{e^2}{r_{12}}\right]\psi=0$$

The associated boundary conditions are given by:

$$\left.\frac{\partial\psi}{\partial r_1}\right|_{r_{10},r_{20},r_{120}}=\left.\frac{\partial\psi}{\partial r_2}\right|_{r_{10},r_{20},r_{120}}=\left.\frac{\partial\psi}{\partial r_{12}}\right|_{r_{10},r_{20},r_{120}}=0\,. \tag{100}$$

And $j = 0,1,2,\ldots$, is the total angular momentum amplitude of the system.

If we put $a=b=c=1$ and $j=0$ in Eq.(98) we get the equation valid for the ground state

$$\delta\int dr_1dr_2dr_{12}\left\{r_1r_2r_{12}\left[\left(\frac{\partial\psi}{\partial r_1}\right)^2+\left(\frac{\partial\psi}{\partial r_2}\right)^2+2\left(\frac{\partial\psi}{\partial r_{12}}\right)^2\right]+r_2\left(r_1^2-r_2^2+r_{12}^2\right)\frac{\partial\psi}{\partial r_1}\frac{\partial\psi}{\partial r_{12}}+\right. \tag{101}$$
$$\left.+r_1\left(r_2^2-r_1^2+r_{12}^2\right)\frac{\partial\psi}{\partial r_2}\frac{\partial\psi}{\partial r_{12}}-2\left[E+2r_{12}\left(r_1+r_2\right)-r_1r_2\right]\psi^2\right\}=0.$$

Executing the variation in Eq.(101) we reobtain the well known Hyleraas equation:

$$\frac{\partial^2\psi}{\partial r_1^2}+\frac{2}{r_1}\frac{\partial\psi}{\partial r_1}+\frac{\partial\psi^2}{\partial r_2^2}+\frac{2}{r_2}\frac{\partial\psi}{\partial r_2}+2\frac{\partial\psi^2}{\partial r_{12}^2}+\frac{4}{r_{12}}\frac{\partial\psi}{\partial r_{12}}+$$
$$+\frac{r_1^2-r_2^2+r_{12}^2}{r_1r_{12}}\frac{\partial^2\psi}{\partial r_1\partial r_{12}}+\frac{r_2^2-r_1^2+r_{12}^2}{r_2r_{12}}\frac{\partial^2\psi}{\partial r_2\partial r_{12}} \tag{102}$$
$$-2\left[E+2\left(\frac{1}{r_1}+\frac{1}{r_2}\right)-\frac{1}{r_{12}}\right]\psi=0.$$

Eq.(102) was in fact Hyleraas starting pointing and apparently he didn't realized the direct connection between the Hamiltonian into the integral equation, Eq.(101), and the corresponding eigenvalue differential equation. He solved Eq. (102) directly, by calculating the integrals of an expansion of the wave function expressed in the new variables

$$\xi = r_1 + r_2,\ \eta = r_2 - r_1,\ \rho = r_{12}, \tag{103}$$

where, for the sake of simplicity he assumed that $\eta \geq 0$, because the electrons are identical.

The expansion was given as $\psi = \varphi(k\xi, k\eta, k\rho)$, where k is a scale factor and $\varphi(\xi,\eta,\rho)$ in an irreducible function defined by:

$$\varphi = e^{-\frac{\xi}{2}} \sum_{n,\ell,m=0}^{\infty} c_{n,2\ell,m} \xi^n \eta^{2\ell} \rho^m . \tag{104}$$

Therefore, the variational problem Eq.(101) becomes the system of equations:

$$\delta\left(k^2 M - kL - \tfrac{1}{4} EN\right) = 0, \tag{105}$$

where the coefficients are given by

$$\begin{aligned} M = \int d\xi d\eta d\rho \Bigg\{ & \rho\left(\xi^2 - \eta^2\right)\left[\left(\frac{\partial\varphi}{\partial\xi}\right)^2 + \left(\frac{\partial\varphi}{\partial\eta}\right)^2 + 2\left(\frac{\partial\varphi}{\partial\rho}\right)^2\right] + \\ & +2\xi\left(\rho^2 - \eta^2\right)\frac{\partial\varphi}{\partial\xi}\frac{\partial\varphi}{\partial\rho} + +2\eta\left(\xi^2 - \rho^2\right)\frac{\partial\varphi}{\partial\eta}\frac{\partial\varphi}{\partial\rho} \Bigg\} \end{aligned} \tag{106}$$

$$L = \int d\xi d\eta d\rho \left(2\xi\rho - \frac{\xi^2 - \eta^2}{4}\right)\varphi^2 \tag{107}$$

$$N = \int d\xi d\eta d\rho \left(\xi^2 - \eta^2\right)\rho\varphi^2 . \tag{108}$$

The original Hylerraas solutions [28] are obtained by using the integration coefficients

$$\int_0^{\infty} d\xi \int_0^{\xi} d\rho \int_0^{\rho} d\eta\, \xi^p \eta^q \rho^r e^{-\xi} = \frac{(p+q+r+2)!}{(q+1)(q+r+2)}, \tag{109}$$

which make possible to determine the c_i coefficients of Eq. (104) through the minimum energy condition

$$\frac{\partial E}{\partial c_i} = 0, \quad i = 0,1,2,\ldots,n. \tag{110}$$

The process results in a infinity homogeneous linear system of equations

$$\tfrac{1}{4}E\frac{\partial N}{\partial c_0}-k^2\frac{\partial M}{\partial c_0}+k\frac{\partial L}{\partial c_0}=0, \tag{111}$$

$$\tfrac{1}{4}E\frac{\partial N}{\partial c_1}-k^2\frac{\partial M}{\partial c_1}+k\frac{\partial L}{\partial c_1}=0,$$

....

$$\tfrac{1}{4}E\frac{\partial N}{\partial c_n}-k^2\frac{\partial M}{\partial c_n}+k\frac{\partial L}{\partial c_n}=0,$$

This system was solved by imposing that the determinant of the system becomes null, which in practice implies the determinant be truncated at a given order of precision. Through this process, Hylerras obtained the Helium atom ground state energy in a very good agreement with experimental measurements [28].

3.4.2. Helium Atom Approximate Solution

We start the calculation [23] by writing the system Hamiltonian from Eq.(12a) in the center-of-mass frame, in the limit of small velocities and neglecting the nucleus motion:

$$H=\frac{p_1^2}{2m_1}+\frac{p_2^2}{2m_1}-e^2\left(\frac{2}{r_1}+\frac{2}{r_2}-\frac{1}{r_{12}}\right). \tag{112}$$

where $r_{12}=r_1+r_2$, and m_1 corresponds to the electron mass. The system total angular momentum is conserved:

$$\vec{J}_{CM}\cong\vec{J}_1+\vec{J}_2 \tag{113}$$

in which the angular momentum of each electron is given by $\vec{J}_a=\vec{r}_a\times\vec{p}_a$.

In order to solve the problem of motion for the system we shall express it as an effective problem for one particle in the electron coordinates, having the other electron coordinates as a parameter. Once the solution has is obtained, we can then vary the coordinates of the second electron and get the equilibrium configuration in which the system energy assumes a minimum value. The problem of motion represented by the constraint $H=E=\text{constant}$ can therefore be written, in atomic units, and considering only effective circular orbits, i.e., assuming that $p_a^2=\frac{j_a^2}{r_a^2}$, as

$$H_1\cong\frac{j_1^2}{2r_1^2}-\frac{Z}{r_1}, \tag{114}$$

where the effective system charge Z is given by

$$Z = 2 + r_1 \left[\frac{2}{r_2} - \frac{1}{r_{12}} - \frac{j_2^2}{r_2^2} \right], \tag{115}$$

which reflects the set of all motion variables of each particle which can have any influence over the possibility of stationary states formation of the charged three-body system.

From the angular momentum action integrals taken into account in Eq.(12b), we conclude that $j = j_1 + j_2$ should be an integer number and, therefore, neglecting internal degrees of freedom due to orbital precessions, we should also have j_1, j_2 as integer numbers. Consequently, being the orbits of circular shape, we obtain for the equilibrium radius r_{1_0} the value

$$r_{10} = \frac{j_1^2}{Z}, \tag{116}$$

from which the binding energy is also determined:

$$H_1 \cong -\frac{Z^2}{2\left(j_1^2 - \alpha^2 Z^2\right)}. \tag{117}$$

The minimization process was conducted using the software *Maple*, in which we have maintained $j_1 = j_2 = 1$ and chosen an initial value for r_{1_0}, and we have then determined the minimum value for H_1 with respect to r_2, from where the value of the point of minimum r_{2_0} results; hence, we have a new value for r_{1_0}. The process was repeated until the input and output values of r_{1_0} converged to the same number.

We have obtained, for the atomic ground-state, the following results for the equilibrium distances:

$$r_{1_0} \cong 0.4166 \text{ au and } r_{2_0} = 0.6000 \text{ au}. \tag{118}$$

As a consequence of the first electron effective charge is $r_{1_0} < r_{2_0}$. This is essentially the total nucleus charge (larger than the second electron effective charge). In the original Bohr model, it follows from the hypothesis that the motion configuration in both radii was the same became unstable against small perturbations. Now it becomes perfectly understandable the supposed Bohr-Sommerfeld model collapse.

We have obtained for the He atom the energy value

$$E_1(He) \simeq -78.38\ eV, \tag{119}$$

with respect to the rest mass, which is only 5% larger than the exact value −79.01 eV. From Eqs. (114) and (115) we can also obtain the binding energy for the He^+ ion ground-state by taking the limit in which r_2 goes to infinite, resulting

$$E_1\left(\mathrm{He}^+\right) \simeq 54.40\ eV, \tag{120}$$

what agrees perfectly with the experimental value and with the one-electron atomic theory [33] and it becomes clear that the small discrepancy that arose in the former item is due to errors introduced by the approximations used to solve the problem analytically. Finally, we obtained for the He atom binding energy, relatively to the He^+ ion ground-state, the following value

$$\Delta E_1 = E_1\left(\mathrm{He}^+\right) - E_1(\mathrm{He}) \simeq 23.98\ eV, \tag{121}$$

which is about 2% smaller than the experimental value 24.57 eV [34,35].

4. Conclusion

It must be considered that the approach used in this work does not have the same precision of QED, but give results in agreement with other methods found in the literature. We expect in future works to apply this method to atomic and molecular system with more than two electrons with lower computational effort than the current procedure of the quantum chemistry.

We hope we have given a first step toward the construction of a mathematical model of the physical reality which makes possible its real visualization, as desired by Einstein, since the beginning of the quantum theory. However, here, we are not trying to create a new theory of hidden variables, as he tried, but to understand the meaning of the variables already existing inside the usual QM within a realistic and deterministic-like conception of the physical nature which is to be applied to atomic and molecular systems.

We have shown in this work that it is possible to obtain the quantization from the continuum time-space geometry in which particles are immersed. The resulting equation of motion is the Hamilton-Jacobi equation, the solutions of which are quantized by Sommerfeld-like action integrals. Alternatively, the conserved momenta can be obtained from an associated Schrödinger-like eigenvalue differential equation. As examples of applications of this method, we have obtained new solutions or reanalyzed old solutions of one, two and three-body problems. The results being in, at least, reasonable agreement with the literature. It is also shown that the new solutions also take into account, through the ordinary orbital angular momentum, the "spin" effects of the electron [21].

Finally, it must be remarked that the well-known difficulties relative to the current versions of the unification problem do not exist in our model. Quantum and relativistic aspects are already unified since its original conception, from which our equations are obtained.

References

[1] Johnson, W. H.; Mohr, P.; Sucher, J. *Relativistic, quantum electrodynamic, and weak interaction effects in atoms : Santa Barbara, CA* 1988; American Institute of Physics: New York, 1989.

[2] Pyykko, P. *Chem. Rev.* 1988, 88, 563-594.

[3] Sapirstein, J. *Rev. Mod. Phys.* 1998, 70, 55-76.

[4] A. Einstein, B. P., N. Rosen *Physical Review* 1935, 47, 777-780.

[5] Bohr, A.; Mottelson, B. R.; Ulfbeck, O. *Found. Phys.* 2004, 34, 405-417.

[6] Griffiths, R. B. *Found. Phys.* 2003, 33, 1423-1459.

[7] Marchildon, L. *Found. Phys.* 2004, 34, 1453-1466.

[8] Birkhoff, G.; Neumann, J. v. *Ann. Math.* 1936, 37, 823-843.

[9] Pais, A. *Phys. Rev. Lett.* 1961, 7, 291-&.

[10] Dixon, G. M. *Division algebras : octonions, quaternions, complex numbers, and the algebraic design of physics*; Kluwer Academic Publishers: Dordrecht ; Boston, 1994.

[11] Stueckelberg, E. C. G. *Helvetica Physica Acta* 1960, 33, 727-752.

[12] Schrödinger, E. *Ann. Physik* 1926, 79, 361-376.

[13] Bohm, D.; Hiley, B. J.; Routledge ;: New York, NY, 1993, p 204-229.

[14] Bohm, D. *Physical Review* 1952, 85, 166-179.

[15] Bohm, D.; Schiller, R.; Tiomno, J. *Nuovo Cimento Supp.* 1955, 1, 48-66.

[16] Goldstein, H. In *Addison-Wesley series in physics*; 2nd ed.; Addison-Wesley: Reading, Mass. ; London, 1980, p 438-498.

[17] Sommerfeld, A. *Ann. Physik* 1916, 51, 1-94.

[18] Schrödinger, E. *Ann. Physik* 1926, 81, 109-139.

[19] Campos, J. A.; Nascimento, D. L.; Cavalcante, D. T.; Fonseca, A. L. A.; Nunes, A. O. C. *Int. J. Quantum Chem* 2006, 106, 2587-2596.

[20] Campos, J. A.; Nascimento, D. L.; Wolf, M.; Fonseca, A. L. A.; Nunes, O. A. C. *THEOCHEM*, 2006, 769, 39-46.

[21] Fonseca, A. L. A.; Nascimento, D. L.; Miranda, D. F. *THEOCHEM*, 2006, 769, 27-31.

[22] Nascimento, D. L.; Fonseca, A. L. A. *Int. J. Quantum Chem* 2003, 95, 274-280.

[23] Nascimento, D. L.; Fonseca, A. L. A. *International Journal of Quantum Chemistry* 2006, 106, 2779-2789.

[24] Nascimento, D. L.; Fonseca, A. L. A.; Portilho, O. *Int. J. Quantum Chem* 2005, 103, 505-508.

[25] Landau, L. D.; Lifshiftz, E. M.; 4th rev. English ed.; Landau, L. D., Ed.; Pergamon: Oxford, 1987, p 74-99.

[26] Morse, P. M.; Feshbach, H. In *International series in pure and applied physics.*; McGraw-Hill: New York ; London, 1953, p 492-674.

[27] Hylleraas, E. A. *Z. Physik* 1931, 71, 739-763.

[28] Hylleraas, E. A. *Z. Physik* 1929, 54, 347-366.

[29] Dirac, P. A. M. *Proc. Roy. Soc. A* 1928, 117, 610-624.
[30] Schiff, L. I. In *International series in pure and applied physics*; 3rd ed.; McGraw-Hill: New York, 1968, p 466-489.
[31] Jaffé, G. *Z. Physik* 1934, 87, 535-544.
[32] Wind, H. *The Journal of Chemical Physics* 1965, 42, 2371-2373.
[33] Karplus, M.; Porter, R. N.; Benjamin/Cummings: Menlo Park, Calif ; London, 1970, p 165-242.
[34] Bergeson, S. D.; Balakrishnan, A.; Baldwin, K. G. H.; Lucatorto, T. B.; Marangos, J. P.; McIlrath, T. J.; O'Brian, T. R.; Rolston, S. L.; Sansonetti, C. J.; Wen, J.; Westbrook, N.; Cheng, C. H.; Eyler, E. E. *Phys. Rev. Lett.* 1998, 80, 3475-3478.
[35] Eikema, K. S. E.; Ubachs, W.; Vassen, W.; Hogervorst, W. *Phys. Rev. A* 1997, 55, 1866-1884.

In: Quantum Chemistry Research Trends
Editor: Mikas P. Kaisas, pp. 205-224
ISBN: 978-1-60021-620-6

Chapter 6

MOLECULAR DYNAMICS SIMULATION OF CHEMICAL PROCESSES IN SOLUTION USING POTENTIALS BASED ON SOLUTE-SOLVENT INTERACTION ENERGY COMPONENTS: APPLICATION TO STRUCTURAL, DIELECTRIC, AND ENERGY PROPERTIES

S. Tolosa Arroyo, J. A. Sansón Martín and A. Hidalgo García
Departamento de Química Física, Facultad de Ciencias, Universidad de Extremadura

Abstract

A molecular dynamics simulation of aqueous solutions at infinite dilution for several molecules was performed using Lennard-Jones 12-6-1 potentials to describe the solute-solvent interactions, and TIP3P to describe the water-water interactions. The Morokuma decomposition scheme of ab initio interaction energies at the SCF level and the ESIE solute atomic charges were used to reproduce the molecular parameters of the solute-water interaction potential. Some improvements, such as including the dispersion component evaluated at MP2 level in the potential function and the extension of this procedure to reactive systems, have been added in order to improve the molecular properties results (solvation energies, dielectric constant, relaxation time, hydration shell, hydrogen bond lifetimes, reaction and activation free energies, etc.). The results showed that the functions that use the EX-PL-DIS-ES interaction model lead to good values of the structural, dielectric, and energy properties when they are compared with those from using other potentials, and with the available theoretical and experimental data.

Introduction

The study, at microscopic level, of chemical reactions and processes in solution is one of the topics of most current interest in quantum chemistry [1-10]. One of the major difficulties in this type of study is the great number of interactions that take place, so that a full quantum mechanical treatment is costly, and recourse is made to approximate methods. Although there exist procedures with an acceptable response to study the solvation of chemical systems, we shall use the classical method of molecular mechanics [11] to calculate the structural and energy properties related to the solvation of molecules [12-13].

In all our studies the description of the solute-solvent interaction will use the Lennard-Jones 12-6-1 analytical function where the molecular parameters come from fitting solute-solvent interaction energies calculated at the *ab initio* level, instead of using the geometric-mean combining rules [14-22]. The solute charges are derived from a fit of the electrostatic component (ES) of the bimolecular solute-water interaction energy, which will henceforth be denoted as ESIE. With respect to the molecular parameters that describe the van der Waals contribution of the LJ(12-6) interaction potential, we employed a fitting procedure based on using different components of the interaction energy – in particular, the repulsion-exchange (EX), polarization (PL), and dispersion (DIS) components – to describe the repulsive and attractive contributions.

Knowledge of the energy curves that guide fluctuations of the solvent is particularly important in the study of chemical processes in solution because it allows one to calculate the activation energy of the process without the necessity of determining the reaction path and the complex structure in the transition state. Outstanding among the different theories proposed is the well-known Marcus relationship [23-26] which describes the free-energy curves of the reactant (R) and product (P) states as quadratic functions of the solvent polarization coordinate. In this scheme, the transition state is associated with the intersection point of the two curves, and the reaction and reorganization energies are obtained directly from relationships between points of these two curves. One of the most commonly employed reaction coordinates is the difference in the solute-solvent interaction energy of a given set of solvent molecules in the presence of the reactant and product structures [27], for which one only needs the potential function that suitably describes this interaction.

A significant test of the quality of a model is to compare the theoretical and experimental properties of the surrounding medium. In this sense, the dielectric constant and the relaxation time of the solvent, determined by simulations, depend on the solute charge distribution in the solution.

The present chapter can be regarded as a review of previous studies [14-22] in which the LJ(12-6-1) potential from the EX-PL-DIS-ES components was proposed to describe molecular solvation in infinitely diluted aqueous solution, the main objective being to validate our model potential for chemical reactions and processes in solution by obtaining structural, thermodynamic, and dielectric properties and by studying energies related with reaction processes (in particular, computing reaction and activation free energies). For this purpose, we chose the solvation of several molecules (acetone, acetamide, acetaldehyde, acetic acid) and the reversible hydration of formaldehyde to obtain methylene glycol, since for these systems there is sufficient theoretical and experimental information available for comparison with our results. We also carried out calculations with other potentials, such as the AMBER [28] force field which applies quantum mechanically derived RESP charges [29] to obtain the electrostatic energy and Lenard-Jones parameters derived from liquid properties [30] to calculate the van der Waals energy.

Formalism, Methods, and Working Conditions

Solute-Solvent Potential

The geometries of the solutes (acetone, acetamide, formaldehyde, acetaldehyde, acetic acid, methylene glycol) and water molecules used in the simulations were obtained at the HF level with the "split valence" basis set 6-31G* of Pople et al. [33,32] using the Gaussian/92 package [33]. A grid of several hundred points was chosen to calculate SCF and MP2 solute(s)-solvent(w) interaction energies. These energies were then fitted by a Lennard-Jones 12-6 potential function that includes, besides van der Waals interactions, an additional Coulomb electrostatic term:

$$U_{sw} = \sum_{ij} \frac{A_{ij}^{sw}}{r_{ij}^{12}} - \sum_{ij} \frac{B_{ij}^{sw}}{r_{ij}^{6}} + \sum_{ij} \frac{q_i^s q_j^w}{r_{ij}}. \tag{1}$$

The net charges on each solute atom q_i^s were obtained using the ESIE procedure, which has been amply described in previous work [14-22]. It can be summarized as fitting the values of the Coulomb electrostatic component of the interaction energy U_{sw}(ES), using the variational scheme of Morokuma and co-worker [34, 35] implemented in the GAMESS package [36], with the expression

$$U_{sw}(ES) = \sum_{ij} \frac{q_i^s q_j^w}{r_{ij}}, \tag{2}$$

where the TIP3P or Mulliken charges of the monomer water q_i^w are assigned previously.

The A_{ij}^{sw} and B_{ij}^{sw} parameters are obtained in a similar way to the q_i^s, but now the energies used in the fits are those that describe the exchange (EX) and polarization (PL) components of the interaction energy at SCF level, and the dispersion (DIS) component related to the MP2 correlation energy [37], i.e.,

$$U_{sw}(EX) = \sum_{ij} \frac{A_{ij}^{sw}}{r_{ij}^{12}}, \tag{3}$$

$$U_{sw}(PL + DIS) = -\sum_{ij} \frac{B_{ij}^{sw}}{r_{ij}^{6}}. \tag{4}$$

The parameters A_{ij}^{sw}, B_{ij}^{sw} and q_i^s obtained with this procedure for the different solutes we shall henceforth denote as ABQ parameters. These parameters have been published in previous papers [14-22].

Molecular Properties Related to Solvation Processes

With the solvation simulated by molecular dynamics, one can then obtain the properties that derive from this process [12,13].

The radial distribution functions g_{sw} is calculated from the probability of finding a pair of atoms at a distance r and its integration gives the coordination number $n_c(r)$ The hydrogen bond half-life τ_{HB} is calculated using a binary-valued function for each pair of molecules (1 if the molecules are bonded, and 0 if they are not bonded) at time *t*.

The solvation Gibbs energy ΔG_{sol} expresses the difference between two states of the system, in particular between the system containing both solute and solvent molecules (λ=0) and the system formed only by solvent molecules (λ=1), i.e., it is found by solving the numerical integral

$$\Delta G_{sol} = G_{\lambda=0} - G_{\lambda=1} = \int_1^0 \langle \frac{dU_{sw}}{d\lambda} \rangle_\lambda \partial\lambda , \tag{5}$$

where the potential at instant λ of the perturbation $U(\lambda)$ is expressed as

$$U_{sw}(\lambda) = (1-\lambda)^k U_0 + \left[1-(1-\lambda)^k\right] U_1, \tag{6}$$

with U_0 being the potential with the original Hamiltonian, U_1 the potential with the perturbed Hamiltonian, and *k* an exponent used to avoid the integrand diverging at λ=1 as a consequence of the solute atoms disappearing in the perturbed state.

The limit of making infinitesimal perturbations at each step of the simulation — the "slow-growth" technique [38, 39] — is frequently used to determine this solvation free energy:

$$\Delta G_{sol} = \sum_{n^\circ step, \lambda=0}^{\lambda=1} U_{sw}(\lambda+1) - U_{sw}(\lambda) \tag{7}$$

The solvation enthalpy ΔH_{sol} is calculated from the interaction of the solute with all the water molecules, using the intermolecular pairwise potential given by Eq.(1). Assuming the ideal-gas approximation for solutes, this energy is

$$\Delta H_{sol} = \langle U_{sw} \rangle - \langle U_{ww} \rangle - RT , \tag{8}$$

where U denotes sums of electrostatic and van der Waals contributions in the whole system (sw) or the pure solvent (ww). This solvation enthalpy can be compared with experimental values determined from the temperature rise by mixing the two substances (ΔT_{exp}) and heating the mixture (ΔT_{cal}) with an electrical energy (W_{el}). The quantity of heat exchanged in the mixing of n_A moles of A component with n_B moles of B component is known as the molar integral mixing enthaphy (ΔH_M^I)

$$\Delta H_M^I = -W_{el} \frac{\Delta T_{exp}}{\Delta T_{cal}} \frac{1}{n_A + n_B}, \tag{9}$$

which is used to calculate the differential mixing enthalpy of component A ($\Delta H_{M,A}^d$) by the expression

$$\Delta H_{M,A}^d = \left(\frac{\partial(\Delta H_M^I)}{\partial x_A} \right) x_B + \Delta H_M^I. \tag{10}$$

The combination of enthalpies of vaporization ΔH_{vap} and solution $\Delta H_{M,A}^d$ gives the experimental enthalpy of solvation ΔH_{sol}

$$\Delta H_{sol} = \Delta H_{M,A}^d - \Delta H_{vap}. \tag{11}$$

Finally, the dielectric constant ε was calculated from the mean squared dipole moment $<M^2>$ of the medium using the "fluctuation formula" [40,41]

$$\frac{(\varepsilon - 1)(2\varepsilon_{rf} + 1)}{3(2\varepsilon_{rf} + \varepsilon)} = \frac{\langle M^2 \rangle}{9\varepsilon_0 k_B TV}, \tag{12}$$

where ε_{rf} is a reference dielectric constant, V the simulation cell volume, k_B Boltzmann's constant, T the temperature, ε_0 the vacuum permittivity, and M the total dipole moment obtained from μ_i the individual dipole moment for each solvent molecule.

$$M = \sum_{i=1}^{N} \vec{\mu}_i. \tag{13}$$

The dielectric constant depends notably on the value taken for ε_{rf} to solve Equation (12), a problem which is usually avoided by setting $\varepsilon_{rf} = \varepsilon$. In this case, Equation (12) can be simplified to

$$\frac{(\varepsilon - 1)(2\varepsilon + 1)}{\varepsilon} = \frac{\langle M^2 \rangle}{\varepsilon_0 k_B TV}. \tag{14}$$

The correlation time τ can be calculated by means of the integral at infinite time of the dipole autocorrelation function $\Gamma_\mu(t)$,

$$\tau = \int_0^\infty \Gamma_\mu(t) \cdot dt\,, \tag{15}$$

where this correlation function is

$$\Gamma_\mu(t) = \frac{\langle \mu(t) | \mu(0) \rangle}{\langle \mu(0) | \mu(0) \rangle}\,, \tag{16}$$

and the relaxation time τ_{rex} by fitting this function in the short-time zone to an exponential form

$$\Gamma_\mu(t) = \Gamma_\mu(0) e^{-t/\tau_{rex}}\,. \tag{17}$$

Molecular Properties Related to Reaction Processes

The hydration of a molecule (reactant, R) to give another different molecule (product, P) can be considered as an example of chemical reactions in an aqueous environment and as a model reaction in physical organic chemistry since it provides a large amount of information about chemical reactivity. The theorical and experimental study of this model reaction provides thermodynamic and kinetic data (e.g., the equilibrium and rate constants).

In order to study reaction processes, it is necessary to define a reaction coordinate which must contain the minima of the reactant and product states, respectively, and the most probable path between them. At a microscopic level, a suitable coordinate choice for reactions in solution is to use any of the solvent coordinates. For example, we can use the differences (ΔE_S) between the solute-water interactions potentials, $U_{sw;S}$, for the reactant (S=R) and for the product (S=P).

$$\Delta E_S = U_{SW;P} - U_{SW;R} \tag{18}$$

Thus, in the MD simulation of the reactant (S = R), we divide the trajectory into N equally separated steps. In each one of these steps the interaction potentials of the solvent with the solute in its reactant and product forms are calculated simultaneously ($U_{sw;R}$ and $U_{sw;P}$). In the same way, the interaction energies are calculated for the product surface (S = P). In both cases, the difference, ΔE_S, fluctuates, and its values are collected in a histogram of the number of times, $n_S(\Delta e)$, that a particular value (Δe) of the macroscopic variable ΔE_S appears in the simulation.

The probability $P_S(\Delta e)$ of finding the system in a particular configuration with an energy Δe can be expressed in a region close to the minimum by a Gaussian type function

$$P_S(\Delta e) = \sqrt{2\pi\left\langle(\Delta E)^2\right\rangle_S}\ \exp\left[\frac{-(\delta\Delta e)^2}{2\left\langle(\delta\Delta E)^2\right\rangle_S}\right], \tag{19}$$

where $\delta\Delta E = \Delta E - \left\langle\Delta E\right\rangle_S$ represents the fluctuations of the energy ΔE in the S state

The free energy $G_S(\Delta e)$ is computed from the normalized probability distribution of the variable Δe:

$$G_S(\Delta e) = -k_B T \ln P_S(\Delta e), \tag{20}$$

which allows one, using a graphic representation (see Figure 1), to obtain the free energies of reaction ΔG, reorganization ΔG_r, and activation $\Delta G^{\#}$.

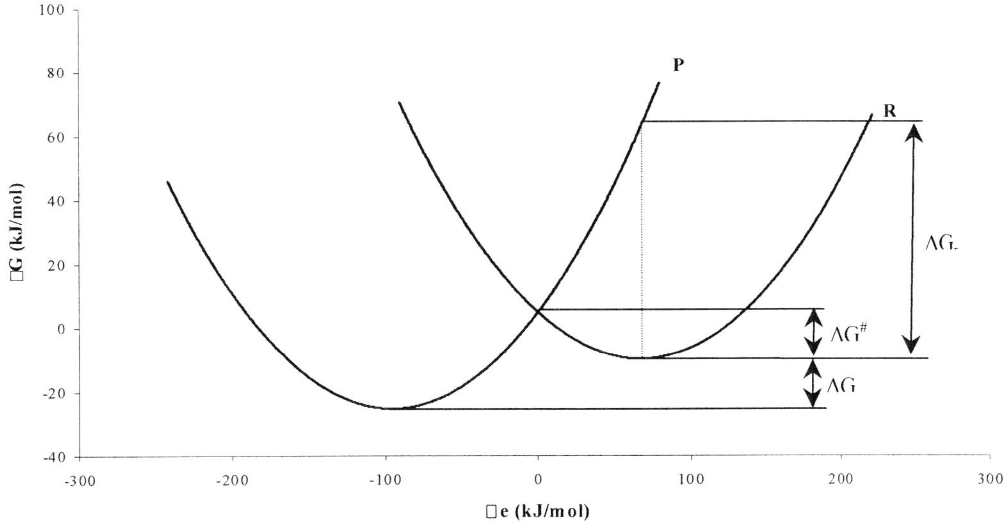

Figure 1. Free energy curves for R and P simulations using a potential function.

Simulation Details

Molecular dynamics simulations of a solute molecule inside a box containing about two hundred solvent molecules were carried out using the KGNMD [42] and/or AMBER [28] programs. The temperature was kept constant at 298 K, and the time considered for the simulation was 1000 ps with time steps of 1 fs. The first 100 ps were taken to reach equilibrium, and the times remaining were used to store the positions required for an adequate evaluation of the molecular properties. For the solvent-solvent interaction we used the TIP3P potential [43], although the MCY potential [44] was also considered with the aim of comparing some properties. In the simulations used to obtain dielectric properties, no cut-off was applied in order to eliminate the dependence of the dipole moment on the cut-off radius. For the solute-water interaction, the aforementioned Lennard-Jones 12-6-1 potential function was used, neglecting at the beginning of the simulation any water molecules that presented an oxygen atom at a distance closer than 1.6 Å from any atom of the solute molecule. This criterion led to some of solvent molecules being left out of the simulation calculations. The electrostatic interactions were treated by the Ewald method [45] and the solute was kept rigid for the calculation of the properties using the SHAKE algorithm [46].

Results and Discussion

Structural Properties

The ABQ molecular parameters that describe the solute-water interaction provide solute-solvent interaction energy maps such as those shown in Figure 2 for the acetamide (R=CH_3, R'=N) as solute molecule. One observes a broad zone with a deep minimum located between the carbonyl oxygen and the hydrogen close to it. This suggests the possibility of a double hydrogen bond with a water molecule of the surrounding medium. One also observes another minimum energy zone near the other hydrophilic hydrogen. On the contrary, the zone describing the hydrophobic group -R lacks the minima typical of hydrogen bonds. The minima become somewhat less pronounced when the potential used in their representation is that obtained from the AMBER parameters. These energy maps represent in practice the electrostatic component of the interaction, since the maps from representing only the LJ(12-6) contribution present hardly any minima with depths greater than -1 kcal/mol.

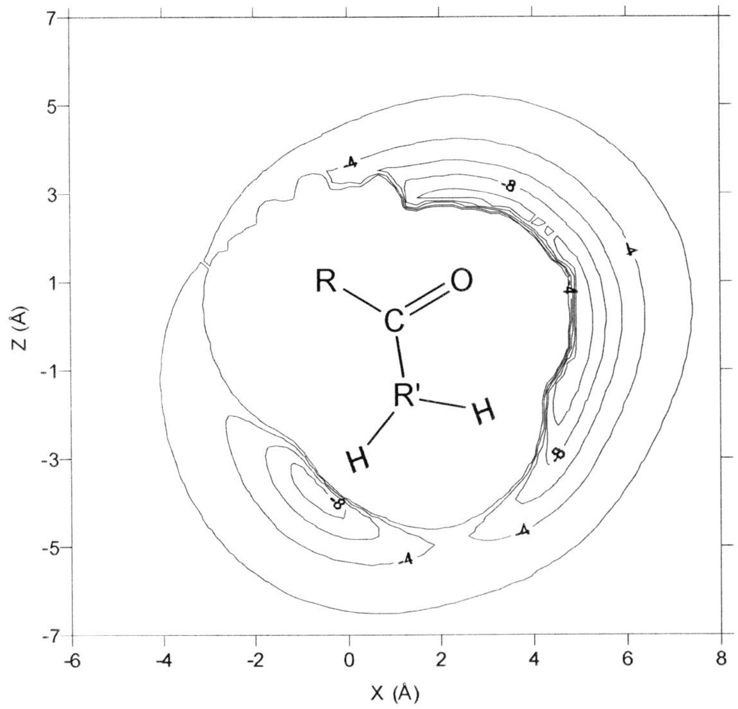

Figure 2. Solute-solvent interaction energy map.

The above energy-map based description is reflected in Figure 3, which shows some of the steps of the simulation of the solute with the water molecules conforming the first solvation shells. One observes water molecules at hydrogen bond distances in the zone of the C=O and $-NH_2$ groups, and the absence of hydration in the zone of the $-CH_3$ hydrophobic group. Also, the hydration shells of this solute are similar with the ABQ and AMBER potentials, there being an average over the course of the simulation of approximately two water molecules at a distance of 2.5 Å from the carbonyl oxygen and amide hydrogen atoms. In the second hydration shell (R≈ 0-5 Å), there appear about 19 water molecules with respect to any solute atom. One also observes in this figure the presence of numerous water molecules linked to each other by hydrogen bonds.

Figure 3. Solvation of the acetamide during the simulation.

Analyzing the radial distribution functions for solute atoms that have the possibility of forming hydrogen bonds with the solvent water (Figure 4), we find for both potentials a maximum at distances that are typical for these bonds (1.85 Å with the function $g_{O\text{-}HW}$ and 1.95 Å with $g_{H\text{-}OW}$). The first minimum of these functions is at around 2.5 Å, leading to a coordination number near 1 for the function $g_{H\text{-}OW}$, and 2 for $g_{O\text{-}HW}$. Comparing these results with those of the aforementioned average structure, we can state that while, in the zone of the carbonyl group the water molecules conforming the first hydration shell are oriented to form hydrogen bonds, in the hydrophilic hydrogen zone not all the solvent molecules are appropriately oriented.

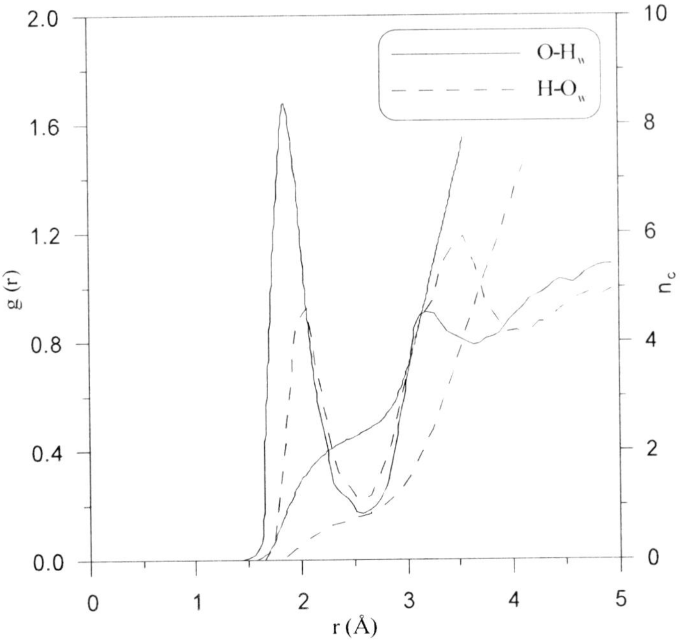

Figure 4. FDR and coordination number for the O and H atoms of the acetamide.

Differences are observed in the intensities of the peaks (slightly lower when using the AMBER parameters), which are greater for the function $g_{O\text{-}HW}$ and thus reflect a stronger hydrogen bond in the zone of the carbonyl group. This leads to half-lives of 0.9 ps and 0.5 ps obtained with our potential for the hydrogen bonds of $O \cdots H_W$ and $H \cdots O_W$, respectively. For the case of the AMBER potential, slightly shorter half-lives are obtained (0.7 and 0.4 ps for the $O \cdots H_W$ and $H \cdots O_W$ bonds, respectively). Given that we obtain a half-life of the $O_W \cdots H_W$ hydrogen bond in the pure solvent of 0.7 ps (1.0 ps by other workers [47]), we can state that our potential is the only one of the two potentials employed that leads to a longer half-life τ_{HB} when the hydrogen-acceptor oxygen has a double bond with another atom.

Thermodynamic properties

Given the charges on atoms of the molecules CH_3-CO-R (with R=H, OH, NH_2) and the capacity to form hydrogen bonds with the water solvent, one could expect an increasing hydrophilic character of the molecules in the following order: acetaldehyde, acetic acid, and acetamide. In Table 1 we show the interaction energies for the ABQ potential. One observes that this potential reflects the expected hydrophilic character of the three solutes.

Table 1. Solute-solvent interaction energies.

Solute	AcH	AcOH	$AcNH_2$
$\langle U_{sw} \rangle^a$	-39.96	-68.74	-71.44

a) Energies in kJ·mol^{-1}

The solute-solvent interaction energy was employed in studying the solvation enthalpy ΔH_{sol} of these solutes. To check the goodness of our ABQ potential with respect to the thermodynamic properties, we compared the results for the solvation enthalpy with the available experimental data and with the results using charges and parameters of the AMBER force field (ff99). Thus, for the case of acetamide, the value obtained with the ABQ potential was ΔH_{sol} = -73.61 kJ/mol (see values in Table 2), which is in good agreement with the experimental -70.55 kJ/mol [48], and improves by almost 3 kJ/mol the result obtained using the AMBER parameters. It is important to note that in Table 2 most of this entalphy comes from the electrostatic contribution of the potential (ΔH_{ES}), although the van der Waals (ΔH_{VDW}) contribution must also be considered.

Table 2. Solvation enthalpy for acetamide.

Energies[a]	AMBER	ABQ	Experimental
ΔH_{ES}	-58.48	-62.28	
ΔH_{VDW}	-18.31	-11.33	
ΔH_{sol}[b]	**-76.79**	**-73.61**	**-70.55[c]; -69.35[d]**

a) In kJ/mol; b) values obtained as $U_{sw} - RT$; c) ref. [48]; c) ref [22]

To compare the theoretical results with experimental data at infinite dilution, we made calorimetric measurements of the mixing enthalpy of different dilute solutions. The value corresponding to infinite dilution, obtained by extrapolation to χ_{AcNH2}=0 of the values presented in Table 3, was $\Delta H^{d}_{M,A}$ =10.95 kJ/mol, in good agreement with the value of 9.74 kJ/mol given by Barone and Giancola [48]. This result leads to a solvation enthalpy of -69.35 kJ/mol (considering an acetamide vaporization enthalpy of 80.30 kJ/mol [48]), in very good agreement with the experimental value of -70.55 kJ/mol [48].

Table 3. Differential mixing enthalpies for dilute solutions of acetamide.

$10^3\ \chi_{AcNH2}$	0.3	0.4	0.5	0.65	0.8	1.0	1.5	2.0
$(\Delta H^{d}_{M,B})^{a}$	14.15	14.86	14.29	16.36	17.12	17.20	17.80	17.99

In kJ/mol.

Another of the applications of the solute-solvent interaction energy is to calculate the free-energy solvation ΔG_{sw} of different solutes. This has been done with the "slow growth" technique (SG) and with the thermodynamic integration method (TI). To check the improvement introduced by our ESIE charges with respect to Mulliken charges, we compared the free energies of solvation, using the ESIE charges and the SG technique, with those provided by experimental studies. The results for the ABQ potential are at 1.2, 4.9, and 3.3 kJ/mol from the experimental values for AcH, AcOH, and $AcNH_2$ solutes, respectively (Table 4). These values lie within the deviations found by other workers according to the solute and method of calculation [49-51]. Larger deviations are obtained in the case of using Mulliken charges, especially in the case of acetamide, as one observes in Table 4.

Table 4. Free solvation energies.

Solute	ΔG_{sol} [a]		
	Q_{MUL}	Q_{ESIE}	Expt.
AcH	-16.10	-13.40	-14.63[e]
AcOH	-35.50	-32.90	-28.01[e]
$AcNH_2$	-47.70	-37.25[b] -39.87[c] -35.19[d]	-40.54[e]

a) Energies in kJ·mol^{-1}; b) obtained with the "slow-growth" technique;
c) obtained with the thermodynamic integration method and ESIE charges;
d) obtained with the thermodynamic integration method and AMBER charges;
e) ref. [53-55].

We also observed that for the acetamide case the TI method leads to better results for the free-energy solvation. The value of the free energy with the ESIE charges was -39.87 kJ/mol, in very good agreement with the tabulated value of -40.55 kJ/mol [52-54], and improving the result of -35.19 kJ/mol obtained with the AMBER(ff99) parameters and the results reported by Shirts et al. [55] using other force fields (-32.60 kJ/mol with AMBER(ff94), -32.98 kcal/mol with CHARMM, and -35.11 kJ/mol with OPLS-AA). This excellent result is also indicative of the improvement in our potential when one includes in the attractive term the components not only of dispersion but also of polarization, and when the ESIE charges and the thermodynamic integration method is used instead of the "slow-growth" technique.

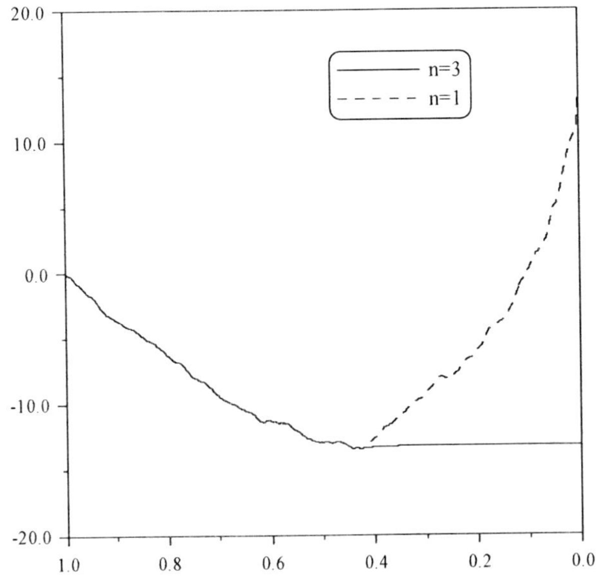

Figure 5. Variation of solvation free energy with the λ parameter for acetaldehyde.

Another aspect that improves the calculations of the free solvation energy using the Free Energy Pertubation method (FEP) is to consider different degrees of perturbation in solute discharge process. This solvation energy improves when a nonlinear perturbation method is used (usually in Equation (6), k=1 from 1.0>λ>0.2 and k=3 from 0.2>λ>0.0). The case of acetaldehyde shown in Figure 5 stands out because is the most hidrophobic solute of those considered. In this case the power n=3 of the perturbation begins to be applied well before the final solute discharge process (λ<0.4) since the van der Waals component surpasses the electrostatic component practically halfway through the discharge. Whatever the case, not incorporating this perturbation exponent leads to results that are very far from the experimental value of -14.63 kJ/mol.

Dielectric Properties

One property related to the relaxation of the aqueous medium considered in the present chapter is the dielectric constant ε. For its determination, we calculated the mean squared dipole moment $<\mu^2>$. The value obtained for the dielectric constant of the pure solvent, which depends strongly on the potential used in the description of the solvent-solvent interactions, was 50.32 with the MCY potential and 87.8 with the TIP4P potential [56], as is seen in the first two lines of Table 5. This last value can be considered to be in full agreement with the experimental value of the dielectric constant at room temperature (ε= 78.3) [57].

Table 5. Dielectric properties of the aqueous medium.

System	Models/Charges	$<\mu^2>^a$	ε	τ^c
Pure water	MCY	9.99 (8.24)^b	50.32 (62.9)^b	8.3
Pure water	TIP4P	17.4	87.8	-
Formaldehyde + Water	MCY/ESIE	9.70 (7.94)	48.87 (61.1)	9.4
	MCY/Mul	9.32 (7.22)	46.97 (58.7)	7.8
	MCY/MEP	9,12 (7.86)	45.98(57.5)	9.3
	MCY/CM2	9.58 (6.01)	48.27 (60.3)	9.3
Formamide + Water	MCY/ESIE	10.32 (8.64)	51.96 (64.9)	5.8
	MCY/Mul	9.72 (7.90)	48.97 (61.2)	8.9
	MCY/MEP	9.32 (6.89)	46.97 (58.3)	9.8
	MCY/CM2	9.03 (7.45)	45.53 (56.9)	8.5
Acetone + Water	MCY/Mul	8.23	41.8	-
	MCY/ESIE	8.20	41.65	-
	TIP4P/Mul	17.5	88.35	-
	TIP4P/ESIE	16.85	85.04	-

a) In D^2; b) values in parentheses are obtained with the correction factor of γ=1.25; c) in ps.

The values of the dielectric constant can be improved by including the correction factor γ obtained as the ratio of the Kirkwood factors for two values of the radius. Here we need to mention the influence of the truncation of the molecular interactions on the dielectric properties of the medium. When a cut-off is applied ($R=R_c$), a number of interactions are neglected which implies a loss of the correlation between the aligned dipoles of the solvent molecules that has to be corrected with the γ factor. The dielectric constant obtained with MCY considering a factor of $\gamma=1.25$ takes the value 62.9, which is in agreement with the values of ε computed for the SPC water model (ε = 72 with Ewald summation and ε = 61 with a reaction-field corrected interaction) in a recent study on the dielectric constant of water using a potential of this type [58]).

In Table 5, one also observes a decrease of the dipole moment with respect to the pure solvent value for the solutions of formaldehyde and acetone described with the ESIE charges due to the greater attraction of the water molecules in the presence of these solutes, and the expected increase in the formamide solute due to the hydrophilic character of the amides. With the other charges, however, the results were not as expected. The dipole moment decreased for the amide even though the solvent molecules are more relaxed. This response is in part due to the increase of the hydrophobic character of the -CH group of the amide when the MEP, MUL, and CM2 charges are employed.

When the working conditions (ρ, T, N) are kept fixed in all the simulations, the dielectric constant only depends on the mean squared dipole moment $<\mu^2>$, so that the above conclusions can be applied directly to this property. Namely, the response obtained with the ESIE charges for the dielectric constant in the presence of formaldehyde or acetone is in consonance with the lesser dielectric character of the solutes (ε= 21 for formaldehyde and ε= 21.4 for acetone [59]). Also, the increase obtained for this property due to the formamide is coherent with the greater values of its dielectric constant (ε= 109 [59]). An unsatisfactory description is again obtained using the Mulliken, MEP, and CM2 charges because in the formamide case there is a decrease in the dielectric capacity of the medium.

These results are also applicable when the TIP4P potential is used. Thus, one observes that the dielectric constant of the water decreases when a hydrophobic solute such acetone is part of the solution. This is the case if the ESIE charges are used to describe the solute-solvent interaction.

Finally, we analyzed the water dielectric relaxation time τ (listed in the last column of Table 5). Again the result of this property depends strongly on the potential used in the solvent-solvent interactions and on the solute considered. Thus, for the pure solvent with the MCY potential a value of 8.3 ps is obtained, in good agreement with the experimental value, τ (25 °C) = 8.1 ps [60]. The presence of small solutes produces significant changes in this time. Thus, a major reduction of the dielectric relaxation time of the medium is observed using the ESIE charges only when the solute is formamide, as a consequence of the solvent molecules having a weaker mutual interaction. For the formaldehyde solution, the hydrophobic character of this solute will strengthen the solvent-solvent interactions, and as a consequence the correlation of the dipole moments will increase. Therefore the time that the solvent molecules need to reorientate in the presence of the solute will be longer. Again, the ESIE (also the MEP and CM2) charges describe this behavior adequately.

Activation and Reaction Energies

The theoretical and experimental study that we shall consider to check the goodness of the ABQ potential is the hydration of formaldehyde to give methylene glycol, a typical example of reactions of nucleophilic addition to the carbonyl group.

The separation and relative orientation of the two solute molecules that form the reactant system (formaldehyde and water) shown in Figure 6, reproduces the minimal energy configuration. Likewise, this figure shows the scheme of the reaction which is a transfer of the water hydrogen (with the rupture of the Ow-H bond) to the carbonyl oxygen to form a diol (O-H), with the formation of a new covalent bond between the water oxygen and the carbon of the aldehyde (C-O).

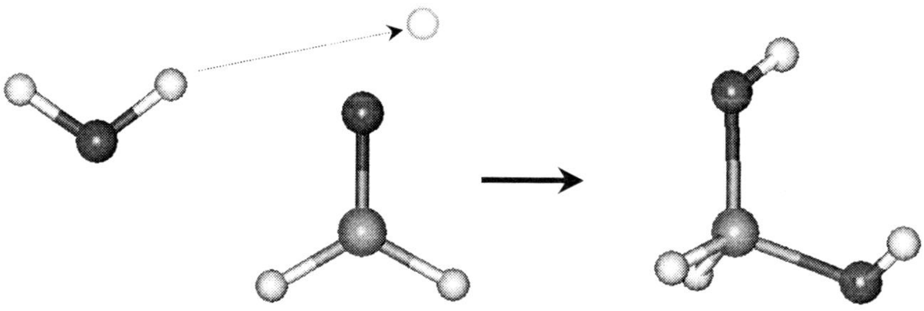

Figure 6. Geometries of the reactant and product systems in the reaction process.

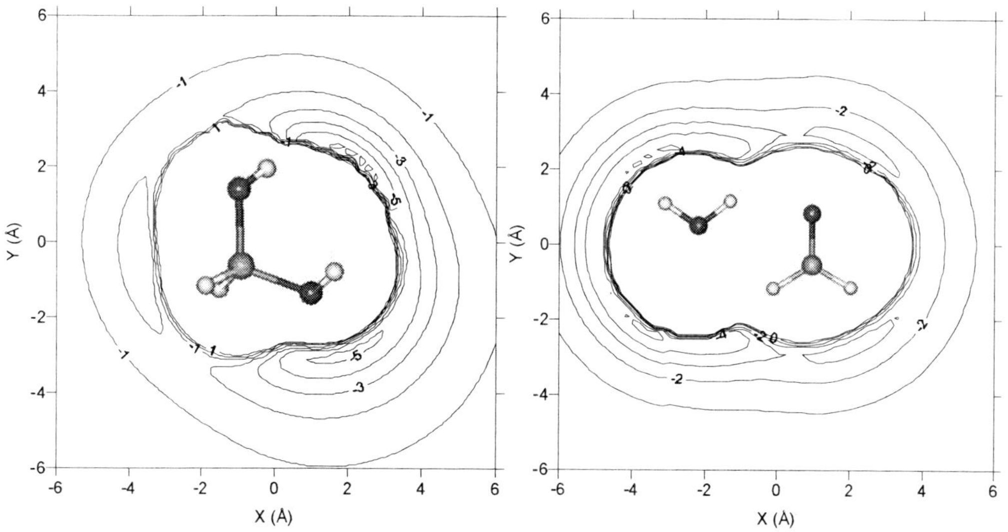

Figure 7. Isoenergy contour map for formaldehyde-water (left) and methylene glycol (right) obtained with the ABQ potential.

Estimation of the interaction molecular parameters shows that the ABQ potential leads to the energy maps showed in Figure 7. These present deeper and more localized minima for the case of methylene glycol.

These iso-energey maps are in consonance with the solute-solvent interaction energies given in Table 6 for the different cases studied. In both cases, the electrostatic (ES) contribution is far greater in absolute value than the van der Waals (VDW) contribution. Comparing the interaction energy in the two systems, one can say that in both potentials the reaction of formation of the methylene glycol is accompanied by an increase in the molecular hydration. This increase is greater when our parameters are used, reflecting the decrease of the solvation enthalpy from -17.5 to -89.0 kJ/mol.

Table 6. Solute-solvent interaction energy components[a].

System	Formaldehyde-water		Methylene glycol	
Potential	ABQ	AMBER	ABQ	AMBER
$U_{s\text{-}w}$(ES)	-17.9	-131.6	-113.7	-150.8
$U_{s\text{-}w}$(VDW)	2.9	19.6	27.2	21.7
$U_{s\text{-}w}$	-15.0	-112.0	-86.5	-129.2
$\Delta H_{s\text{-}w}{}^{b}$	-17.5	-114.5	-89.0	-131.7

a) In kJ/mol; b) obtained as $U_{s\text{-}w} - U_{w\text{-}w}$ -RT, where $U_{w\text{-}w}$ = -10.566*N_{water}.

To calculate the reaction and activation energies, we build the R and P free-energy curves (obtained from their respective simulations and corrected according to the work of Tachiya [61] where the vertical ΔG free-energy difference must be equal to the Δe energy difference). Only when our potential is used are the curves corresponding to the products somewhat deeper than that of the reactant. This leads to slightly exothermic reactions with a small activation barrier (see Figure 1 as a prototype of these curves)

The results obtained for the reaction and activation free energies are given in Table 7, together with the reorganization energy of the solvent (ΔG_r). From an inspection of this table, it appears that only when the calculations are carried out with the ABQ potential do the results of the different free energies show an exothermic process with a low activation barrier and smaller energy cost to reorient the molecules of the solvent in the course of the reaction.

Table 7. Free energies associated with the reaction process.

Potential[a]	AMBER	ABQ
ΔG	1.12	-16.05
$\Delta G^{\#}$	23.24	3.70
ΔG_r	63.91	29.38

In kJ/mol.

If one bears in mind that the free energy of this reaction at 298 K is -17.7 kJ/mol when obtained from the experimental equilibrium constant $k_{eq}(T) = e^{3769/T-5.494}$ [60] and -19.4 kJ/mol when obtained from ultraviolet spectrophotometry [63], one can conclude that the results obtained with our electrostatic and van der Waals parameters describe the energetics of this reaction better than when the AMBER parameters are used in the LJ(12-6-1) potential function.

Likewise, if one analyzes the values of the activation free energy in the methylene glycol formation, and compare them with those from experimental studies, one observes an appreciable discrepancy between the two results. Thus, in the work of Schecker and Schulz [64], where they obtain a temperature dependence of the rate constant given by $k_v(T) = 2.04{\cdot}10^5{\cdot}e^{-2936/T}$ for the aqueous hydration of formaldehyde, or the work of Winkelman [62] with $k_v(T) = 7800{\cdot}e^{-1913/T}$, the free energy of activation is about 62.7 kJ/mol, very far from any of the theoretical values presented in this chapter. This makes us think that some of the solvent water molecules also participate in the reaction mechanism, catalyzing the process and lowering the activation barrier considerably. Thus, the activation energy required to dehydrate the methylene glycol and to obtain formaldehyde (the inverse reaction to that considered in the present chapter) was computed to be 19.8 kJ/mol with our potential and 22.1 kJ/mol with the AMBER parameters. Both values come closer to the activation barrier of 16.5 kJ/mol given by Meijer and Sprik in an ab initio molecular dynamics study of this inverse reaction in an acid medium [65], with our model again being the one that better reproduces this energy.

Conclusions

The use of simple potentials of the type LJ(12-6-1) to describe the solute-solvent interactions in the molecular dynamics simulation of molecular solvation and chemical reactions in aqueous media leads to acceptable results when the interaction parameters are chosen appropriately. Thus, when one uses the A_{ij}^{sw} and B_{ij}^{sw} parameters obtained from the EX-DIS-PL components of the interaction energy and the ESIE charges, the results for the free energy of reaction are in very good agreement with experiment. Moreover, the free energy of activation obtained with our model approaches that of this reaction catalyzed by acids. Also, the perturbational FEP method and the slow-growth technique lead to good values of the solvation free energy ΔG_{sol}, as long as —especially for not very hydrophilic solutes— a nonlinear dependence of the energy U_{sw} on the perturbation parameter λ is taken into account in the final solute discharge process. Nevertheless the application of the thermodynamic integration method improves the solvation free energy even more. Also, the results of the dielectric constant were improved by appropriately modifying the solvent-solvent interaction with the TIP4P potential. The dielectric constant and the relaxation times obtained with the EX-DIS-PL-ES procedure agreed well with those from other similar studies, thus confirming the goodness of the solute-solvent interaction model employed in this chapter. Likewise, the structural properties, such as the FDR functions and the hydration shells, are very well described with our potential, even improving the results of the half life of the hydrogen bonds calculated with the AMBER potential.

Acknowledgement

This research was sponsored by the Consejería de Infraestructuras y Desarrollo Tecnológico de la Junta de Extremadura (project GRU06007).

References

[1] Cramer, C.J.; Truhlar, D.G. *Structure and Reactivity in Aqueous Solution.* Am. Chem. Soc.: Washington D.C., 1994.
[2] Warshel, A. *Computer Modeling of Chemical Reactions in Enzymes and Solutions.* Wiley & Sons: New York, NY, 1991.
[3] Jorther, J.; Levine, R.D.; Pullman, B. *Reaction Dynamics in Cluster and Condensed Phases.* Kluwer Acad. Pub.: London, GB, 1999.
[4] Moreau, M.; Turq, P. *Chemical Reactivity in Liquids. Fundamental Aspects.* Plenum: New York, NY, 1988
[5] Honeisel, C. *Theoretical Treatment of Liquids and Liquid Mixtures.* Elsevier: New York, NY, 1993.
[6] Simkin, B.Y.; Sheikhet, I.I. *Quantum Chemical and Statistical Theory of Solutions. A Computational Approach.* Ellis Horwood: Madrid, SP, 1995.
[7] Politzer, P.; Murray, J.S. *Quantitative Treatments of Solute-Solvent Interactions.* Elsevier: Amsterdam, NL, 1994.
[8] Tapia, O.; Bertrán, J. *Solvent Effects and Chemical Reactivity.* Kluwer: Dordrecht, GE, 1996.
[9] Truhlar, D.G.; Isaacson, A.D.; Garret, B.C. In *The Theory of Chemical Reaction Dynamics*; Baer, M.; CRC Press: Boca Raton, FL, 1985; Vol. 4
[10] Muller, A.; Ratjczak, H.; Junge, W.; Diemann, E. *Electron and Proton Transfer in Chemistry and Biology*; Elsevier: New York, NY, 1992.
[11] Alder, B.J.; Winwright, T.E. *J. Chem. Phys.* 1959, 31, 459.
[12] Haile, J.M. *Molecular Dynamics Simulation. Elementary Methods*; Wiley: New York, NY, 1987.
[13] Allen, M.P.; Tildesley, D.J. *Computer Simulations of Liquids*; Clarendon Press: Oxford, GB, 1987.
[14] Tolosa, S.; Sansón, J.A.; Hidalgo, A. *Chem. Phys.* 2001, 265, 207
[15] Tolosa, S.; Sansón, J.A.; Hidalgo, A. *Chem. Phys. Lett.* 2002 357, 279
[16] 16 .Tolosa, S.; Sansón, J.A.; Hidalgo, A. In *Recent Research Developments in Chemical Physics*; Pandalai, S.G.; Transworld Research Network: Trivandrum, IN, 2002; vol. 2.
[17] 17 .Tolosa, S.; Sansón, J.A.; Hidalgo, A. *Chem. Phys.* 2003, 293, 193
[18] Tolosa, S.; Sansón, J.A.; Hidalgo, A. *J. Solution Chem.* 2005, 34, 407
[19] Tolosa, S.; Sansón, J.A.; Hidalgo, A. *Chem.Phys.* 2005, 315, 76
[20] Tolosa, S.; Sansón, J.A.; Hidalgo, A. *Molecular Simulation* 2005, 31, 549
[21] Tolosa, S.; Sansón, J.A.; Hidalgo, A. *Chem. Phys.* (in press, 2006)
[22] Tolosa, S.; Sansón, J.A.; Hidalgo, A. *J Phys.Chem.A.* (submitted for publication, 2006)
[23] Marcus, R.A. *J. Chem. Phys.* 1956, 24, 966
[24] Marcus, R.A. *J. Chem. Phys.* 1956, 24, 979.
[25] Marcus, R.A. *J. Phys. Chem.* 1968, 72, 891

[26] Marcus, R.A.; Sutin, N. *Comments Inorg. Chem.*, 1986, 5, 119
[27] Carter, E.A.; Hynes, J.T. *J. Phys. Chem.*, 1989, 93, 2184
[28] Case, D.A.; Darden, T.A.; Cheatham III, T.E.; Simmerling, C.L.; Wang, J.; Duke, R.E.; Luo, R.; Mert, K.M.; Wang, B.; Pearlman, D.A.; Crowley, M.; Brozell, S.; Tsui, V.; Gohlke, H.; Mongan, J.; Hornak, V.; Cui, G.; Beroza, P.; Schafmeister, C.; Caldwell, J.W.; Ross W.S.; Kollman, P.A. *AMBER 8*, University of California, San Francisco, (2004).
[29] Shirts, M.R.; Pitera, J.W.; Swope, W.C.; Pande, V.S. *J. Chem. Phys.* 2003, 119, 5740
[30] Jorgensen, W.L.; Tirado-Rives, J.J. *J.Am. Chem. Soc.*, 1988, 110, 1657
[31] Ditchfield, R.; Hehre, W.J.; Pople, J.A. *J. Chem. Phys.* 1971, 54, 724
[32] Hehre, W.J.; Ditchfield, R.; Pople, J.A. *J. Chem. Phys.* 1972, 56, 2257
[33] Frisch, M.J.; Trucks, G.W.; Head-Gordon, M.; Gill, P.M.W.; Wong, M.W.; Johnson, B.G. ;Schlegel, H.B.; Robb, M.A.; Replogle, E.S.; Gomperts, R.; Andres, J.L.; Raghavachari, K.; Binkley, J.S.; Gonzalez, C.; Martín, R.L.; Fox, D.J.; DeFrees, D.J.; Baker, J.; Stewart, J.J.P.; Pople, J.A. *GAUSSIAN-92 Revision D.3*, Gaussian Inc, Pittsburgh PA, 1992
[34] Morokuma, K. *J. Chem . Phys.* 1970, 19, 553
[35] Kitaura, K.; Morokuma, K. *Int. J. Quantum Chem.* 1976, 10, 325.
[36] Dupuis, M.; Spangler, D.; Wendoloski, *GAMESS Program QG01* J. National resource for Computations in Chemistry Software Catalog, University of california: Berkeley, CA, 1980
[37] Moller, C.; Plesset, M.S. *Phys. Rev.* 1934, 46, 618
[38] Kollman, P. *Chem. Rev.* 1993, 93, 2395
[39] Basch, P.A.; Singh, U.C.; Langrdge, R.; Kolmann, P. *Science* 1987, 236, 564
[40] Matyushov, M.V.; Landayi, B.M. *J Chem. Phys.* 1999, 110, 994
[41] Neumann, M.; Steinhauser, O.; Pauley, G.S. *Mol. Phys.* 1984, 52, 97
[42] Corongiu, G. *IBM Corporation center Scientific and Engineering Computation*, department 48B/428, New york
[43] Mahoney, M.W.; Jorgensen, W.L. *J. Chem. Phys.* 2000, 20, 8910
[44] Matsuoka, D.; Clementi, E.; Yoshimina, M. *J. Chem. Phys.* 1976, 64, 1351
[45] Ewald, P., *Ann. Phys.* 1921, 64, 253
[46] Ryckaert, P.; Ciccotti, G.; Berendsen, J.J.C. *J. Comp. Phys.* 1977, 23, 237
[47] Barthel, J.; Bachhuer, K.; Buchner, R.; Hetzenauer, H. *Chem. Phys. Lett.* 1976, 165, 369
[48] Barone, G.; Giancola, C. *Pure & Appl. Chem.* 1990, 62(1), 57
[49] Hawkins, G.D.; Cramer, C.J.; Truhlar, D.G. *J.Phys. Chem.* 1996, 100, 19824
[50] McDonald, N.A.; Carlson, H.A.; Jorgensen, W.L. *J. Phys. Org. Chem.* 1997, 10, 563
[51] Still, W.C.; Tempczyk, A.; Hawley, R.C.; Hendrickson, T. *J. Am. Chem. Soc.* 1990, 112, 6127.
[52] Cabani, S.; Paolo, G.; Mollica, V.; Lepori, L. *J. Solution. Chem.* 1981, 10, 563
[53] Hine, J.; Mookerjee, P.K. *J.Org.Chem.* 1975, 40, 292
[54] Pearson, R.G. *J. Am. Chem. Soc.* 1986, 108, 6109
[55] Shirts, M.R.; Pitera, J.W.; Swope, W.C.; Pande, V.S. *J. Chem. Phys.* 2003, 119, 5740
[56] Jorgensen, W.J.; Chandraseichar, J.; Madura, J.D.; Impey, R.W.; Klein, M.L. *J. Chem. Phys.* 1983, 79, 926
[57] Barthel, J.; Bachhuber, K.; Buchner, R.; Hetzenauer, H. *Chem. Phys. Lett.* 1990, 165, 369
[58] Heinz, T.N.; van Gusteren, W.F.; Hünenberg, P.H. *J. Chem. Phys.* 2001, 115, 1125

[59] Lide, D.R. *Handbook of Chemistry and Physics*; CRC. Press, Boca Raton, FL, 1992
[60] Barthel, J.; Bachhuber, K.; Buchner, R.; Hetzenauer, H. *Chem. Phys. Lett.* 1990, 165, 369
[61] Tachiya, M. *J. Chem. Phys.* 1989, 93, 7050
[62] Winkelman, J.G.M.; Voorwinder, O.K.; Ottens, M.; Beenackers, A.A.C.M.; Janssen, L.P.B.M. *Chemical Engineering Science*, 2002, 57, 4067
[63] Zavitsas, A.A.; Coffiner, M.; Wiseman, T.; Zavitsas, L.R. *J. Phys. Chem.*, 1970, 74, 2746
[64] Schecker, H.G.; Schulz, G. *Zeitschrift für Physikalische Chemie Neue Folge* 1969, 65, 221
[65] Meijer, E. J.; Sprik, M. *J. Am. Chem. Soc.* 1998, 120, 6345

In: Quantum Chemistry Research Trends
Editor: Mikas P. Kaisas, pp. 225-238
ISBN: 978-1-60021-620-6

Chapter 7

THE THEORETICAL STUDY OF THE INTERACTIONS BETWEEN THE 1, 3-BUTANEDIOL AND THE HYDROGEN PEROXIDE

Zheng-yu Zhou[1,2], Hai-tao Zhang[1] and Yun Shi[1]
[1]Department of Chemistry, Qufu Normal University,Shandong, Qufu 273165, People's Republic of China
[2]State Key Laboratory Crystal Materials Shandong University, Shandong, Jinan 250100, People's Republic of China

Abstract

In this work, we report a theoretical study on the interactions between the 1,3-butanediol and the chiral hydrogen peroxide. The complexes formed between two isolate chiral hydrogen peroxide (R and S) and 1,3-butanediol have been investigated by using B3LYP method at different basis set levels. Four pairs of equilibrium structures have been determined, one of which are one-hydrogen bonded structure and others are cyclic double–hydrogen bonded structures at the largest basis level. The optimized geometric parameters, interaction energies and chirodiastatic energy for various isomers at different levels are estimated. The infrared spectrum frequencies and IR intensities for various complexes are reported.

Introduction

The structures of many compounds in the nature, as none-superposable right- or left- handed mirror image, are enantimorph. This character is defined as Chirality. Chirality is one of the essence characters of the nature. The biological function of biomolecules, such as nucleic acid, protein and enzyme, are almost chiral. Intermolecular interactions between two isolated chiral molecules M and P (where each also can be R or S) are inherently asymmetric, leading to chirally discriminating molecular recognition and binding. Chiral discrimination forces and energies are very important for understanding and determining the mechanisms of chiral recognition, enatioselective reactions and a multitude of chirally discriminating effects in chemistry and biology[1-5]. There is a large body of experimental work based on the existence of these so-called chirally discriminating interactions; the interaction energy ΔE between two molecules, both of which can exist in chiral R and S enantiomers, depends on their relative handedness. For a M...M' complex, the interaction energies for the interactions between these two homochiral partner are identical, $\Delta E_{hom} \equiv \Delta E(R...R') = \Delta E(S...S')$; this is also true for the two possible heterochiral interactions $\Delta E_{het} \equiv \Delta E(R...S') = \Delta E(S...R')$. The difference of these energies $\Delta E_{chir} \equiv \Delta E_{hom} - \Delta E_{het}$ is the energy term responsible for chiral discrimination.[6] In this work, we only discussed the complexes formed between the 1,3-butanediol and the chiral hydrogen peroxide, so the Chirodiastatic energies motioned below are calculated as $\Delta E_{chir} = \Delta E(R...R') - \Delta E(R...S')$.

Hydrogen peroxide is not only the smallest chiral O-H hydrogen donor molecule but also the smallest O-containing chiral proton-acceptor. The equilibrium geometry of HOOH exhibits helical chirality, the enantiomes are denoted M and P (see Figure 1). In fact, HOOH is not a protypical chiral molecule, since in the gas phase it interconverts between the (M)- and (P)-HOOH enantiomers. Hydrogen peroxide in low temperature matrics has been studied by a few groups,[7-9] and the monomer spectrum has been published by pettersson et al. [9] Goeble et al.[10] have studied the complex between hydrogen peroxide and dimethyl ether and shown that hydrogen peroxide forms a single, relatively strong hydrogen bond to the other oxygen. Engdahl et al. have studied the complex between water and hydrogen peroxide[11] and hydrogen peroxide dimmer.[12] To the best of our knowledge, no study has been reported on the interactions between the 1,3-butanediol and the chiral hydrogen peroxide. In this work, the chiral hydrogen bonds between the 1,3-butanediol and the chiral hydrogen peroxide have been studied. The 1,3-butanediol also has two O-H bonds. It is not only a proton acceptor but also a proton donor. And there exists an intermolecular hydrogen bond between the H1 atom and O2 atom. So this is a good model to study the chiral hydrogen bond.

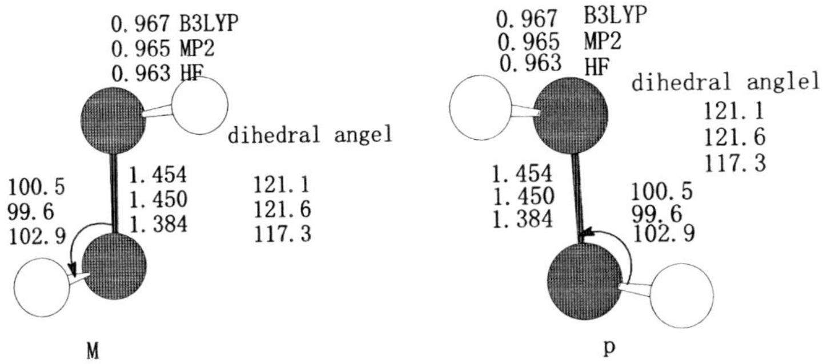

Figure 1. The optimized geometries of hydrogen peroxide at the B3LYP, MP2 and HF/6-311++G(d,p) basis levels, respectively. Bond lengths are in Å and angles are in degrees.

Recently, density functional theory (DFT) has been accepted by the ab initio quantum chemistry community as a cost-effective approach for the computation of molecular structure, vibrational frequencies, and energies of chemical reactions. Many studies have shown that molecular structures and vibrational frequencies calculated by DFT methods are more reliable than MP2 methods.[13-15] We thus report geometry optimization, calculated binding energies and chirodiastaltic energies of these complexes for a variety of basis sets. The role of basis set size and basis set superposition error effects are analyzed in detail. In addition, the vibrational frequencies of the monomer and the stationary complexes are calculated; the intramolecular frequencies and their shifts due to the complex formation are analyzed.

Computational Methods

In the present paper, the popular hybrid density functional B3LYP method, namely Becke's three-parameter nonlocal exchange functional[16] with the nonlocal correlation functional of Lee, Yang and Parr[17] has been used throughout the study, which had been shown to be effective at accurately predicting structure and energies.

For hydrogen bonding, it is expected that both diffuse and polarization functions may be necessary in the basis sets, we thus analyze the separate influence of the diffuse and polarization functions. The geometry optimization of the hydrogen peroxide dimer has been carried out using B3LYP method with the 6-31G, 6-31G(d), 6-31+G(d) and 6-31++G(d,p) basis set along with analytic vibrational frequency calculations .

After obtaining these optimized structures, the calculated binding energies obtained with all the above theoretical procedures are then corrected for the basis set superposition error (BSSE).[18] This is done using the counterpoise method.[19] All calculations are performed using the Gaussian98 program.[20]

Results and Discussion

Structure of Hydrogen Peroxide Monomer

We optimized the H_2O_2 monomers by using B3LYP, MP2 and HF method. The fully optimized geometries for the H_2O_2 monomers (M and P) are depicted in Figure1. The two chiral structures are completely same in bond lengths, angles and molecular energy. The only different between them is dihedral angle, ∠HOOH in M structure is 121.13°, 121.62° and 117.36° whereas in P structure is -121.13°, -121.62° and -117.36 for B3LYP, MP2 and HF/6-311++G(d,p) level, respectively. So the two structures are mirroring image each other.

Structure of 1,3-Butanediol Monomer

Ryderg electron transfer spectroscopy (RET) had been used to determine the dipole-bound (DB) electron affinity (EA) of 1,3-butanediol by S.Carles.[21] In their works, a series of ab initio calculations also has been carried out. They calculated the energies of four possible configurations, which differ in terms of thee orientation of the OH...OH hydrogen-bonded link with respect to the molecular axis and in terms of the shape of the C1-C2-C3-C4 backbone (linear or 'L' shape). They found that the configurations with linear backbone are more stable than those with 'L' shape backbone. On the basis of their discussion, by using B3LYP method we calculated the two configurations with linear backbone, configuration A and B (Figure 2). Then these configurations were fully optimized by using B3LYP method at 6-31G(d), 6-31++G(d,p), 6-311++G(d,p)and 6-311++G(2d,2p) level. All these calculation show that configuration A is more stable than configuration B. All the calculated energies of the two configuration of 1,3-butanediol are presented in Table 1.

Table 1. B3LYP Energies of the Two Conformers of 1,3 Butanediol at 6-31G(d), 6-31++G(d,P), 6-311++G(d,P) and 6-311++G(2d,2P) Level. Values in the Paragraph Are Corrected for Zero-point Vibrational Energy (KJ/mol).

	B3LYP/6-31G(d)	B3LYP/6-31++G(d,P)	B3LYP/6-311++G(d,P)	B3LYP/6-31++G(2d,2P)
Configuration A	-810955.2	-811064.3	-811255.7	-811285.4
	(-810579.9)	(-810691.2)	(-810883.4)	(-810722.7)
Configuration B	-810953.8	-811064.1	-811255.3	-811285.2
	(-810578.6)	(-810691.1)	(-810883.1)	(-810722.6)
ΔE	-1.460041	-0.138889	-0.367045	-0.193237
	(-1.26024)	(-0.89267)	(-0.33869)	(-0.107646)

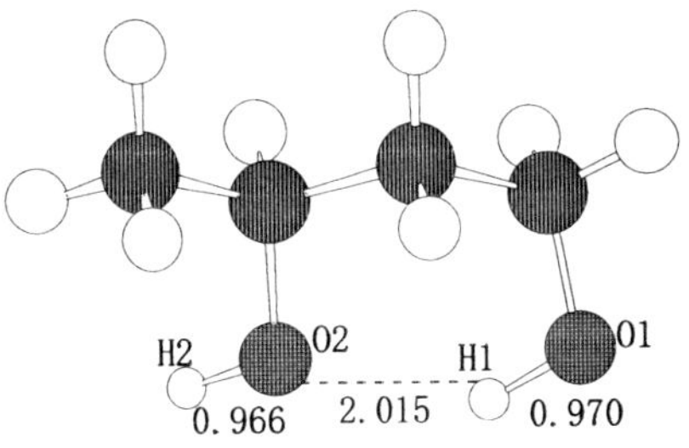

configuration A

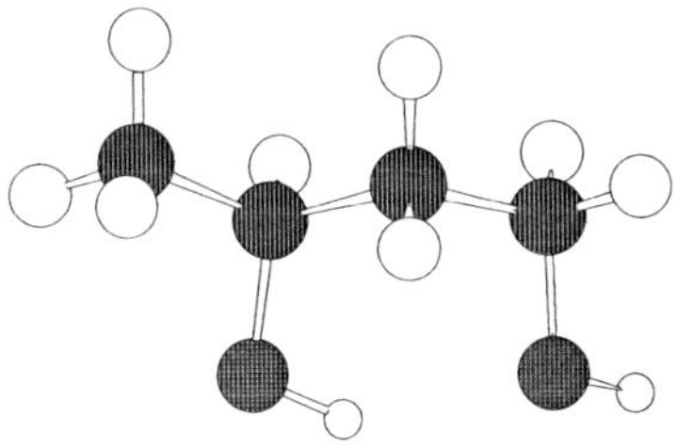

configuration B

Figure 2. The two configurations of 1,3-butanediol optimized at the B3LYP/6-31++G(d,p) basis levels, respectively. Bond lengths are in Å.

Structure of 1,3-Butanediol-Peroxide Complex

At all the levels of B3LYP method employed here, 8 stable structures (four eight-member ring structures, two five-member ring structure and two single hydrogen bond structures) were optimized. The structures of all the complexes are shown in Figure 3. All the complexes

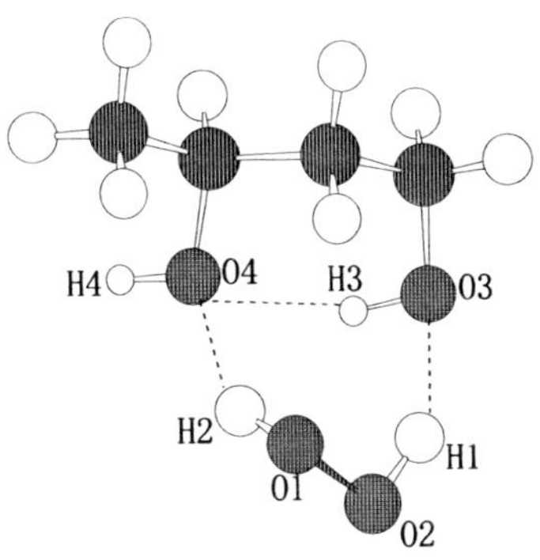

BH1

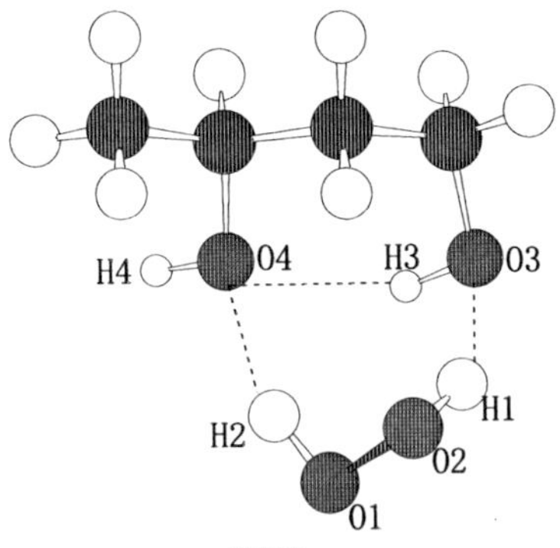

BH2

Figure 3. Continued on next page.

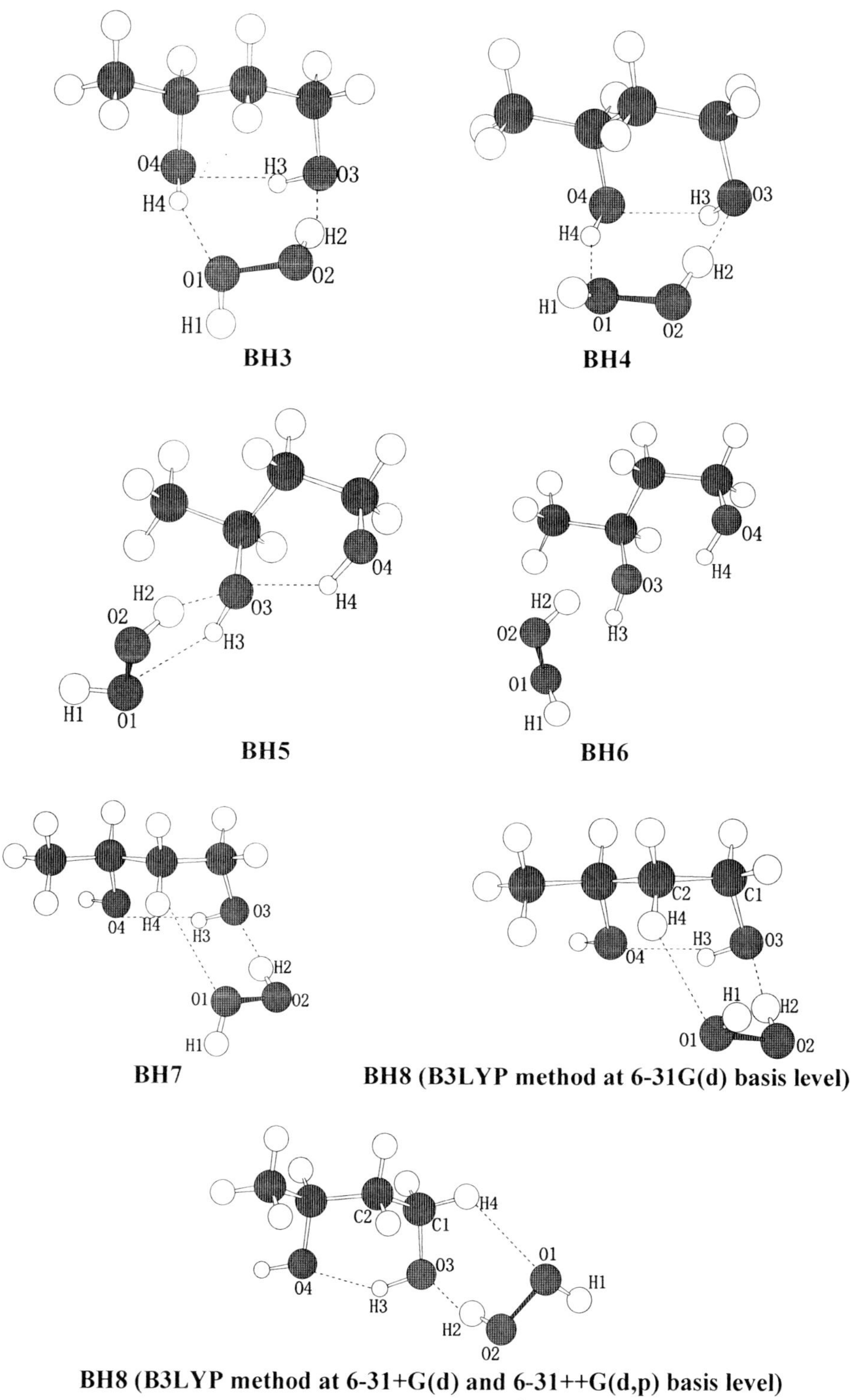

Figure 3. The configurations of 1,3-butanediol-peroxide optimized at the B3LYP/6-31++G(d,p) basis levels, respectively.

(from BH1 to BH8) have been positively identified for real minimum energy structures without imaginary frequency. The most interesting geometrical parameters of these structures are listed in Table 2, Table 3, Table4 and Table 5, respectively.

Table 2. The Optimized Geometric Parameters of BH1 and BH2 Using B3LYP Method at Different Basis Set Levels[a].

		B3LYP/6-31G	B3LYP/6-31G(d)	B3LYP/6-31+G(d)	B3LYP/6-31++G(d,p)
BH1	$R_{O4\text{-}H2}$	1.956	2.014	2.047	2.085
	$R_{O4\text{-}H3}$	1.938	1.997	1.055	2.067
	$R_{O3\text{-}H1}$	1.853	1.942	1.977	1.990
	A1[b]	148.5	149.2	151.4	149.8
	A2[c]	134.7	135.1	130.9	130.9
	A3[d]	144.7	145.6	146.1	146.5
	ψ[e]	81.7	82.7	79.1	79.7
BH2	$R_{O4\text{-}H2}$	1.910	1.993	2.032	2.051
	$R_{O4\text{-}H3}$	2.029	2.052	2.099	2.114
	$R_{O3\text{-}H1}$	1.867	1.929	1.962	1.986
	A1	150.1	147.7	150.7	151.2
	A2	131.6	133.3	129.5	129.2
	A3	144.3	147.5	148.9	148.1
	ψ[e]	-85.7	-85.8	-80.5	-80.9

[a]Distance in Å, angles in degree. [b]A1 is O4H2O1. [c]A2 is O4H3O3. [c]A3 is O3H1O2 [e]ψis the dihedral angel of HOOH.

Table 3. The Optimized Geometric Parameters of BH3 and BH4 Using B3LYP Method at Different Basis Set Levels[a]

		B3LYP/6-31G	B3LYP/6-31G(d)	B3LYP/6-31+G(d)	B3LYP/6-31++G(d,p)
BH3	$R_{O1\text{-}H4}$	1.844	1.927	2.004	2.032
	$R_{O3\text{-}H2}$	1.634	1.764	1.771	1.767
	$R_{O4\text{-}H3}$	1.814	1.864	1.941	1.961
	A1[b]	154.8	159.3	156.5	155.8
	A2[c]	167.8	169.8	170.1	169.7
	A3[d]	137.1	139.6	135.4	134.7
	ψ[e]	137.6	106.6	113.3	113.8
BH4	$R_{O1\text{-}H4}$		1.927	2.0354	2.063
	$R_{O3\text{-}H2}$		1.763	1.774	1.769
	$R_{O4\text{-}H3}$		1.887	1.955	1.980
	A1		159.4	156.3	155.6
	A2		169.4	169.9	170.1
	A3		137.9	134.1	133.1
	ψ[e]		-151.9	-127.4	-128.2

[a]Distance in Å, angles in degree. [b]A1 is O4H4O1. [c]A2 is O2H2O3. [c]A3 is O4H3O3 [e]ψ is the dihedral angel of HOOH.

Table 4. The Optimized Geometric Parameters of BH5 and BH6 Using B3LYP Method at Different Basis Set Levels[a]

		B3LYP/6-31G	B3LYP/6-31G(d)	B3LYP/6-31+G(d)	B3LYP/6-31++G(d,p)
BH5	$R_{O1\text{-}H3}$		2.057	2.250	2.256
	$R_{O3\text{-}H2}$		1.919	2.049	2.063
	$R_{O3\text{-}H4}$		1.992	2.049	2.063
	A1[b]		128.4	118.5	118.9
	A2[c]		143.9	147.1	146.6
	A3[d]		137.2	135.1	134.9
	ψ[e]		-121.3	-124.6	-124.9
BH6	$R_{O1\text{-}H3}$	2.005	2.023	2.219	2.221
	$R_{O3\text{-}H2}$	1.833	1.920	1.908	1.929
	$R_{O3\text{-}H4}$	1.922	1.978	2.027	2.041
	A1	125.4	129.9	119.7	120.5
	A2	143.9	142.9	146.3	145.9
	A3	139.1	138.3	136.2	136.2
	ψ[e]	144.2	115.1	118.9	119.3

[a]Distance in Å, angles in degree. [b]A1 is O1H3O3. [c]A2 is O2H2O3. [c]A3 is O3H4O4 [e]ψ is the dihedral angel of HOOH.

Table 5. The Optimized Geometric Parameters of BH7 and BH8 Using B3LYP Method at Different Basis Set Levels[a]

		B3LYP/6-31G	B3LYP/6-31G(d)	B3LYP/6-31+G(d)	B3LYP/6-31++G(d,p)
BH7	$R_{O4\text{-}H3}$	1.784	1.886	1.934	1.945
	$R_{O3\text{-}H2}$	1.627	1.773	1.771	1.759
	$R_{O1\text{-}H4}$	2.495	2.537	2.779	2.762
	A1[b]	141.8	142.4	139.2	138.9
	A2[c]	169.5	172.8	173.6	172.8
	ψ[e]	154.68	110.4	116.4	116.4
BH8	$R_{O4\text{-}H3}$		1.878	1.942	1.942
	$R_{O3\text{-}H2}$		1.781	1.782	1.773
	$R_{O1\text{-}H4}$		2.543	2.840	2.778
	A1		141.8	138.4	138.9
	A2		174.3	172.7	171.9
	ψ[e]		-121.3	-113.1	-112.3

[a]Distance in Å, angles in degree. [b]A1 is O4H3O3. [c]A2 is O3H2O2 [e]ψ is the dihedral angel of HOOH

Surveying the calculated results for the different methods at different basis set level is that changes in the monomer geometries upon complexation are relatively minor. For the monomer peroxide, complex formation induces a small elongation of the O-H bond and a very small contraction of the O-O bond. The maximum bond length change is less than 0.06

Å at 6-31++G(d,p) basis level. As to 1,3-butanediol, due to the formation of the hydrogen bonding, the O-H bond and the C-O bond are prolonged; of which, the O-H bond is elongated less than 0.03 Å and other bond length change is less than 0.015 Å.

BH1 (formed by peroxide (P) and 1,3-butanediol) and BH2 (formed by peroxide (M) and 1,3-butanediol) are eight-member ring made up of two hydrogen of peroxide and two oxygen of 1,3-butanediol. In BH1 the O4-H3 bond (the intramolecular hydrogen bond distance of 1,3-butanediol) is shortened, while in BH2 it is elongated. The length of this bond is 1.938 Å, 1.997 Å, 2.055 Å and 2.067 Å, respectively in BH1 and 2.029 Å, 2.052 Å, 2.099 Å and 2.114 Å, respectively in BH2 for B3LYP method at different level. At 6-31G level the difference of the length is about 0.091 Å, but it is only about 0.047 Å at 6-31++G(d,p) level. As Table 2 shows that compared with the lengths of other intermolecular hydrogen bonds, in BH2 this intramolecular hydrogen bond is relatively weak. In BH1 the length of the O4-H2 bond is 1.956 Å, 2.014 Å, 2.047 Å and 2.085 Å, respectively at different level but the length of the corresponding bond in BH2 (O3-H1 bond) is 1.910 Å, 1.993 Å, 2.032 Å and 2.051 Å, respectively, which shows that this hydrogen bond of BH2 is stronger than that of BH1. But the variation of the length of the hydrogen bond at different basis set is less than the intramolecular hydrogen bond, at 6-31G level the difference of the length is about 0.046 Å, and it is 0.054 Å at 6-31++G(d,p) level. In BH 1, $R_{O3\text{-}H1}$ is 1.853 Å, 1.942 Å, 1.977 Å and 1.990 Å, respectively, but it is 1.867 Å, 1.929 Å, 1.962 Å, and 1.986 Å, respectively in BH2 at different level, which denote that this hydrogen bond in BH2 is stronger than that in BH1 too. Furthermore, the cyclic arrangement results in more bent hydrogen bonds, for A1 the deformation from linearity being 30.2°(28.8°), while the deformation of A2 is 33.5° and 31.9° for BH1 and BH2 at the largest basis level, respectively. The values of A1 in the two structures are similar to each other, and the maximum difference at different level is less than 1.6 degree. It can be seen from Table 3 that for A1 there is little difference caused by the using diffuse functions and polar functions in both two structures. As to A2 and A3, high basis levels may get a comparatively low value. It is very interesting to see that for A2 and A3 in the two structures at 6-31+G(d) and 6-31++G(d,p) level we almost get the same value. The dihedral angle of HOOH is 79.7° for BH1 and –80.7° for BH2 at the largest level. Compared with the parameters of hydrogen peroxide monomer, the distortions are obviously, which should attribute to the cyclic arrangement. It seems that when the diffuse functions and polarization functions are used, the dihedral angle of HOOH doesn't vary significantly.

As to BH3 (formed by peroxide (P) and 1,3-butanediol) and BH4 (formed by peroxide (M) and 1,3-butanediol), they are also cyclic conformers, in which hydrogen peroxide accepted a proton from the 1,3-butanediol group and donated a proton to the 1,3-butanediol group. There are two hydrogen bonds, O1-H4 bond and O3-H2 bond in the two structures. $R_{O1\text{-}H4}$ is 2.032 Å and $R_{O3\text{-}H2}$ is 1.767 Å for BH3 at the largest basis level, while in BH4 the two bonds is 2.063 Å and 1.769 Å, respectively, which is a little longer than those of BH3, hence, the interaction energy of this structure will be higher than that of BH3. The cyclic arrangement also results in more bent hydrogen bonds, the deformation from linearity being 24.2°(24.4°) for A1, while the deformation of A2 is 33.5° and 31.9° for BH3 and BH4 at the largest basis level, respectively. At B3LYP/6-31++G(d,p) level as Table. 2 shows that the differences of the hydrogen bonds caused by the by different basis set levels are more significant than those of the angles. The B3LYP/6-31++G(d, p) calculations which yield the A1 of 155.8° and A2 of 169.7° are in good agreement with the B3LYP/6-31+G(d) values of 155.6° and 170.1°. The intramolecular hydrogen bond is strengthened in the two structures and from Table 2 it can be seen that the hydrogen bond in BH3 is stronger than that in BH4. But from 6-31G to 6-31++G(d, p) level, R_{O4-H3} and R_{O1-H4} is increased by 0.147 Å and 0.188 Å, respectively. We can not find steady conformer of the BH4 at 6-31G level, which shows that the polarization functions are very important in getting the accurate conformer of complexes formed by chiral hydrogen peroxide and other molecules. The dihedral angle of HOOH is 113.8° for BH3 and −128.2° for BH4 at the largest level. Compared with the parameters of hydrogen peroxide monomer, the dihedral angle of HOOH has been increased, which should attribute to the intermolecular repulsion of other parts of 1,3-butanediol. It has been shown in Table 2 that at low basis level the values of the dihedral angle of HOOH are inaccurate, but when one diffuse function and one polarization function are used the values are very closed to those at 6-31++G(d,p) level.

BH5 (formed by peroxide (M) and 1,3-butanediol) and BH6 (formed by peroxide (P) and 1,3-butanediol) are 5-number rings, in which the peroxide accepts one proton from the 1,3-butanediol group and donated a proton to it. We cannot get the configuration of BH5 at 6-31G basis level, which is due to the lack of the polarization function in the basis set. In BH5 and BH6, the O3-H2 bond is 2.063Å and 1.929Å at the largest basis level, while the O1-H3 bond is 2.256 Å and 2.221Å, respectively, which denote that in the two structures the O3-H2 bond is much stronger than the O1-H3 bond and the intermolecular hydrogen bonds in BH5 is relatively weak. Form Table 4 it can be found that in the both two structures the intramolecular bonds are elongated and the bond in BH5 are weaker than that in BH6. For A1 the deformation from linearity being 81.1°(79.5°), while the deformation of A2 is 33.4° and 34.1° for BH5 and BH6 at the largest basis level, respectively. And the dihedral angle of HOOH is 116.4° for BH5 and −112.3° for BH6 at the largest level. Compared with the parameters of hydrogen peroxide monomer, the distortions of A1, A2 and the dihedral angle of HOOH are not obviously, it is due to the peroxide only formed hydrogen bonds with one of the O-H bond of the 1,3-butanediol and there is no cyclic arrangement as other complexes. Form Table 4 it is not difficult to find that if we ignore the result at 6-31G level for BH6, there is little difference among the results calculated at different basis set levels for A1, A2 and the dihedral angle of HOOH.

BH7 (formed by peroxide (P) and 1,3-butanediol) and BH8 (formed by peroxide (M) and 1,3-butanediol) are also cyclic structures. There is only one classical hydrogen bond in BH7 and BH8, namely O3-H2 bond. And there also exist a weakly intermolecular hydrogen bond, O1-H4 bond, in the two structures. In Figure.2, we also line out O1-H4 bond in BH7 and BH8 respectively, because this bond play a very important role in determining the configurations of BH8 at different basis set levels. At 6-31G level, we cannot find steady configuration for BH8, which is due to the lack of the polarization function at this basis level. When it is optimized at a same initial structure, at 6-31G(d) and other basis levels, two different configurations of BH8 have been found. In the two structures there exists one hydrogen bond between the H2 atom and the O3 atom, but in the structure calculated at 6-31+G(d) level the O1 atom formed one hydrogen bond with H4 atom, while in the structure calculated at 6-31++G(d,p) level the O1 atom formed one hydrogen bond with H5 atom. Maybe the two weak hydrogen bonds is the most determinant of forming the different configurations. The diffuse functions and the polarization functions play an important role again, and in this case without proper basis set the right configuration even cannot be achieved. The length of the O4-H3 bond is 1.945 Å and 1.942 Å at 6-31++G(d,p) level for BH7 and BH8 respectively. Compared with other intermolecular hydrogen bonds in other structures, there is little difference between the two bonds; at 6-31G(d) level, the difference between the length of the O1-H4 bond for the two structures is neglectable, which should attribute to that the peroxide only formed hydrogen bonds with one of the O-H bond of the 1,3-butanediol and there is no circularity in them. The O4-H3 bond is 1.945 Å and 1.942 Å for the two structures at 6-31++G(d,p) level, which shows that the intramolecular bonds in them are strengthened. Because there is no the cyclic arrangements, the A2 at different basis levels are almost linearity. Similar to BH5 and BH6, there is little distortion on the dihedral angle of HOOH.

Form the discussion above it can be found that in the structures with cyclic arrangements the distortions of the bond angle and dihedral angle of HOOH are obviously But for the structures with one hydrogen bond, the influence is neglectable. With the diffuse functions and polarization functions are considered the bond angle is decreased, but the length of the hydrogen bond is increased. In some case, without polarization functions we even cannot get a steady configuration. For BH8, at different level two different configurations have been achieved, which means that in further calculations work on larger system, if accurate calculations work are wanted to be performed, at least the 6-31G(d,p) basis level is supposed to be used.

Interaction Energies

Interaction energies are calculated for the hydrogen peroxide-1,3-butanediol hydrogen bond by taking the energy difference between the fragments and the complex. To correct the basis set superposition error (BSSE), the counterpoise (CP) method [20] is employed. The zero-point vibrational energy (ZPVE) corrections are also applied in the present case.

Table 6. Interaction Energies of 1,3-butanediol - Peroxide Complex (kJ/mol)

	B3LYP/6-31G	B3LYP/6-31G (d)	B3LYP/6-31+G (d)	B3LYP/6-31++G (d,p)
BH1	-58.4 (-49.6) (-46.1)	-57.1 (-34.8) (-41.0)	-34.7 (-31.1) (-25.3)	-30.9 (-28.8) (-25.2)
BH2	-59.3 (-47.7) (-47.6)	-52.9 (-34.2) (-42.4)	-35.7 (-32.2) (-26.0)	-33.2 (-30.9) (-24.1)
BH3	-86.2 (-73.4) (-73.2)	-71.8 (-53.9) (-59.7)	-51.6 (-48.9) (-40.3)	-48.1 (-45.9) (-37.7)
BH4		-67.6 (-49.9) (-56.2)	-46.4 (-44.6) (-36.2)	-43.2 (-41.1) (-33.8)
BH5		-40.7 (-23.7) (-32.1)	-28.6 (-24.8) (-21.7)	-26.6 (-24.3) (-20.1)
BH6	-50.9 (-36.3) (-41.1)	-42.9 (-27.6) (-34.0)	-31.3 (-27.1) (-23.8)	-29.1 (-26.5) (-22.1)
BH7	-64.2 (-54.6) (-55.6)	-47.7 (-36.0) (-40.0)	-39.4 (-35.5) (-32.4)	-37.9 (-35.7) (-31.0)
BH8		-46.5 (-36.9) (-39.1)	-39.4 (-35.1) (-31.9)	-37.5 (-35.3) (-30.7)

Values in the First Parenthesis Are Results with Correction for Basis set superposition Error And in the Second Parenthesis Are Corrected for Zero-point Vibrational Energy.

To analyze in more detail the role of basis set size effects on the binding energy between hydrogen peroxide and 1,3-butanediol, we show all the results in Table 6, which gives a detailed analysis of the binding energy obtained with B3LYP method at several basis set. The numbers in the first parenthesis are corrected for zero-point vibrational energy and in the second parenthesis are results with correction for basis set superposition error using the counterpoise method of Boys and Bernadi.[20] As expected, basis set sensitivity exists. The interaction energy computed with B3LYP using the minimal basis set 6-31G is much higher. As the basis set enlarged, the computed values decrease and converge smoothly. All the ZPVE corrected energies are smaller than the uncorrected ones and all of the ZPVE corrected energies calculated at 6-31+G(d) 6-31++G(d,p) levels are smaller than the BSSE corrected ones at the same basis level. Moreover, at 6-31+G(d) 6-31++G(d,p) levels the correction is not sensitive to the basis sets. The general importance of including BSSE corrections in calculated binding energies has been well documented in the literature. From the Table 6 we can see, the magnitude of BSSE are decreasing with the basis set enlarged, when the diffusion and polarization functions are considered, especially for the 6-31++G(d,p) basis set using B3LYP method, the inclusion of BSSE correction has minor importance to the binding energy.

From Table 6, we can see that the relative stability order of the eight structures is BH3> BH4> BH7> BH8> BH2> BH1> BH6> BH5. It is easy to understand the stability of BH3 derives from the formation of a pair of hydrogen bonds between the hydrogen peroxide and 1,3-butanediol due to the strongest interaction (the shortest hydrogen bond distance). The reason why BH3 is more stable than BH4 is that in BH4 the distance between H1 atom and of 1,3-butanediol is shorter than that in BH3. At the same time, the instability of BH5 is by reason of the weakest interaction, though it is also a cyclic structure. In BH7 and BH8 the H1 atom distance 1,3-butanediol, so there is little caused by the chirality of hydrogen peroxide. For the other structures, we can also estimate the stability by the interaction energy and the lengths of the intermolecular hydrogen bonds.

Table 7. Chirodiastatic Energy of 1,3-butanediol - Peroxide Complex (kJ/mol)

	B3LYP/6-31G	B3LYP/6-31G(d)	B3LYP/6-31+G(d)	B3LYP/6-31++G(d,p)
BH1- BH2	0.87(-1.82)(1.51)	0.86(-0.51)(1.44)	0.97(1.11)(0.72)	2.23(2.10)(-1.1)
BH3- BH4		-4.23(-3.99)(-3.51)	-5.22(-4.28)(-4.08)	-4.91(-4.88)(-3.93)
BH5- BH6		2.21(3.86)(1.89)	2.74(2.25)(2.14)	2.56(2.23)(2.03)
BH7-BH8		-1.11(0.90)(-0.94)	0.03(-0.49)(-0.48)	-0.33(-0.36)(-0.07)

Values in the First Parenthesis Are Results with Correction for Basis set superposition Error And in the Second Parenthesis Are Corrected for Zero-point Vibrational Energy.

To analyze the difference caused by the chiral structures of hydrogen peroxide, we show the chirodiastatic energy for the four pairs complexes at different level in Table 7. When diffuse functions and polarization functions are added most of the chirodiastatic energies calculated by using ZPVE data are smaller than the uncorrected ones and most of the ZPVE corrected energies are smaller than the BSSE corrected ones at the same basis level. We also find that in BH7 and BH8 when the BSSE data are considered we even get the opposite conclusions, which mainly because at different level there is a different configuration of BH8.

Infrared Spectrum

Vibrational spectroscopy is one of the most useful experimental tools for study of the H-bonded clusters, so the information on calculated harmonic vibrational frequencies can be useful. The values for both vibrational frequencies and IR intensities of the monomers and the eight complexes are list in Table 8 at the B3LYP/6-31++G(d, p) level. Since there are little shifts at other frequencies of 1,3-butanediol, we have only discussed the frequencies involved in the intermolecular hydrogen bond. Since the frequency shifts are relatively stable with respect to variation of basis set, it can estimate the IR spectrum for the complex by combining the observed fundamental vibrational frequency of its moieties and the frequency shift in Table 8.

Table 8. Frequencies And IR intensities of Monomers And Complexes at B3LYP/6-31++G(d,p) Level, Vibrational Frequencies (ν) in cm^{-1}, IR Intensities (I) in km/mol.

Hydrogen Peroxide			BH1	BH2	BH3	BH4	BH5	BH6	BH7	BH8
Expt	ν (I)	Assignment	ν (I)	ν (I)	ν (I)	ν (I)	ν (I)	ν (I)	ν (I)	ν (I)
371	371 (221)	HOOH bend out of plane	734 (138)	710 (169)						
877	945 (1)	OO stretch	944 (0.5)	943 (0.1)	943 (6)	942 (6)	940 (25)	940 (2)	944 (3)	942 (3)
1266	1301 (96)	HOOH rock in plane					1320 (70)	1327 (70)	1331 (68)	1338 (58)
1402	1445 (0.1)	HOOH bend in plane	1448 (35)	1431 (52)	1571 (13)	1562 (11)	1528 (13)	1522 (11)	1527 (8)	1560 (19)
3599	3767 (60)	HOOH asym stretch	3594 (185)	3587 (185)						
3608	3768 (11)	HOOH sym stretch	3671 (171)	3662 (211)						
1,3-butanediol										
3756 (127)		O3-H3 stretch	3751 (112)	3764 (95)	3648 (180)	3655 (172)	3760 (179)	3754 (193)	3703 (181)	3704 (183)
3827 (22)		O4-H4 stretch	3817 (27)	3818 (28)	3680 (256)	3690 (230)	3773 (43)	3769 (50)	3830 (27)	3830 (27)

The HOOH frequency of the bend in plane mode is found to reduce for all structures considered here (1448, 1431, 1571, 1562, 1528, 1522, 1527 and 1560 cm^{-1} respectively). Expect in BH1 and BH2, the frequencies of the stretch mode in other complexes are all blue shifted. The frequency of the O-O stretching mode shows very slight change on all of the structures, which shows that the forming of the complexes has little effect on this vibrational mode. The HOOH frequencies of the bend out of plane mode that has disappeared in other complexes, it only has been found in BH1 and BH2, which mainly because the hydrogen peroxide forms the intermolecular hydrogen bond with only one OH bond of 1,3-butanediol. The HOOH asymmetric stretch mode and HOOH symmetric stretch mode only appear in BH1 and BH2, it maybe is also due to the vibration of H atom in hydrogen peroxide is restricted to the forming of the intermolecular hydrogen bond. The HOOH rock in plane mode vanishes in the four eight-member ring structure which mainly because the forming two intermolecular hydrogen bond in them. For the vibrational modes of 1,3-butanediol in the complex, like that of H_2O_2, the most of the stretching frequencies associated with the hydrogen bond undergo red shift compared to the free monomer. The reason is that the formation of the hydrogen bond weakens the O-H bond.

With respect to IR intensities, they are all IR-active and most of them have large intensities. These predicted IR spectral characteristics might be of great interest in the analysis of the experimental spectral features. It is more difficult to predict accurate shifts in absorption intensities. For this system, there is an extremely large increase in the intensity of the stretching vibrational modes of the hydrogen donor. From the results presented in Table 8, it can be seen that the hydrogen bonded OH stretching mode intensities are enhanced largely in every hydrogen bond complexes. For example, the HOOH bend in plane mode intensities varied from 8km/mol to 52 km/mol for the eight structures, and it is notable that the largest change of this mode intensity occurs in BH2 (to 53 km/mol, approximately 500 times bigger than that of the mode in the monomer.), which is due to the strongest interaction. For the O-O stretch mode, the intensities have decreased in BH1 and BH2, which should be due to the forms of hydrogen bond restricted the vibration of O-O bond; while in other structures it have increased, which mainly because there is no restriction on the O-O bond. In a word, owing to the formation of the hydrogen bond, the force constants of the bond involved in the hydrogen-bonding reduce and the frequencies are red-shifted. At the same time, the increasing of the change of the vibrational dipole moments results in the enhancement of the IR intensities.

Conclusions

The hydrogen bond interaction of 1:1 complex between hydrogen peroxide and 1,3-butanediol has been analyzed by B3LYP method employing different basis set levels. Eight structures are considered, one of which are one-hydrogen bonded structure and others are cyclic double–hydrogen bonded structures at the largest basis level. In the eight structures, BH3 is the most stable at all levels and BH5 is the most unstable, it is due to the weakest interaction. For other complexes, their stabilities depend on the values of the interactions between hydrogen peroxide and 1,3-butanediol. Moreover, the infrared spectrum frequencies, IR intensities and the vibrational frequency shifts are reported. It can be found that the configurations of the complexes play an important role in determining the IR intensities and the vibrational frequency shifts. The stretching frequency associated with the hydrogen bond undergoes a shift to a lower frequency compared to the free monomer and there is an extremely large increase in the intensity of the stretching vibration of the hydrogen donor.

Acknowledgements

This work is supported by the Natural Science Foundation of Shandong Province (Z2002F01) the State Key Laboratory Foundation of Crystal Material and the National Natural Science Foundation of China (2967305).

References

[1] Craig, D. P.; Power, E. A.; Thirunamachaandran,T. *Proc. R. Soc. London, Scr. A* 1971, 322, 65.
[2] Craig D. P.; Mellor, D. P. *Top. Curr. Chem.* 1976, 63, 1.
[3] Salen,L.; Chapuisat, X. G Segal,. P.; Hiberty,C.; Minot, C.; Leforestier, C.; Sautet, P. J. *Am. Chem. Soc.* 1987, 109, 2887.
[4] Collet, A. *Problems and wonders of Chiral Molecules* (Akademiai Kiado, Budapest, 1990)
[5] Craig, D. P.; Thirunamachandran, T. *Theor. Chem. Acc.* 1999, 63, 1.
[6] Portmann, S. *J. Chem. Phys.* 2000, 113, 6577.

[7] Catalano, E.; Sanborn, R. *J. Chem. Phys.* 1963, 38, 2273.
[8] Lannon, J. A.; Vederame, F. D.; Anderson, R. W. *Jr. J. Chem. Phys.* 1997, 90, 2212.
[9] Pettersson, M.; Touminen, S.; Rasanen, M. *J. Phys. Chem. A* 1997,104, 1166.
[10] Goebel, J.; Ault, B.; Del Bene, J. *J. Phys. Chem. A* 2000 104, 2033.
[11] Engdahl, A.; Nelander, B. *Phys. Chem. Chem. Phys.* 2000 2, 3967.
[12] Engdahl, A.; Nelander, B.; Karlstrom,G. *J. Phys. Chem. A* 2001,105, 8393.
[13] Liu, R. F.; Dennis, R. T.; Jefferey, A.C.; Panla, R. M. *J. Phys. Chem.* 1996, 100 3430
[14] Zhou, X. F.; Liu, R. F. *Spectrochim. Acta. Part A.* 1997, 53, 259.
[15] Nwobi, O.; Higgins, J.; Zhou, X. F. Liu, R. F. *Chem. Phys. Lett.* 1997, 272,155.
[16] D. Becke, *J. Chem. Phys.* 1993, 98, 5648.
[17] Lee, W. T. Yang, R. G. Parr, Phys. Rev. B 1988, 37, 785.
[18] Vanduijneveldt, F. B.; Van Duijneveldt-Vande Rijdt, J. G. C. M.; Van Lenthe, J. H. *Chem. Rev.* 1994, 94, 1873.
[19] Boys, S. F.; Bernadard, F. *Mol. Phys.* 1970, 19, 553.
[20] Frisch, M. J.; Trucks, G.W.; Schlegel, H. B.; Gill, P. W. M.; Johnson, B.G.; Robb, M.A.; Cheeseman, J. R.; Keith, T. A.; Petersson, G. A.; Montgomery, J. A.; Raghavamery, K.; Al-Laham, M. A.; Zakrezewski, V. G.; Ortiz, J. V.; Foreman; J. B.; Cioslowski, J.; Stefanov, B. B.; Nanayakkara, A.; Challacombe, M.; Peng, C. Y.; Ayala, P. Y.; Chen, W. Wong, M. W.; Andres, J. L.; Replogle, E. S.; Gomperts, R.; Martin, R. L.; Fox, D. J.; Binkley, J. S.; Defrees D.J.; Baker, J.; Stewart, J. P.; Head-Gordon, M.;, C. Gonzales, J.A. Pople, *Gaussian 98, Revision A.l,* Gaussian, Inc., Pittsburgh, PA, 1998.
[21] Carles, S.; Desfrancois, C.; Schermann, J. P.; Jalbout, A. F.; Adamowicz Ludwik *Chem. Phys. Lett.* 2001, 334, 374.

In: Quantum Chemistry Research Trends
Editor: Mikas P. Kaisas, pp. 239-251
ISBN 978-1-60021-620-6

Chapter 8

A NEW METHOD FOR APPROXIMATING THE COULOMB POTENTIAL GENERATED BY THE PRODUCT OF TWO 1S SLATER ORBITALS

***J. E. Pérez*[1]*, O. E. Taurian*[1]*, J. C. Cesco*[2] *and F. S. Ortiz*[1]**
[1]Departamento de Física, FCEFQyN, Universidad Nacional de Río Cuarto;
Ruta Nac. 36 Km 601, X5804BYA Río Cuarto, Argentina
[2]Instituto de Matemática Aplicada, Universidad Nacional de San Luis,
CONICET; Avda. Ej. de los Andes 950, 1er piso,
D5700HHW, San Luis, Argentina

Abstract

The present chapter is a study in the evaluation of the Coulomb Potential (CP) generated by the product of two Slater Type Orbitals (STO). This study is motivated by the recent developements carried out in our research group, see refs. [1,2,3]. The methos presented in these references open several possibilities. Here, the efforts are directed to obtain approximated values of a parameter called Q, similar to that defined in [1], and also, to improve the evaluations in the intermediate auxiliary integrals which are analogous to those defined in [2]. The CP is represented as a series whose terms involve integrals which are functions of the approximated value for Q. This parameter is determined in two differents ways: as an appropriated average and considering a model of the error of truncation of the series expansion. In both cases the value of CP corresponds to the truncated series evaluated at these particular values of the parameter. The interest in this approach resides in the fact that it is possible to perform all the necessary integrals in the complex plane, similarly as it is done in ref. [3]. This step is performed by using rational functions to approximate the nonoscilating parts of the integrands. The error observed in the obtained CP values can be explained from the errors involved in the rational functions used. The test calculations are performed for the ethylene molecule.

A. Introduction

Recently, our research group has developed a methodology to approximate multicenter integrals (MI) involving 1s STO, see references [1-3]; from now on these references will be denoted by I, II and III, respectively. Each one of the articles I, II and III, presents a theoretical framework from which a numerical procedure is implemented to approximate the multicenter bielectronic integral on 1s STOs. Based on these works, it is presented here a complimentary methodology to improve the evaluation of the intermediate integrations involved in the approximants of MIs; the direct numerical integrations are replaced by the integration in the complex plane, this is accomplished by replacing the non-oscilating functions of the integrands by rational functions. This turns out to be a very efficient algorithm because the rational approximants, which depend only on two 1s STOs, have a very simple pole structures.

In order to simplify the formulation of the new methods, without loosing generality, it is only considered the electric potential (EP) generated by the product of two 1s STOs, i.e. the three-center integral. This kind of MI was not considered previously in I-II-III. In the present chapter, two different ways for approximating the EP are explored, both proposals are based on an adapted version of the truncated series expansion of the Spherical Bessel function $j_0(pr)$ developed in II, see eq. (10) there. The approximation of $j_0(pr)$ in the MI by its truncated series, permits the decoupling of variables r and those involved in p, in subsequent integrations.

Besides of the decoupling of variables in the integration mentioned above, another important reason for using the expansion of $j_0(pr)$ is the introduction of an arbitrary positive parameter, called q in II. In II, it is shown that if q is taken as an average of p, for example see eq. (9) in II, it turns out that the first term is the principal contribution to MI however the remainder is a non negligible correction; here, these properties are used to approximate MIs representing the EP.

One of the algorithm to be developed, follows a close analogy to the algorithm given in II. Firstly, q is calculated as an averaged of p and it is introduced in the truncated series. Each one of the terms in the truncated expansion representing the MI is an integral containig oscilating functions over an infinite interval, which is integrated with similar procedures to those given in III, replacing the non oscilating part in the integrands by rational approximants, this permits o carry out the integrations using the Residues Theorem based technique in the complex plane.

The other algorithm to be presented, is based on the theorem proved in I, i.e. there exists a unique Q such that $F(Q) = MI$; in this equation, the series expansion of $j_0(pr)$ is introduced in MI, in order to take advantage of the arbitrariness of the parameter q, it is taken equal to Q. This permits to truncate the series retaining only the first two terms in the expansion and, in this way, to obtain an approximated value of Q, denoted by Q_a. Then, in the evaluation of EP the series is developped up to the third term, it should be noticed that the second term is equal to zero because the definition of Q_a.

The important point in these two procedures is the introduction of the racional approximants in all the integrals, involved in the calculation, by performing them in the complex plane. This permits to improve the computational efficiency. The criteria that permit the calculation of the rational approximants are deduced from the relevant properties of the approximated functions and, they are discussed and proved in detail.

The chapter is organized as follows, in *Method* it is described in detail the two algorithms, in *Results* it is shown the numerical trend that the new method can give, ending with some comments and final considerations in *Conclusion*.

B. Method

In subsection $B.1$ the theoretical expression for the electric potential to be aproximated in the chapter, taken from ref. [4], is introduced. In subsection $B.2$ the series expansion of the function $j_0(pr)$, taken from II is introduced in MI, and it is presented at this point because it is used in both algorithms. In subsection $B.3$ one of the approximated expression for the EP is obtained; with the aim of making this subsection clearer, some of the needed results are given in $APPENDIXES$ 1 and 2, respectively. In subsection $B.4$, the other procedure to get an expression for the EP is given in detail.

B.1. The Electric Potential

It is well known [4], that the EP at $\vec{R}$, generated by the product of two 1s STOs (located at $\vec{R}_1$ and $\vec{R}_2$, respectively) is given by:

$$\Phi(\vec{R}) = \int dv \frac{\phi_1(\vec{r}-\vec{R}_1)\phi_2(\vec{r}-\vec{R}_2)}{|\vec{r}-\vec{R}|}$$

$$= CN \int_0^{\infty} dr \int_0^1 dx U(x,r) j_0(p(x,\vec{R})r) \tag{1}$$

where

$$\phi_i(\vec{r}-\vec{R}_i) = (\alpha_i^3/\pi)^{0.5} exp(-\alpha_i|\vec{r}-\vec{R}_i|) \tag{2}$$

$$U(x,r) = x(1-x)K(Z_{1_2}(x,r)) \tag{3}$$

$$K(z) = (\pi/2)^{0.5} exp(-z)\left[1+\frac{3}{z}+\frac{3}{z^2}\right]/z^3 \tag{4}$$

$$Z_{12}(x,r) = |\vec{R}_1-\vec{R}_2|\left[r^2x(1-x)+\alpha_1^2x+\alpha_2^2(1-x)\right]^{0.5} \tag{5}$$

$$p(x,\vec{R}) = \left|\left[(1-x)\vec{R}_1+x\vec{R}_2\right]-\vec{R}\right| \tag{6}$$

$C = 4\alpha_1\alpha_2|\vec{R}_1-\vec{R}_2|^5$, $N = \frac{(\alpha_1\alpha_2)^{1.5}}{\pi}$ and $j_0(z) = \frac{sin(z)}{z}$.

B.2. The Series for $j_0(pr)$

Following a close anlogy to the procedure given in II and for a positive parameter q such that $Min[p(x,\vec{R})] < q < Max[p(x,\vec{R})]$, where $p(x,\vec{R})$ is given in eq. (6), it follows that

$$j_0(p(x,\vec{R})r) = j_0(\frac{p(x,\vec{R})}{q}qr) = j_0(qr) -$$

$$-\frac{j_1(qr)}{2(qr)}r^2[p(x,\vec{R})^2-q^2] + \frac{j_2(qr)}{8(qr)^2}r^4[p(x,\vec{R})^2-q^2]^2 + \ldots \tag{7}$$

where $p(x,\vec{R})$ is given in eq. (6) and $j_n(z)$ are the Spherical Bessel functions [5].

B.3. Algorithm 1

The EP given in eq. (1), transforms into the expression given in eq. (8) by using the eq. (7) with $q = Q_2$,

$$\Phi(\vec{R}) = CN \int_0^\infty dr j_0(Q_2 r) \int_0^1 dx U(x,r) -$$
$$-\frac{CN}{2} \int_0^\infty dr \frac{j_1(Q_2 r)}{(Q_2 r)} r^2 \int_0^1 dx U(x,r)[p(x,\vec{R})^2 - q^2] +$$
$$+\frac{CN}{8} \int_0^\infty dr \frac{j_2(Q_2 r)}{(Q_2 r)^2} r^4 \int_0^1 dx U(x,r)[p(x,\vec{R})^2 - q^2]^2 + \dots \tag{8}$$

where

$$Q_2 = \left[\frac{\int_0^1 dx U(x,0) p(x,\vec{R})^2}{\int_0^1 dx U(x,0)}\right]^{0.5} \tag{9}$$

and C, N, $p(x,\vec{R})$ and $U(x,r)$ are defined in subsection $B.1$; with

$$j_1(z) = \frac{sin(z)}{z^2} - \frac{cos(z)}{z} \tag{10}$$

$$j_2(z) = (\frac{3}{z^3} - \frac{1}{z}) sin(z) - \frac{3}{z^2} cos(z). \tag{11}$$

The eq. (8) can be rewriten as

$$\Phi(\vec{R}) = \int_0^\infty dr j_0(Q_2 r) Ip(r,0) -$$
$$-\frac{1}{2Q_2^2} \int_0^\infty dr j_1(Q_2 r)(Q_2 r) \Big[Ip(r,1) - Ip(r,0) Q_2^2\Big] +$$
$$+\frac{1}{8Q_2^4} \int_0^\infty dr j_2(Q_2 r)(Q_2 r)^2 \Big[Ip(r,2) - 2Ip(r,1) Q_2^2 + Ip(r,0) Q_2^4\Big] + \dots \tag{12}$$

where

$$Ip(r,n) = CN \int_0^1 dx U(x,r) p(x,\vec{R})^{2n}. \tag{13}$$

The eq. (12) yields aproximants to $\Phi(\vec{R})$ by truncating the series. Such a procedure is based on the methodology delineated in II and, consequently, it follows the same qualitative behaviour observed in the approximants to the bielectronic integrals, i.e. being Q_2 an average of $p(x,\vec{R})$ the first term in eq. (8) is the principal contribution and the remaining ones are corrections. The aditional simplifing step proposed in this chapter, is the replacement of the functions $Ip(r,n)$ by rational functions, with the main aim of performing all the integrations over r in the complex plane. In the *Appendix* 1 it is shown that the functions $Ip(r,n)$ given in eq. (13), can be approximated by

$$Ap(r,n) = \frac{Ip(0,n)}{1 + a2(n) r^2 + a4(n) r^4} \tag{14}$$

where the constants $a2(n)$ and $a4(n)$ are determined from the criteria given in the *Appendix* 2, i.e. they can be choosen positive. When the functions $Ap(r,n)$ are introduced in the place of $Ip(r,n)$ in eq. (13), the integration on r can be performed straightforwardly by using the Residues Theorem in the complex plane, see [6] and III. In *Appendix* 3 are discussed the details of the integration in the complex plane for obtaining the auxiliary functions used in eq. (15), which is the approximant for the EP only considering terms up to the third in eq. (12):

$$\Phi(\vec{R}) \approx A0(Q_2) + A1(Q_2) + A2(Q_2) \tag{15}$$

where

$$A0(Q_2) = \int_0^\infty dr j_0(Q_2 r) Ap(r,0) = \frac{1}{2}\mathbf{Im}\Big(\int_{-\infty}^{\infty} dr \frac{exp(iQ_2 r)}{Q_2 r} Ap(r,0)\Big) \tag{16}$$

$$\begin{aligned} A1(Q_2) &= -\frac{1}{2Q_2^2}\int_0^\infty dr j_1(Q_2 r)(Q_2 r)\Big[Ap(r,1) - Ap(r,0)Q_2^2\Big] = \\ &= -\frac{1}{2Q_2^2}\frac{1}{2}\mathbf{Im}\Big(\int_{-\infty}^{\infty} dr \frac{exp(iQ_2 r)}{Q_2 r}\Big[Ap(r,1) - Ap(r,0)Q_2^2\Big]\Big) + \\ &+ \frac{1}{2Q_2^2}\frac{1}{2}\mathbf{Re}\Big(\int_{-\infty}^{\infty} dr exp(iQ_2 r)\Big[Ap(r,1) - Ap(r,0)Q_2^2\Big]\Big) \end{aligned} \tag{17}$$

$$\begin{aligned} A2(Q_2) &= \frac{1}{8Q_2^4}\int_0^\infty dr j_2(Q_2 r)(Q_2 r)^2\Big[Ap(r,2) - 2Ap(r,1)Q_2^2 + Ap(r,0)Q_2^4\Big] = \\ &= \frac{1}{8Q_2^4}\frac{1}{2}\Big[3\mathbf{Im}(A21(Q_2)) - \mathbf{Im}(A22(Q_2)) - 3\mathbf{Re}(A23(Q_2))\Big] \end{aligned} \tag{18}$$

$$A21(Q_2) = \int_{-\infty}^{\infty} dr \frac{exp(iQ_2 r)}{Q_2 r}\Big[Ap(r,2) - 2Ap(r,1)Q_2^2 + Ap(r,0)Q_2^4\Big] \tag{19.1}$$

$$\begin{aligned} A22(Q_2) = \int_{-\infty}^{\infty} dr (Q_2 r) exp(iQ_2 r)\Big[Ap(r,2) - \\ -2Ap(r,1)Q_2^2 + Ap(r,0)Q_2^4\Big] \end{aligned} \tag{19.2}$$

$$A23(Q_2) = \int_{-\infty}^{\infty} dr exp(iQ_2 r)\Big[Ap(r,2) - 2Ap(r,1)Q_2^2 + Ap(r,0)Q_2^4\Big] \tag{19.3}$$

B.4. Algorithm 2

By analogy to the procedure defined in I, see eqs. (9) and (11) in I, it follows that there exists a unique $Q(\vec{R})$ such that

$$\Phi(\vec{R}) = CN\int_0^\infty dr j_0(Q(\vec{R})r)\int_0^1 dx U(x,r) \tag{20}$$

where $\Phi(\vec{R})$ and $U(x,r)$ are given in eqs. (1) and (3), respectively. The eq. (20) can be writen as

$$CN\int_0^\infty dr\int_0^1 dxU(x,r)\Big[j_0(p(x,\vec{R})r)-j_0(Q(\vec{R})r)\Big]=0. \tag{21}$$

Introducing in eq. (21) the series for $j_0(p(x,\vec{R})r)$ given in eq. (7) with $q=Q(\vec{R})$, it is obtained

$$0=-\frac{CN}{2}\int_0^\infty dr\frac{j_1(Q(\vec{R})r)}{Q(\vec{R})r}r^2\int_0^1 dxU(x,r)\Big[p(x,\vec{R})^2-Q(\vec{R})^2\Big]+$$

$$+\frac{CN}{8}\int_0^\infty dr\frac{j_2(Q(\vec{R})r)}{(Q(\vec{R})r))^2}r^4\int_0^1 dxU(x,r)\Big[p(x,\vec{R})^2-Q(\vec{R})^2\Big]^2+\ldots \tag{22}$$

Reteining only the first term in eq. (22) it is obtained the eq. (23), which defines an approximated value for $Q(\vec{R})$, denoted by Q_a:

$$0=-\frac{CN}{2}\int_0^\infty dr\frac{j_1(Q_ar)}{Q_ar}r^2\int_0^1 dxU(x,r)\Big[p(x,\vec{R})^2-Q_a^2\Big]=$$

$$=-\frac{1}{2Q_a}\int_0^\infty drj_1(Q_ar)(Q_ar)\Big[Ip(r,1)-Ip(r,0)Q_a^2\Big]. \tag{23}$$

The rational approximants of eq. (14), when introduced in eq. (23) give the eq. (24), which can be solved directly because it is expressed in terms of elementary functions,

$$0\approx-\frac{1}{2Q_a}\int_0^\infty drj_1(Q_ar)(Q_ar)\Big[Ap(r,1)-Ap(r,0)Q_a^2\Big]=A1(Q_a). \tag{24}$$

where it has been used the same notation of subsection $B.3$ for all the quantities in eq. (24). Once the eq. (24) has been solved for Q_a, the approximant for the EP is

$$\Phi(\vec{R})\approx A0(Q_a)+A2(Q_a) \tag{25}$$

with identical notation used in eq. (15), with the difference that in eq. (25) the term $A1(Q_a)$ is absent because it is zero because eq. (24).

C. Results

In TABLE 1 the numerical trends for the Algorithms 1 and 2 respectively, are shown; these values correspond to the 1s STO of the hydrogens in ethylene molecule, see FIGURE 1 for the numerations.

Table 1. The orbital exponents of the 1s STO are $\alpha(H) = 1$, the bond lengths are $d(C - H) = 2.0409$ and $d(C = C) = 2.5323$, the bond angles are 120°; see ref. [7]. All the quantities are given in atomic units. The notation "R.A.-N.I." means that the corresponding values are evaluated by using Rational Approximants and direct Numerical Integrations, respectively; the direct Numerical Integration performed, follows the steps defined in the numerical evaluations given in II. The $\Phi(\vec{R}_k)$ values are obtained following the algorithm used in I for evaluating the exact multicenter molecular integral there. The number of points to perform the minimum squared calculations is $N = 20$.

H (i)H (j)k	Q	$\Phi(R_k)$	Q_2	Eq. (16) R.A.-N.I.	Q_a R.A.-N.I.	Eq. (25) R.A.-N.I.
12 6	2.2722	0.065564	2.4027	0.0660378-0.066515	2.4431-2.4346	0.065801-0.066326
12 3	4.2447	0.038174	4.3325	0.0382997-0.038339	4.3345-4.3343	0.038298-0.038228
13 2	3.1224	0.021535	3.1659	0.0215496-0.021638	3.1965-3.1872	0.021532-0.021626
13 6	1.6453	0.033001	1.8093	0.0350588-0.035350	1.9451-1.9378	0.034039-0.034413
14 6	3.6436	0.086767	3.6396	0.0868608-0.086770	3.6428-3.6420	0.086804-0.086770
14 2	4.9475	0.064282	4.9662	0.0643649-0.064296	4.9660-4.9665	0.064365-0.064296

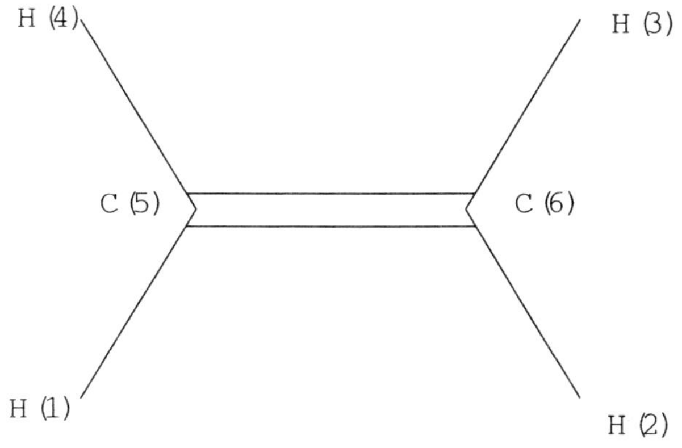

Figure 1. Sketch of the ethylene molecule with the corresponding numerations used in TABLE 1.

D. Conclusion

Within the framework of refs. [1,2,3,4], in this chapter, the simplest model for approximating the CP three-center molecular integral involving 1s STO, is developed. In this model, the inherent series expansions are truncated retaining only three terms and, the functions $Ip(r,n)$ which appear in each one of these terms, are replaced by rational approximants possesing the simplest pole structures.

The numerical test performed with the Algorithm I is in agreement with that of II, up to the third order, both approaches give the same number of significant digits for the MI. This is remarkable because the series in II, which represent bielectronic integrals, have two functions $Ip(r,n)$ in each term. The number of significant digits obtained, up to the third order, is conserved when the approximants are used; this is a consequence of the fact that, the approximants have an overall mean squared root error with respect to original functions, of about 0.001 or better.

The Algorithm 2, for approximating Q, differs from the one developed in I; and it is worth mentioning that it is designed on the basis of a decoupling in the variables of integration, a property that has some advantages which will be mentioned in the last paragraph.

The precision obtained is a consequence of the simplifying hypothesis made and some care must be exercised in extending the number of terms in the series. As the order of truncation in the series is greater, the replacement of $Ip(r,n)$ by $Ap(r,n)$, can make that the numerator not to be balanced in all the integrals, with undesirable asymptotic errors. This behaviour contrast with that of the four-center bielectronic integrals, where each term has two $Ip(r,n)$-like functions in the integrand, as mentioned in the above second paragraph. The adition of more terms must be accompained by raising de degree of the polynomial in $Ap(r,n)$, e.g. going to 6 or 8, perhaps analyzing a double pole structure, etc. This constitutes one of the fascinating perspectives of this chapter.

Finally, it is stressed that the present methodology is based on handling the functions $Ip(r,n)$ that depend on two orbitals and this is a very important property originated in the decoupling of the variables of integration. In this respect, when the Algorithms 1 and 2, presented in this chapter, be applied to the four-center bielectronic integrals, they will decouple the four indexes in pair, with a great gain in computational efficiency.

Appendix 1

The following items give the properties of functions $Ip(r,n)$ that make possible the adaptation of $Ap(r,n)$ as approximants.

$A1.a-$ The first partial derivative with respect to r of the function $Ip(r,n)$ is given by:

$$\frac{\partial}{\partial r}Ip(r,n) = \int_0^1 dx x(1-x)p(x,\vec{R})^{2n}K^{(1)}(Z(x,r))\frac{\partial}{\partial r}Z(x,r)$$

$$= \int_0^1 dx x(1-x)p(x,\vec{R})^{2n}K^{(1)}(Z(x,r))\frac{D_{12}rx(1-x)}{Z(x,r)} \qquad (A1.1)$$

where C, N, $K(z)$, $Z(x,r)$ and $p(x,\vec{R})$ are given in subsection $B.1$; $K^{(j)}(z)$ is the $j-th$ derivative of $K(z)$ and $D_{12} = |\vec{A}_1 - \vec{A}_2|$. In eq. (A1.1) the integrand is negative because

$K^{(1)}(Z(x,r)) < 0$, as can be seen deriving eq. (4); so that $\frac{\partial}{\partial r} Ip(r,n) < 0$ for $r > 0$. Also, from eq. (A1.1) it follows that $\frac{\partial}{\partial r} Ip(r,n) = 0$ at $r = 0$. It is worth mentioning that these properties are also satisfied by $Ap(r,n)$, because the coefficients $a2(n)$ and $a4(n)$ are positive, see *Appendix* 2.

*A*1.*b*– By using analogous arguments to those of Section 4 in III, it can be shown that the asymptotic behaviour of $Ip(x,r)$ for large r is bounded by $[const/r^{3.5}]$. From this and the fact that $Ip(r,n)$ is an even function of r, it results that THE DEGREE 4 OF THE EVEN POLYNOMIAL IN EQ. (14) IS THE SIMPLEST CHOICE.

*A*1.*c*– In this item is given a function, denoted by $G(r,n)$ and defined in (A1.5), which is needed for demostrating that $a4(n) > 0$; such a function is an upper bound of $Ip(r,n)$. From the eqs. (A1.2), (A1.3) and (A1.4):

$$Ip(r,n) = CN \int_0^1 dx x(1-x) p(x,\vec{R})^{2n} K(Z(x,r)) \tag{A1.2}$$

$$Z(x,r) > D_{12}[r^2 x(1-x)]^{0.5} \tag{A1.3}$$

$$Z(x,r) > Z(x,0) \tag{A1.4}$$

it follows that

$$Ip(r,n) < CN \int_0^1 dx x(1-x) p(x,\vec{R})^{2n} \frac{exp(-Z(x,0))}{[D_{12}[r^2 x(1-x)]^{0.5}]^3} \cdot$$

$$\cdot \left[1 + \frac{3}{Z(x,0)} + \frac{3}{Z(x,0)^2}\right] =$$

$$= \frac{g(n)}{r^3} = G(r,n) \tag{A1.5}$$

for $r > 0$.

Appendix 2

In this appendix, it is demostrated that the coefficients $a2(n)$ and $a4(n)$ can be selected positive.

*A*2.*a*– The coefficients $a2(n)$ in eq. (14), are determined from the condition:

$$\frac{\partial^2}{\partial r^2} Ip(r,n)\Big|_{r=0} = \frac{\partial^2}{\partial r^2} Ap(r,n)\Big|_{r=0} \tag{A2.1}$$

which, together with eqs. (13), (14) and (A1.1), it gives

$$a2(n) = -0.5 \frac{\frac{\partial^2}{\partial r^2} Ip(r,n)\Big|_{r=0}}{Ip(0,n)} =$$

$$= -\frac{0.5CN}{Ip(x,0)} \int_0^1 dx x(1-x) p(x,\vec{R})^{2n} K^{(1)}(Z(x,0)) \frac{D_{12} x(1-x)}{Z(x,0)} \tag{A2.2}$$

The resulting $a2(n)$ FROM EQ. (A2.2) IS POSITIVE, because the function $K(z)$ is decreasing, i.e. $K^{(1)}(Z(x,0)) < 0$ in eq. (A2.2).

$A2.b-$ The coefficients $a4(n)$ in eq. (14) are determined from a minimum squared procedure, i.e. minimizing $F(a4(n)) = \sum_{i=1}^{N}\left[\frac{Ip(r_i,n)}{Ip(0,n)} - \left(1+a2(n)r_i^2+a4(n)r_i^4\right)^{-1}\right]^2$ with respect to a4(n), it is obtained

$$0 = \sum_{i=1}^{N}\left[\frac{Ip(r_i,n)}{Ip(0,n)} - \left(1+a2(n)r_i^2+a4(n)r_i^4\right)^{-1}\right].$$

$$.r_i^4\left(1+a2(n)r_i^2+a4(n)r_i^4\right)^{-2}. \qquad (A2.3)$$

After some rearrange, the eq. (A2.3) transforms into

$$\sum_{i=1}^{N}\left[\frac{Ip(r_i,n)}{Ip(0,n)}\right]r_i^4\left(1+a2(n)r_i^2+a4(n)r_i^4\right)^{-2} =$$

$$= \sum_{i=1}^{N} r_i^4\left(1+a2(n)r_i^2+a4(n)r_i^4\right)^{-3} \qquad (A2.4)$$

In eq. (A2.3), the set of points $[r_i]$ with $i = 1...N$ is chosen so that they are equally distributed in the interval $(0, L_n]$, the L_n are determined from the equation:

$$Ip(L_n,n) = 0.01Ip(0,n), \qquad (A2.5)$$

i.e. L_n is the point where the the function $Ip(r,n)$ has decreased to 1 percent of its maximun value. The eq. (A2.4) involves the following two functions

$$F1(N,n;x) = \sum_{i=1}^{N}\left[\frac{Ip(r_i,n)}{Ip(0,n)}\right]r_i^4\left(1+a2(n)r_i^2+xr_i^4\right)^{-2} \qquad (A2.6)$$

$$F2(N,n;x) = \sum_{i=1}^{N} r_i^4\left(1+a2(n)r_i^2+xr_i^4\right)^{-3} \qquad (A2.7)$$

where the variable x is in the place of a4(n). For fixed values of N and n, both functions $F1$ and $F2$ respectively, are decreasing for $x \in (0,\infty)$, as can be seen deriving with respect to x; consequently, they reach their corresponding maxima at $x = 0$. There exists a positive solution $a4(n)$ in eq. (A2.4), i.e. $F1(N,n;a4(n)) = F2(N,n;a4(n))$, as a consequence of the following properties: a2.I) $F1(N,n;x) > F2(N,n;x)$ fro x sufficiently large; a2.II) $F2(N,n;0) > F1(N,n;0)$. The property a2.I results from their asymptotic, however, the property a2.II requires a demostration, which is given in what follows.

The proof of property a2.II comes from the following inequality

$$Ip(r,n) < \frac{Ip(0,n)}{1+a2(n)r^2} \qquad (A2.8)$$

for every $r > 0$. The proof of eq. (A2.8) is based on te principle of induction applied to the following succession, denoted by $[v_j]$ for $j = 1, 2...$, given in eq. (A2.9),

$$Ip(v_j, n) = \frac{Ip(0,n)}{1 + a2(n)v_{j-1}^2}. \qquad (A2.9)$$

The proof comes from the following three steps:
Step 1.There exists r_c such that

$$\frac{Ip(0,n)}{1 + a2(n)r^2} > G(r,n) \qquad (A2.10)$$

for $r > r_c$; eq. (A2.10) is satisfied because both functions $\frac{Ip(0,n)}{1 + a2(n)r^2}$ and $G(r,n)$ are decreasing, but $G(r,n)$ is unbounded with faster decreasing that $\frac{Ip(0,n)}{1 + a2(n)r^2}$, consequently, in this region (A2.8) is satisfied. So that, there exist v_1 and v_0 both bigger than r_c, satisfying eq. (A2.9) and

$$v_1 < v_0. \qquad (A2.11)$$

The eq. (A2.11) is a fundamental property that $[v_j]$ must satisfy.
Step 2. It is assumed that, for a given entier k, it is true that

$$v_k < v_{k-1}. \qquad (A2.12)$$

for v_k and v_{k-1} satisfying eq. (A2.9).
Step 3. Now it must be proved that $v_{k+1} < v_k$, for v_{k+1} and v_k satisfying eq. (A2.9):

$$Ip(v_{k+1}, n) = \frac{Ip(0,n)}{1 + a2(n)v_k^2}. \qquad (A2.13)$$

Considering the hypothesis in Setp 2, it follows that:

$$\frac{Ip(0,n)}{1 + a2(n)v_k^2} > \frac{Ip(0,n)}{1 + a2(n)v_{k-1}^2} = Ip(v_k, n). \qquad (A2.14)$$

The eqs. (A2.14) and (A2.13) and, the decreasing property of $Ip(r,n)$, imply that $v_{k+1} < v_k$, which completes the proof.

Appendix 3

In this appendix it is shown the procedure for obtaining the auxiliary functions $A0(z)$, $A1(z)$, $A2(z)$ (and $A21(z)$, $A22(z)$ and $A23(z)$), given in eqs. (16), (17) and (18) (and (19.1), (19.2) and (19.3)) respectively. The structure of poles of the functions $Ap(r,n)$ is very simple, because they are consequence of the roots of $[1 + a2(n)r^2 + a4(n)r^4]$ with a2(n) and a4(n) both positive; the simple poles are located on the vertices of rectangles centered around the origin with the edges paralel to the axis, or they can be located on the vertical axis, two in the upper part and the other two in the lower part of the complex plane respectively. The double poles that may arise will not be considered explicitly, although their treatment is similar.

Function $A0(z)$. The contour of integration in the complex plane, denoted by C_0, which gives eq. (16), is composed of four parts C_{0i} with $i = 1..4$: C_{01} is the oriented line on the horizontal axis defined by the interval $[-d, -\epsilon]$ going from $-d$ to $-\epsilon$; C_{02} is the semicircle of radius ϵ centered in the origin lying in the upper part of the complex plane going clockwise; C_{03} is the oriented line on the horizontal axis defined by the interval $[\epsilon, d]$ going from ϵ to d; and C_{04} is the semicircle of radius d centered in the origin lying in the upper part of the complex plane going counter clockwise. From the properties of the integration in the complex plane, see ref. [6], it follows that:

$$\oint_{C_0} du \frac{exp(izu)}{zu} Ap(u,0) = i2\pi \sum_{j=1}^{2} R\Big[\frac{exp(izu)}{zu} Ap(u,0), u_{0j}\Big] \qquad (A3.1)$$

where $z > 0$ and the set $[u_{0j}$ for $j = 1, 2$ is formed with the simple poles of $Ap(r, 0)$ having positive imaginary parts. Partitioning the left side of eq. (A3.1), it is obtained:

$$\int_{C_{01}} du \frac{exp(izu)}{zu} Ap(u,0) + \int_{C_{02}} du \frac{exp(izu)}{zu} Ap(u,0) +$$

$$\int_{C_{03}} du \frac{exp(izu)}{zu} Ap(u,0) + \int_{C_{04}} du \frac{exp(izu)}{zu} Ap(u,0) =$$

$$= i2\pi \sum_{j=1}^{2} R\Big[\frac{exp(izu)}{zu} Ap(u,0), u_{0j}\Big] \qquad (A3.2)$$

In eq. (A3.2) by leting that $d \to \infty$ and $\epsilon \to 0$, each term transforms into

$$\int_{-\infty}^{0} du \frac{exp(izu)}{zu} Ap(u,0) - i\pi \frac{Ip(0,0)}{z} +$$

$$\int_{0}^{\infty} du \frac{exp(izu)}{zu} Ap(u,0) + 0 =$$

$$= i2\pi \sum_{j=1}^{2} R\Big[\frac{exp(izu)}{zu} Ap(u,0), u_{0j}\Big]. \qquad (A3.3)$$

The eq. (A3.3) gives

$$\int_{-\infty}^{\infty} du \frac{exp(izu)}{zu} Ap(u,0) = i\pi \frac{Ip(0,0)}{z} +$$

$$+ i2\pi \sum_{j=1}^{2} R\Big[\frac{exp(izu)}{zu} Ap(u,0), u_{0j}\Big], \qquad (A3.4)$$

from which the auxiliary function $A0(z)$, given in eq. (16), can be obtained.

Function $A1(z)$. The contour of integration C_0 defined for $A0(z)$ is useful for evaluating the first term in eq. (17):

$$\int_{-\infty}^{\infty} dr \frac{exp(iQ_2 r)}{Q_2 r} \left[Ap(r,1) - Ap(r,0)Q_2^2\right] \qquad (A3.5)$$

taking into account the necessary changes coming from the different poles $[u_0 j]$ and $[u_1 j]$ of the functions $Ap(r, 0)$ and $Ap(r, 1)$, respectively. In this way, the evaluation of the term (A3.5) in eq. (17) is completely analogous to that involved in eq. (A3.4). The other term in eq. (17):

$$\int_{-\infty}^{\infty} drexp(iQ_2 r) \left[Ap(r,1) - Ap(r,0)Q_2^2\right] \qquad (A3.6)$$

requires a simpler contour than C_0 because the absence of the simple pole at the origin in the integrand in eq. (A3.6); so that, a contour like C_0 without C_{02} and joining the C_{01} and C_{03} paths, permits the evaluation of the term (A3.6) giving the corresponding part in eq.(17).

Function $A2(z)$. The function $A2(z)$, defined in eq. (18) through the functions $A21(z)$, $A22(z)$ and $A23(z)$, is evaluated by using the contours mentioned in relation to $A0(z)$ and $A1(z)$, again taking into account the corresponding poles of $Ap(r, n)$.

Acknowledgments

The authors greately acknowledge to the Editor for the opportunity of participating in the present book. Grants from SeCyT-UNRC and IMASL-CONICET-UNSL are acknowledged.

References

[1] Pérez J. E., Cesco J. C., Taurian O. E., Ortiz F. S., Rosso A. E., Denner C. C., Giubergia G. O.; *Int. J. Quantum Chem.* 2004, Vol 99, 70.

[2] Pérez J. E., Cesco J. C., Taurian O. E., Ortiz F. S., Rosso A. E., Denner C. C., Giubergia G. O.; *Int. J. Quantum Chem.* 2005, Vol 102, 1056.

[3] Cesco J. C., Pérez J. E., Denner C. C., Giubergia G. O., Rosso A. O.; *Applied Numerical Mathematics* 2005, Vol 55, 173.

[4] Shavitt I., Karplus M., *J. Chem. Phys.* 1965, Vol 43, 398.

[5] Abramowitz M., Stegun I. Eds., *Handbook of Mathematical Tables*, Dover: New York, 1972.

[6] Weinberger H. F., *Partial Differential Equations*, Blaisdell Publishing Company, New York, 1965.

[7] Pople J. A., Beveridge D. L., *Approximate Molecular Orbital Theory*, Mc Graw-Hill Inc, USA, 1970.

INDEX

A

B

C

F

G

H

N

O

P

Q

R